S

AIR POLLUTION CONTROL ENGINEERING

AIR POLLUTION CONTROL ENGINEERING

Basic Calculations for Particulate Collection

Second Edition

WILLIAM LICHT

Department of Chemical and Nuclear Engineering
University of Cincinnati
Cincinnati, Ohio

MARCEL DEKKER, INC. **New York and Basel**

Library of Congress Cataloging-in-Publication Data

Licht, William
Air pollution control engineering: basic calculations for particulate collection / William Licht. -- 2nd ed.

Includes bibliographic references and index.
ISBN 0-8247-7898-7
1. Gases--Purification--Mathematical models. 2. Dust--Removal--Mathematical models. I. Title.
TD884.5.L52 1988 87-33225
628.5'3--dc19 CIP

MARCEL DEKKER, INC.

270 Madison Avenue, New York, New York 10016

Current printing (last digit):

10 9 8 7 6 5 4 3 2 1

PRINTED IN THE UNITED STATES OF AMERICA

Preface to the Second Edition

Since the appearance of the first edition research has continued to produce an improved understanding of the performance of particulate collectors; hence improvements and refinements in the mathematical models that describe that performance are needed. The increasing importance of collecting finer particles and those of hazardous materials has largely motivated such research.

Some models are now so sophisticated as to be embodied in complex computer simulation programs. No attempt is made to present these, as the emphasis of the book remains the basic principles underlying the models and calculations that can be done at the desk.

For this revision the chapters in Part II have been largely rewritten to incorporate the new knowledge into the models and design procedures. Although the fundamental principles covered in Part I have not, for the most part, been altered, this material has also been carefully reconsidered and updated. A set of problems has been added to each chapter as a further aid to using the book as a text, particularly for self-directed study. Answers to selected problems are given.

The first edition has been used as a text in a number of courses. I have used Chapters 2, 4, and 5 in a course on the fundamentals of small particle behavior, and Chapters 1, 3, 6, 7, 8, and 9 in a following course on design of particulate control systems. I am grateful to the several hundred students who have used it, for their painstaking study of it as well as for their comments and suggestions.

I appreciate the interest in the first edition shown by professional colleagues, especially those who have raised questions and made suggestions in regard to its use, and I hope that this new edition will continue to be useful in both study and work. I am especially grateful to my wife, Helen, not only for her steadfast support but also for her painstaking proofreading of the entire manuscript. The responsibility for any errors in content or presentation, however, remains solely my own.

William Licht

Preface to the First Edition

This book is concerned with how to make some basic calculations involved in air pollution control engineering. Specifically it has to do with the cleaning of gases by the removal of particulate matter.

The Federal legislation embodied in the Clean Air Act of 1963, the Air Quality Act of 1967, and the Clean Air Amendments of 1970, has resulted in the setting of ambient air quality standards and of emission regulations by governmental agencies, on the basis of protecting public health and welfare. The enforcement of these laws by the government at local, state, and national levels has become a dominant factor in the design of present day particulate collection systems. It has become necessary to do an increasingly better job of particulate collection.

In meeting this need the engineering profession is faced with two major problems: to improve its "tools" for this kind of design work, and to teach many practicing engineers (as well as students) the specialized technical skills required. The response has involved (a) research to elucidate scientific principles related to forces acting upon suspended particles and their movement, and (b) development of more precise and detailed engineering application of these principles to analyzing, synthesizing, and optimizing the performance of devices and systems. As a result, there has accumulated a large body of technical information scattered through many scientific and engineering publications: journals, patents, government reports, trade literature, etc. There has also been an outpouring of educational programs: formal degree programs at the universities, and intensive continuing education courses given by professional societies and by the U.S. Environmental Protection Agency.

This book attempts to present the relevant body of information in a unified approach to basic principles involved and their applications. It is devoted to presentation of basic (scientific) principles and basic (engineering) calculations involved in the design of particulate collection systems. It seeks to provide this information in such a way as may be useful not only in formal courses, but also to the practicing engineer previously untrained in this area and confronted with a need to teach himself how to work in it.

This is not a book devoted to descriptions of equipment or to specific problems of application to certain industries. It is intended only to show how to do fundamental calculations which underly all kinds of collection problems. It assumes some familiarity with equipment, hardware, and control techniques.

The most important design tool is an adequate mathematical model of the collection process. This model must do two things: predict how successful the collection will be for a given design and set of operating conditions, and predict the energy required at the same time. The model can then be used to explore alternative designs or systems, to optimize a design, and to provide a basis for estimation of the cost of the system.

Construction of a mathematical model for a given type of particle collector usually involves a fairly intricate combination of underlying scientific principles. As more knowledge becomes available models may become more accurate, but also more sophisticated and complex. There is great urgency to speed up this process in order to meet the goals set by the Clean Air Act.

A presentation of the calculational tools falls naturally into two parts. The first covers the basic scientific principles. This includes the methods of describing particulate systems, an overview of the fundamental concepts of collection, and the physics of particle behavior. These topics are presented in Part I, Chapters 2 through 5.

Part II deals specifically with the modelling process. It gives the best current models for cyclones, filter, electrostatic precipitators, and scrubbers in Chapter 6 through 9. Hopefully it also shows how to proceed in building models for other devices which may be developed.

A mixture of metric (S.I.) and traditional English units is used, because this is the state of affairs today. While we are on our way to adoption and widespread use of the S.I. System, many customary values in the old system remain in use and must be understood. Where these are appropriate they are retained, without apology.

The author is indebted to a number of colleagues for ideas and fruitful discussions, especially to Prof. Wm. E. Ranz of the University of Minnesota (where the manuscript first began to develop), to Profs. J.N. Pattison and C.W. Gruber of the University of Cincinnati, and to Dr. R. Lee Byers of the Exxon Research and Engineering Co. He is also grateful for several classes of students on whom these ideas could be tested. And to his wife, Helen, for insuring the presence of an atmosphere of tranquillity which is so essential to work of this kind.

William Licht

Contents

AIR POLLUTION
CONTROL ENGINEERING

1
Particulate Collection in Air Pollution Control

I. AIR QUALITY STANDARDS

The Clean Air amendments of 1970 to the Clean Air Act of 1963 of the U.S. Congress provide that the Administrator shall promulgate national primary and secondary ambient air quality standards for each type of pollutant. The Act defines: "National primary ambient air quality standards shall be ambient air quality standards the attainment and maintenance of which...are requisite to protect the public health." And "National secondary ambient air quality standards shall specify a level of air quality the attainment and maintenance of which...is requisite to protect the public welfare from any known or anticipated adverse effects associated with the presence of such air pollutant in the ambient air." Protection of the public welfare has been taken to mean protection against effects on soil, water, vegetation, materials, animals, weather, visibility, and personal comfort and well-being.

Pursuant to this requirement, the Administrator of the Environmental Protection Agency (EPA) on April 30, 1971 promulgated the following ambient air quality standards for particulate matter.

Primary standard:

-75 micrograms (TSP) per cubic meter, annual geometric mean

-260 micrograms (TSP) per cubic meter, as a maximum 24-hour averaged concentration not to be exceeded more than once a year.

Secondary standard:

-60 micrograms (TSP) per cubic meter, annual geometric mean

-150 micrograms (TSP) per cubic meter, as a maximum 24-hour averaged concentration not to be exceeded more than once a year

These standards are expressed in terms of the amount of "total suspended particulate (TSP)" matter as collected from the air by a "high-volume" sampler, which effectively collects particulate matter up to a nominal diameter of 25 to 45 micrometers (μm) .

Congress further amended The Clean Air Act in 1977 to direct the Administration to periodically review and revise (if necessary) these standards. Accordingly in 1984, the EPA proposed to replace TSP by a new indicator which includes the amount of only those particles with an aerodynamic diameter (see Chapter 4) smaller than or equal to a nominal 10 μm (referred to as PM_{10}). This size was selected because particles finer than 10 μm are those most likely to be inhaled and remain in the lungs to the greatest extent.

The proposed new standards are:

Primary standard:

-value to be selected between 50 and 60 microgram (PM_{10}) per cubic meter, annual arithmetic mean

-value to be selected between 150 to 250 micrograms (PM_{10}) per cubic meter?

Secondary standard:

-value to be selected between 70 and 90 microgram (PM_{10}) per cubic meter, annual arithmetic mean.

A method of sampling and measuring PM_{10} in the ambient air was also proposed. As of mid-1986 these proposals have not been finalized, but it seems very likely that they will be adopted in some form.

The Clean Air Amendments of 1970 further require that each state must devise and establish an implementation plan by which it intends to attain the ambient air quality standards, not only for particulate matter but for all kinds of pollutants. A state is free to adopt more stringent ambient air quality standards than those promulgated as national standards, but may not adopt less stringent standards. Existing state implementation plans will also need to be revised if and when the PM_{10} standards are adopted.

II. EMISSION REGULATIONS

The only practical way to insure attainment of the legally specified ambient air quality standards is to place a suitable limitation upon the amount of each pollutant which may be emitted from each source where it arises. Thus, a necessary part of any air pollution control implementation plan is a set of emission regulations which may be applied to individual sources. The 1970 Amendments provide that the Adminstrator shall promulgate Federal standards of performance for new stationary sources, and for moving sources. States may, again, adopt emission regulations more stringent, but not less stringent, than those set forth by the Federal standards of performance.

The relationship between the individual emissions of a given pollutant from the multiplicity of sources within an air quality region, and the resulting average concentration of that pollutant in the ambient air is obviously a complex one. It must involve a reasonably accurate and complete listing of all the sources (their location, height of stack, rate of emission etc.), known as an "emission inventory", as well as local meteorological data covering the direction and velocity of winds, presence of inversion, etc. However our knowledge of this overall relationship is still imcomplete. Various dispersion models have been developed and are applied with computer calculations in order to estimate the effect of a given source upon the air quality of a region.

Because of the uncertainties in the quantitative relationship between the strength of individual point source emissions and overall regional ambient air quality, it is difficult to determine appropriate standards for emission regulations. Hence the 1970 Amendments state that

> The term 'standard of performance' means a standard for emissions of air pollutants which reflects the degree of emission limitation achievable through the application of the best system of emission reduction which (taking into account the cost of achieving such reduction) the Administrator determines has been adequately demonstrated.

Acting under these provision of the law, the administrator promulgated the first set of Standards of Performance for New Stationary Sources (NSPS), on December 23, 1971 relative to particulate matter, as follows:

A. FOSSIL FUEL-FIRED STEAM GENERATORS

No owner or operator...shall cause the discharge into the atmosphere of particulate matter which is:

1. In excess of 0.10 lb per million B.T.U. heat input (0.18 g. per million cal.) maximum 2-hour average.
2. Greater than 20 percent opacity, except that 40 percent opacity shall be permissible for not more than 2 minutes in any hour. (Except where the presence of uncombined water is the only reason for failure to meet these requirements).

B. INCINERATORS

No owner or operator...shall cause the discharge into the atmosphere of particulate matter which is in excess of 0.08 gr/s.c.f. (0.18 g/MN^3) corrected to 12 percent CO_2, maximum 2-hour average.

C. PORTLAND CEMENT PLANTS

1. No owner or operator...shall cause the discharge into the atmosphere of particulate matter from the kiln which is:

(a) In excess of 0.30 lb/ton of feed to the kiln (0.15 kg/metric ton), maximum 2-hour average.

(b) Greater than 10 percent opacity (except where the presence of uncombined water vapor is the only reason for failure to meet this requirement).

(c) No owner or operator...shall cause the discharge into the atmosphere of particulate matter from the clinker cooler which is:

(1) In excess of 0.10 lb/ton of fee/ to the kiln (0.050 Kg/metric ton), maximum 2-hour average.

(2) 10 percent opacity or greater.

(d) No owner or operator...shall cause the discharge into the atmosphere of particulate matter from any affected facility other than the kiln and clinker cooler which is 10 percent opacity or greater.*

Subsequently, Federal standards of performance regarding emission of particulate matter have been listed for over 60 types of stationary sources including asphalt concrete plants; catalyst regenerators in petroleum refineries; furnaces in secondary lead smelters, secondary brass and bronze smelters; electric arc furnaces and basic oxygen furnaces in iron and steel mills; sludge incinerators for sewage treatment plants; air tables and thermal

*Federal Register, Dec. 23, 1971.

dryers in coal cleaning plants; Knoft pulp mills; grain elevators; stationary internal combustion engines; many types of chemical manufacturing plants; solvent dry cleaning; beverage can coating; wool fiberglass insulation production; and many more. Details are published from time to time in the Federal Register, or may be obtained from local air pollution control agencies. Additional sources are continually added to the listing as time goes on.

The above regulations are quoted as examples of the type of compliance which is required for new sources. In addition states and cities will have emission regulations for existing sources which may differ in form of statement and in stringency from the Federal standards. No doubt some will be revised as a better knowledge is obtained of how emission is related to abmient conditions.

Obviously, it is of the greatest importance that the designer of any particulate collection system or control device learn the emission regulations which apply to the location where the control is to be used. He would be well-advised to design not only to be well within the present regulations but also to anticipate that an even higher degree of collection may be necessary in the future to meet more stringent standards of performance for emissions. He must also become thoroughly familiar with the methods of stack sampling and emission measurement which are to be used in testing whether his equipment is functioning within compliance of the regulations. Some standard test method is usually specified by the appropriate authority, along with the emission regulations.

The air quality standards and the emission regulations are at present always stated in terms of total mass of particulate being emitted per unit time, and usually also in terms of the opacity of the emission stream. No cognizance is taken of the fact that the mass will usually be composed of particles of a wide range of sizes, and perhaps of different composition as well. The coarser particles in an aerosol account for the large proportion of its total mass. Hence a collection system which is designed to do a good job on the removal of coarser particles, can usually be made to meet the regulations.

However, very fine particles, i.e. especially of the order of 1 μm or smaller in size, are those which cause the greater health hazard. Collection of such small particles is a more difficult problem, and one to which increasing attention is being given, as witness the proposal for PM_{10} standards mentioned above. It may be that in time emission regulations will be expressed in terms of particle size. The design of more sophisticated collectors will be required. This can only be achieved on the basis of a thorough understanding of certain fundamental principles, which are presented in this book.

III. SELECTION OF COLLECTION SYSTEM

There are many kinds of particulate collection devices available on the market. They utilize a number of different principles for accomplishing removal of particles from a gas stream. The differences result mainly from the nature of the force which is applied to a particle in order to collect it. The possible number of fundamentally different forces is not great. Either singly or in certain combinations they consitute the basic mechanisms by which collection is achieved. But these mechanisms are not all successful to the same degree on particles of all sizes.

The selection of an appropriate type of collector for a given particulate collection duty must be based upon a knowledge of collector performances which may be anticipated (or predicted) under various operating conditions. There are advantages and disadvantages accompanying the utilization of any particular type of collector embodying certain collecting forces or mechanisms. With regard to performance in terms of collection effectiveness, energy requirements, initial cost, operating cost, ease of installation and maintenance, etc. each of the collectors available finds its most appropriate type of application.

Table 1 is a comprehensive check list of items which must be considered in the design of a system to meet the needs of any air pollution control problem, gaseous or particulate in nature.

TABLE 1 Check list of considerations in the design of an air pollution control system

---

I. Survey of the Problem (Existing; Potential)

A. Type of Harmful Effect (Anticipated or Present)

1. Illness and health effects on humans
2. Toxicity to plants and/or animals
3. Damage to material
   a. Corrosion, deterioration, etc.
   b. Soiling
4. Reduction of visibility
5. Odor
6. Annoyances
7. Nature of complaints received

B. Properties of the gas

1. Pollutants present
2. Composition (analysis)
3. Temperature
4. Pressure
5. Humidity and Dewpoint
6. Chemical equilibria among components

C. Flow Rate

1. Quantity of gas to be treated
2. Steadiness of flow; possible fluctuations and start-up, shut-down

D. Particulates

1. Liquid or solids
2. Size distribution of particles
3. Amount of loading of particulates, mass/volume of gas
4. Physical and chemical properties
   a. Composition
   b. Density
   c. Shape
   d. Corrosiveness
   e. Abrasiveness
   f. Electrical properties
   g. Hygroscopicity
   h. Agglomeration
   i. Adsorption
   j. Combustibility

Table 1 (continued)

E. Location of plant

1. Meteorology factors of area
2. Plant and adjacent property, terrain
3. Stacks
   a. Height
   b. Location
4. Existing plant vs. Site Selection for new plant

F. Degree of Pollution Control Required

1. Local Standards and Regulations
2. Tolerable Levels; toxic materials
3. Future requirements of control system
4. Methods of testing and measurement
5. Public image of plant and company

G. "Excursions" of conditions

1. Known intermittent operations
2. Accidental releases

II. Solving the Problem

A. Process Changes

B. Alternative Materials in process

C. Selection of Methods of Control of Emissions

1. Gaseous removal
2. Particulate removal
3. Wet or dry system
4. Single system or combination
5. Efficiency

D. Existing control system (if any) and possible adaptation or retrofit

E. Use of Pilot Units

F. Consideration of other pollution (E.g. water, or solid waste) which may occur in consequence

III. Design of system

A. Size of Equipment and dimension specifications; retrofit to existing plant

Table 1 (continued)

B. Space needed for control unit and auxiliary equipment (fans, ducts, disposal, pumps, power supply etc.)

C. Materials of Construction

Corrosion allowance

D. Power requirements and pressure drop

E. Controls needed

1. Temperature
2. Humidity
3. Flows
4. Other

F. Maintenance needs

1. Ease of maintenance; preventive maintenance
2. Time required for maintenance - schedule

G. Flexibility of System - Future needs

H. Safety aspects; hazards and controls

I. Availability of standard equipment, or need for special design

J. Cost Estimates

K. Optimizing Design

L. Alternatives

M. Installation permit or "variance"

IV. Performance

A. Testing of installed equipment

B. Operating permit or "variance"

C. Monitoring regular operation

Obviously a great deal of practical as well as theoretical information will be needed to determine the optimum design of control system for each problem.

It is not the purpose here to present the descriptive details of hardware available. It is assumed that the reader possesses the minimum knowledge of this kind which is needed to understand the application of theoretical principles. There are a number of excellent references which may be consulted for a general survey of types of particulate collectors and their typical performance, costs and other features. Some of these are listed in Table 2.

TABLE 2 General references on particulate collection

The following references deal especially with the practial aspects, such as descriptions of typical equipment, considerations in selecting equipment, typical performance data and cost data. More specific references dealing with the theory of individual collectors will be found in the respective chapters.

1. S. Calvert, H. M. Englund, ed., Handbook of Air Pollution Technology, John Wiley & Sons Inc., New York (1984).

   Covers all aspects of air pollution and its control in topical chapters written by individual experts.

2. L. Theodore, A. J. Buonicore, ed., Air Pollution Control Equipment, Prentice-Hall, Inc., Englewood Cliffs N.J. (1982).

   Selection, design, operation, and maintenance of all types of equipment.

3. A. Stern, ed., Air Pollution - Vol. IV, 3rd ed., Academic Press, New York, (1977).

   Deals generally with control concepts of all kinds, and especially of problems in specific industries.

4. W. Strauss, Industrial Gas Cleaning, 2nd ed., Pergamon Press, Oxford (1975).

   Detailed presentation of theories and equipment for all kinds of gas cleaning.

5. J. M. Marchello and J. J. Kelly, ed., Gas Cleaning for Air Quality Control, Marcel Dekker, Inc., New York (1975).

   General survey of methods, equipment, and application.

6. Air Pollution Engineering Manual 2nd Edition, Chap. 4 — J. A. Danielson, Editor. Publication AP-40 Environmental Protection Agency, Research Triangle Park, N.C. (1973).

7. Compilation of Air Pollutant Emission Factors, 2nd edition, Publication AP-42 U.S. Environmental Protection Agency, Research Triangle Park, N.C. (1973).

Continually up-dated with supplementary data sheets.

8. L. S. Chaput, J. Air Poll. Cont. Assoc., 26: 1055 (1976). "Federal standards of performance for new stationary sources of air pollution - a summary of regulations."

9. R. B. Neveril, J. V. Price, and K. L. Engdahl, Capital and Operating Costs of Selected Air Pollution Control Systems, J. Air Poll. Cont. Assn., 28: 829, 963, 1069, 1171, (1978).

Detailed data for cost estimation.

10. G. J. Celenza, Designing Air Pollution Control Systems, Ch.E. Progress, 66, no. 11, pp. 31-40, Nov. 1970.

Good general introduction to equipment and its selection.

11. C. J. Stairmand, Selection of Gas Cleaning Equipment, Filtration and Separation, 7, pp. 53-63, Jan./Feb. 1970.

Good general discussion from British practice.

2
Characterization of Particles and Aerosols

No discussion of the collection of aerosol particles can be undertaken without adequate means available to describe the relevant properties of individual particles, and of the group of particles constituting an aerosol. The most important properties of an individual particle are its size, shape and density. All particles having a specified size, shape and density constitute a "grade" of particles, to be designated as the ith grade. Other related properties of interest are the surface area, volume, mass and chemical composition of a particle.

The assembly of particles making up an aerosol must be described in terms of its overall concentration (e.g. total mass/volume or total number/volume) and of the distribution of grades by number-, volume-, or mass-fraction within each grade.

The techniques by which individual particles are examined, and by which a size- (or more accurately a grade-) distribution analysis is performed, constitute an important and difficult branch of physical measurements. For a detailed discussion of methods, the reader is referred to sources listed in the references [1,2,3,4]. This chapter is devoted primarily to the interpretation of the results of such measurements and their application to the problems of particulate collection.

I. DESCRIPTION OF INDIVIDUAL PARTICLES

The basic tool for examining particles is microscopy, either of the optical or electronic variety. It reveals that solid particles making up most dusts and fumes are usually very irregular in shape, but occasionally correspond to a regular crystalline state. Liquid particles present in mists and sprays tend to be spherical, although they may be more or less distorted in larger sizes. A concise summary of the range of sizes found for many materials is given in Fig. 1. An excellent descriptive source is "The Particle Atlas" by McCrone [5] which is a comprehensive collection of photomicrographs of particles. This can be of great assistance in the identification of the nature and source of particles of unknown origin.

The irregular shapes observed tend to fall into one of three general classes:

(a) Isometric (granular, modular) - in which all three overall dimensions of the particle are roughly of the same magnitude;
(b) Flat (flaky) - in which there are much greater lengths in two dimensions than in the third, e.g. platelets, scales, leaves, etc.
(c) Needlelike (acicular, fibrous) - in which there is a much greater length in one dimension than in the other two, as for instance in prisms or fibers.

Examination of a particle under the microscope only reveals the two-dimensional outline which happened to fall at random into the plane of observation. The thickness of the particle may be inferred indirectly. Other methods of observation measure the total volume, total mass, or total surface of a particle, its mechanical behavior (motion) under the application of certain forces (gravitational, centrifugal, electrostatic), or the way in which particles scatter light.

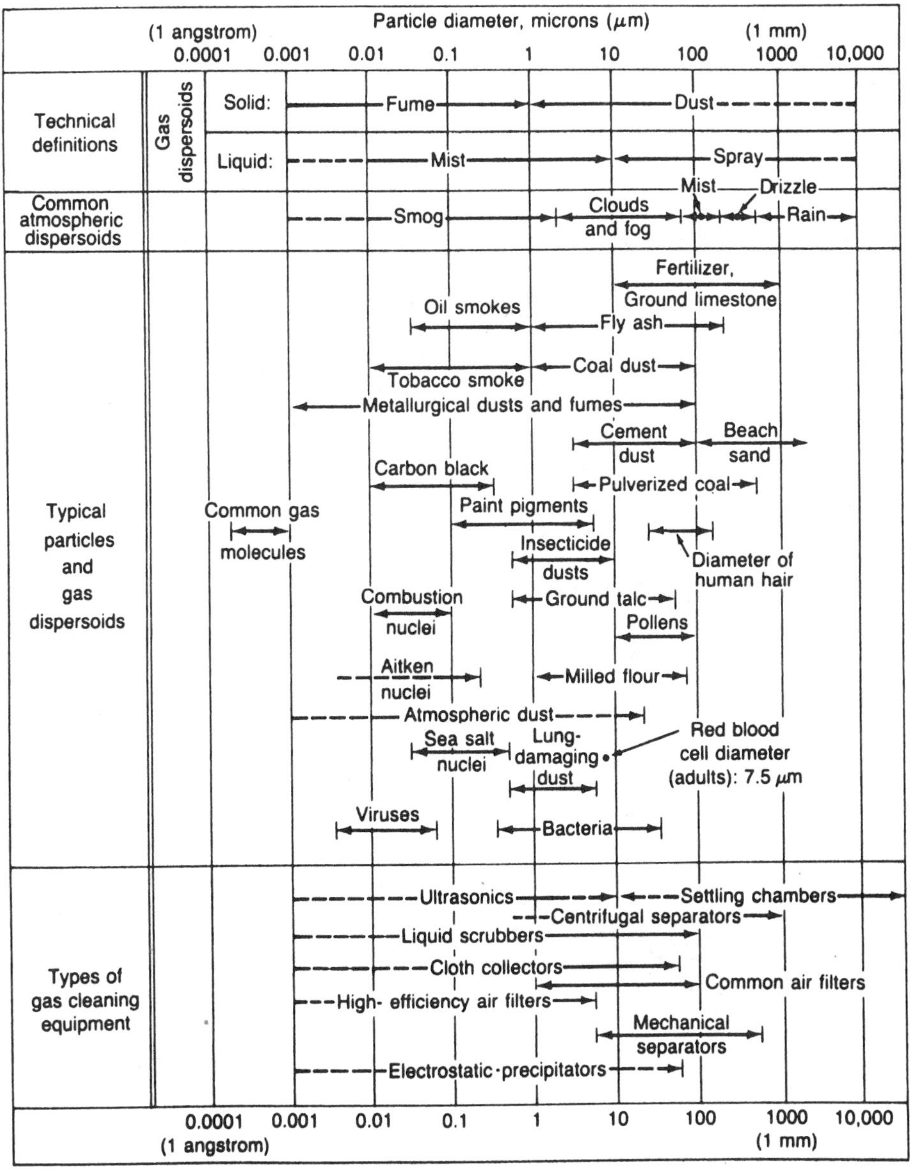

FIG. 1 *Characteristics of particles*
Sources: *C. David Cooper, F.C. Alley "Air Pollution Control: a Design Approach." PWS Publishers, Boston 1986, as adapted from C.E. Lapple, SRI Journal 5, 94 (1961)*
Reproduced by permission of both sources.

These measurements are of some property of the particle which is in some way related to its "size". A numerical value for the "size" is then determined on the basis of some sort of "equivalent behavior", usually equivalent to the behavior of a spherical particle under similar conditions.

For example, the rate of settling in air under gravity may be measured for some irregular particles. Then a "size" number may be assigned to each particle on the basis of calculation of the diameter of a spherical particle of the same density which would settle at the same speed as that measured. For the purpose of discussing the movement of such a particle under the influence of gravity, the diameter so calculated may be referred to as the "effective size" of the particle. This kind of calculation is explained in Chapter 4.

A. *PARTICLE SIZE AND SHAPE*

With these concepts in mind, some definitions of particle "size" in common use may be stated:

"Projected area diameter" - the diameter of a circle having the same area as the image of the particle projected parallel to the plane of the microscope view.

"Equivalent volume diameter" - the diameter of a sphere having a volume equal to that of the particle.

"Sedimentation diameter" (free-falling diameter) - the diameter of a sphere of equal density, having the same settling velocity as the particle in a specified fluid.

"Aerodynamic diameter" - the diameter of a sphere of unit density (1 gm/cm^3) having the same settling velocity in air as the particle (discussed in Chapter 4).

"Drag diameter" - the diameter of a sphere having the same resistance to motion (drag force) as the particle in a fluid of the same viscosity and at the same velocity (discussed in Chapter 4).

Other "diameters" are defined with reference to the particle image under the microscope as indicated by the sketch:

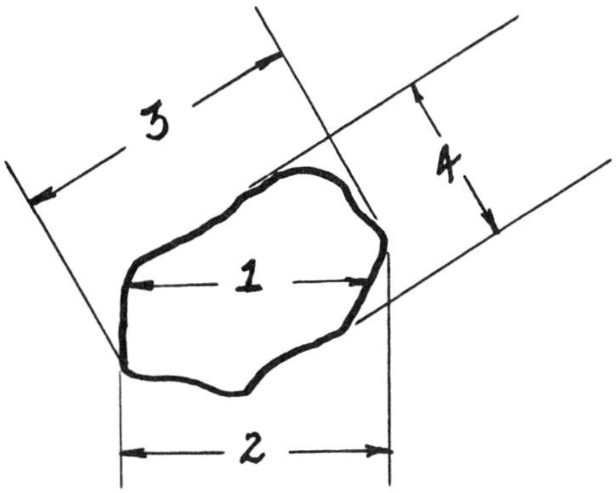

(1) Martin's diameter, divides projected area into equal parts along line parallel to base of field of view.
(2) Feret's diameter, must be longest dimension along line parallel to base of field of view.
(3) Longest diameter, without regard to direction.
(4) Shortest diameter, without regard to direction.

Which of all of these possible "diameters" to use in a given case depends upon the application for which the particle behavior is being considered. Even in the case of a spherical particle, not all of these diamters are the same. In most aerosol work it is customary to assume the particles are spheres for simplicity. For different types of particle collectors, however, different "diameters" may be most appropriate to use. This point will be discussed further in later chapters.

Another method of describing a real irregular particle is in terms of the size of some simple geometrical shape which will behave in the equivalent manner. The simple geometrical shapes employed are:

Sphere:	diameter = d
Cube:	side = d
Cylinder:	length = ℓ, diameter = d
Ellipsoid of revolution:	polar radius = p, equatorial radius = r

$$\beta = \frac{p}{r} \quad \begin{cases} \text{Prolate} & \beta > 1 \\ \text{Oblate} & \beta < 1 \\ \text{Sphere} & \beta = 1 \end{cases}$$

An isometric particle may be approximated by either a sphere or cube, a disk-like particle by either a cylinder with ℓ/d very small or an oblate ellipsoid with $\beta \to 0$, and a needlelike particle by a cylinder with large ℓ/d or prolate ellipsoid with $\beta \to \infty$. The system of ellipsoids may be used to cover many cases, but the geometrical calculations for them tend to become rather complicated.

The particle silhouette shape as seen in two dimensions may be represented by a method of Fourier transforms. This consists of finding the particle outline and its center of gravity, then measuring the distance from the centroid to the circumference at equal angular intervals of rotation around the centroid. The relationship between the radial distance to the periphery and the angle of rotation is then expressed in terms of Fourier Coefficients. Details are given by Beddow [6].

B. *OTHER PARTICLE PROPERTIES*

The surface, volume and mass of a particle may be related to an appropriately defined diameter d_p, such that

$$\text{surface} = \alpha_s\, d_p^2$$

$$\text{volume} = \alpha_v\, d_p^3$$

$$\text{mass} = \alpha_v \rho_p\, d_p^3$$

where α_s and α_v are the surface and volume shape coefficients respectively.

The "specific surface" is usually defined as surface per unit volume (occasionally per unit mass) and is given as

$$\Omega = \alpha_s/\alpha_v d_p = \alpha_{s,v}/d_p$$

$$\Omega = 6/d_p \text{ for sphere } (d_p\text{=diameter}), \text{ or cube } (d_p\text{=side}), \; \alpha_{s,v}=6$$

$$\Omega = \frac{2}{\ell}\left(1 + \frac{2\ell}{d}\right) \text{ for cylinder} \qquad \alpha_{s,v} = 2[d/\ell + 2]$$

The "sphericity" of a particle is defined by Wadell [7] as: ϕ_s = surface area of a sphere having the same volume as the particle ÷ surface area of the particle. The value of ϕ_s is always less than one. In general $\phi_s = 4.836\ \alpha_v^{2/3}/\alpha_s$. For a cube $\phi_s = 0.806$. For a cylinder $\phi_s = 2.62\ (\ell/d)^{2/3}/(1 + 2\ \ell/d)$ and for $\ell/d = 10$, $\phi_s = 0.579$. Some experimentally measured values for ϕ_s are given in Table 1.

For some purposes e.g. catalysis, adsorption, etc., the question of what constitutes the "surface" of a particle requires closer study. These phenomena occur at a size scale of the order of magnitude of atomic and molecular dimensions. The presence of surface imperfections, active sites, and other situations at this size scale may make the effective surface quite different from and far greater than the gross geometric surface which is related to drag forces, light scattering, aerodynamic capture, and other phenomena. These are characterized by a surface interaction on

TABLE 1 *Values of sphericity*

Material	ϕ_s
Sand	0.600, 0.681
Iron catalyst	0.578
Bituminous coal	0.625
Celite cylinders	0.861
Broken solids	0.63
Sand	0.534-0.628
Silica	0.554-0.628
Pulverized coal	0.696

Source: Ref. [8]

the size scale of the particle dimension itself which is usually at least several orders of magnitude larger than atomic or molecular dimensions.

Particles frequently have a tendency to agglomerate, with the resulting clump being "the particle" whose behavior is to be studied or controlled. The presence of clumps can usually be detected by microscopic examination. Some sort of appropriate dimension can be assigned to the clump acting as a single particle. Care must be taken in using such size values, for the apparent density of the agglomerate will be much less than that of the material of which the individual particles are composed. The clump may become deagglomerated again in a stream of gas flowing through collecting equipment.

The chemical composition of the material of the particle determines its density as well as a number of other properties of importance in the collection process. among these are its hardness, tendency to absorb moisture (become sticky), electrical resistivity, and tendency to acquire and hold electrical charges, all of which have consequences which may be of great practical importance in the collection process.

II. SIZE DISTRIBUTION IN AEROSOLS

In order to make calculations such as will be discussed in Chapter 3 for the collection of any given aerosol, it is necessary to know the grade-distribution (or "size distribution") expressed say in terms of mass-fractions m_i expressed as a function of d_p:

$$m_i = f(d_{p_i}) \qquad \text{where } \sum m_i = 1$$

Experimental methods are needed for determining the mass or number of particles of each size grade in a representative sample of the aerosol. Such methods are described in detail in references [1,2,3,4 and 9]. The size values for d_{p_i} will usually be expressed in terms of some simple geometrical shape having "equivalent behavior" in the sense of the measuring technique employed.

In particulate collection work it is generally desired to measure the size distribution of particles as they exist in a flowing gas. Sampling and measuring techniques are desired which do not alter the particles in any way. Thus it is best not to collect particles in a liquid or on a solid medium and then examine them, although sometimes this has to be done. The sampling and sizing methods most useful for aerosols are listed by Sparks [9] as given in Table 2.

For the purpose of calculating the total emission by weight from a collector, an accurate knowledge of the grade distribution of the coarser particles is sufficient, because these account for the overwhelming proportion of the total weight of many aerosols. However, since most of the bad health effects on humans and animals arise from the inhaling of very fine particles (e.g., less than 10 um), greater attention must be given to keeping these out of the ambient air. Hence greater importance must be attached to more accurate measurement of grade-distribution in the very fine size range, as in the PM_{10} range mentioned in Chapter 1.

It is frequently handy to characterize an entire aerosol by some kind of "average" or "mean" particle size. There are many ways (listed below) in which an average or mean value may be defined. One must be careful to use a mean value which correctly

TABLE 2 *Techniques for measuring particle size distribution in aerosol*

Technique	Particle diameter range	Theory discussed
Cascade impactors	0.3 - 15 μm	Chap. - 5
Cyclones	0.3 - 15 μm	Chap. - 6
Light scattering	0.5 - 10 μm	
Diffusion batteries	0.01 - 0.1 μm	Chap. - 4
Mobility analyzer	0.001 - 0.1 μm	Chap. - 4,5

Source: Ref. [9]

realtes to the behavior of the aerosol under the particular circumstances being considered.

A. *RAW DATA FOR GRADES*

An examination of an aerosol sample by any of the usual techniques will give size-distribution information in terms of size intervals or ranges. That is raw data will be in the form of the number of particles, or the total mass of particles, falling within a specified range of sizes. Alternatively the total number (or mass) of particles smaller than (or larger than) a specified size may be given. Very rarely (such as when counting glass beads under the microscope) will it be possible to identify individual particles by exact sizes, and to tabulate the number of particles found to be of each precise size. Data of the first kind may be displayed graphically as a histogram, in which the intervals of size-range need not all be equal. Figure 2 is a typical illustration, the raw data for which are tabulated in Example 1.

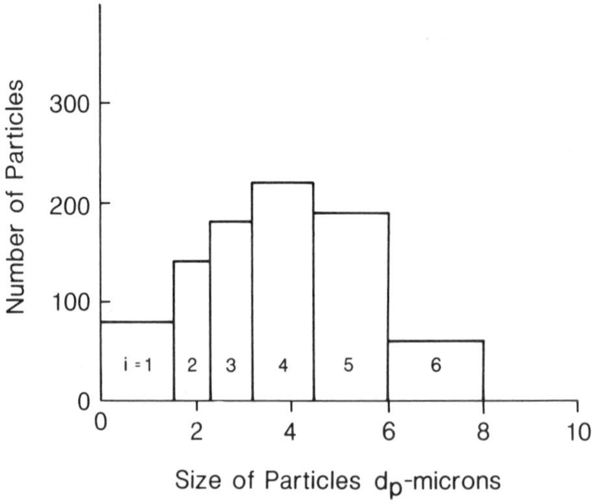

FIG. 2. *Histogram of particle count - Example 1.*

In general, it is reasonable to expect the size-distribution to indicate a modal (most frequently occurring) value, or perhaps occasionally a bi-modal form. If the raw data do not clearly indicate any mode, the data may be suspect as inadequate or inaccurate and further checking of them is in order. It is also reasonable to expect that there will be relatively few very large particles, so that the mode is not likely to be at either extreme end of the total size range.

B. *FREQUENCY AND CUMULATIVE FRACTIONS*

The number, or weight, values displayed by the raw data may be transformed into (a) fractional, or frequency, values, or (b) cumulative values. This will be developed first in terms of number data, and the corresponding calculations for weight data presented as an example below.

Let

n_i = number of particles observed in i^{th} interval

$\sum n_i$ = total number of particles of all sizes observed

Then: (a) Fractional or frequency values:

$$f_i = n_i / \sum n_i = \text{fraction of particles in } i^{th} \text{ interval} \tag{2.1}$$

where $\sum f_i = 1$. Thus, f_i also represents the frequency of occurrence of particles in the ith interval.

Or: (b) Cumulative values:

$$F_j = \frac{\sum_{i=1}^{i=j} n_i}{\sum_{i=1}^{n} n_i} \qquad F_j = \sum_{i=1}^{j} f_i \qquad \text{and } F_n = 1 \tag{2.2}$$

F_j = fraction of all particles which are smaller than the largest size in the jth interval, called the "cumulative-less-than" fraction. Similarly, a "cumulative-greater-than" fraction may be defined. The raw data, especially in terms of weights, may sometimes be obtained in the cumulative form also.

The raw data of the particle count may be recalculated into frequency values and plotted against the particle size representing the mid-point of each interval. This is shown on Fig. 3. There is then a temptation to draw a smoothed curve through these points. This temptation should be resisted, because such a curve would be misleading. It would imply that at any point in the total range of particle sizes one may take a size interval to be as small as desired, and that for this interval

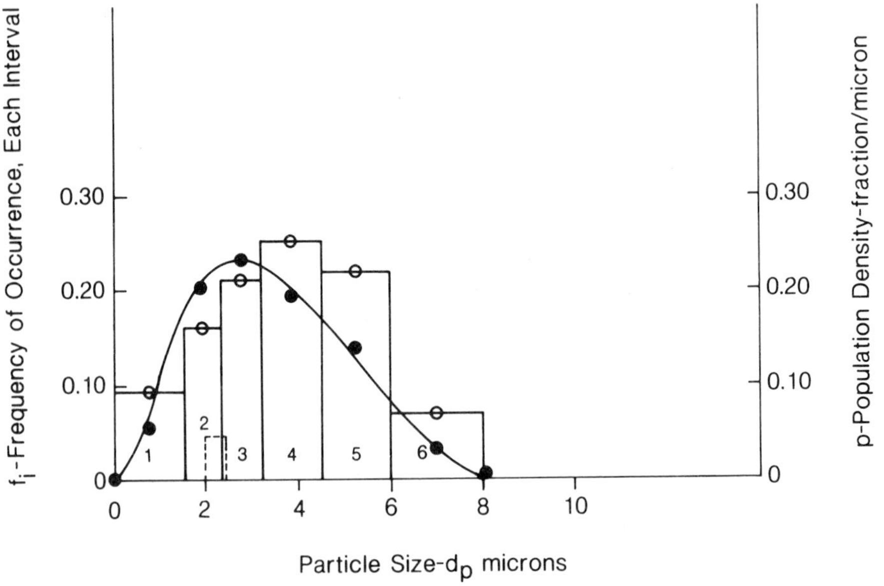

FIG. 3. *Relation between frequency of interval occurance and population density, Example 1. Bar chart: calculated discrete values of f_i; curve: population density = p.*

around the selected point, the curve would give a correct value for the frequency. Obviously this is not the case, for the smaller the interval chosen, the smaller the value of f must be. The only meaning that could be given to a smoothed curve drawn through the histogram points would be that the ordinates could represent the frequency of occurrence of particles of the same size interval as measured, but around other midpoints as might be read on the curves. But if the original raw data were not determined for size intervals of equal width, then this would not apply.

Two important questions then arise:

(a) How can we determine the frequency of occurence of particles lying within any size-interval other than those actually counted?

(b) In the limiting case, can we determine the frequency of occurrency of particles of a specific size?

These questions are best explored through the use of the cumulative-less-than function F and its graph.

The raw data are used to calculate the F-values, as defined above, and these are each plotted against the particle size value which is the upper limit of each size interval counted. This is shown in Fig. 4. Here a smooth curve may be drawn without question and its meaning is clear. If we take any two points, designated as (a) and (b), upon this curve we define an interval of particle sizes d_{p_a} to d_{p_b}. The difference between the F values $(F_a - F_b)$ is clearly equal to the value of f for this interval. The calculation may also be represented in terms of the slope of the F-curve thus:

$$f_{a-b} = F_a - F_b = \int_{F_b}^{F_a} dF = \int_{d_{p_b}}^{d_{p_a}} \frac{dF}{dd_p} \times dd_p = \int_{d_{p_b}}^{d_{p_a}} p\ dd_p \qquad (2.3)$$

The F-curve also makes it easy to determine the number median particle size (NmD), which is the value of d_p for which F = 0.50. Half of the particles are larger and half smaller than this size.

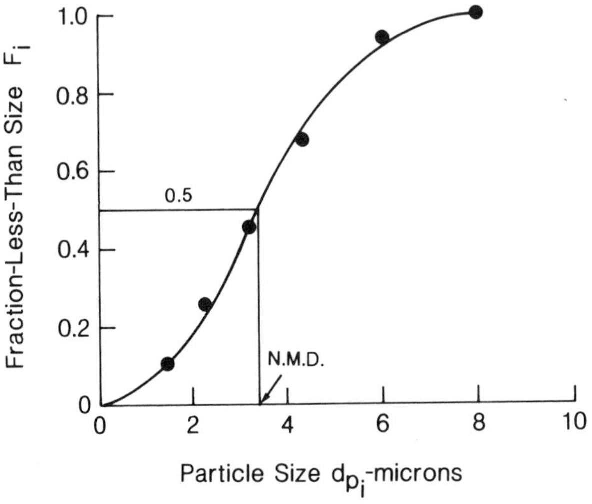

FIG. 4. *Cumulative fraction less-than particle size, Example 1. M.M.D. = Number median diameter = 3.3 microns.*

C. *DENSITY FUNCTIONS*

1. *Population density* Equation (3) implies the definition of a useful function $p = \frac{dF}{dd_p}$, called "population density," having dimensions of (size^{-1}). It is a continuous function of particle-size, and from the definitions it is evident that:

$$F = \int_0^{d_p} p.dd_p \qquad \text{and} \qquad \int_0^{\infty} p.dd_p = 1 \tag{2.4}$$

$$\frac{dp}{dd_p} = \frac{d^2F}{dd_p^2} \tag{2.5}$$

In the limits as $d_p \to o$, or $d_p \to \infty$, f must approach zero, hence $p \to o$ and the F-curve must be S-shaped with a point of inflection. This point of inflection occurs at a particle size for which p is a maximum i.e., where

$$\frac{d^2F}{dd_p^2} = \frac{dp}{dd_p} = 0 \tag{2.6}$$

The p-function may be constructed by measuring the slope of the F-curve, or approximately by calculating the ratio $(F_a - F_b)/(d_{p_a} - d_{p_b})$ to represent the slope at the point $(d_{p_a} + d_{p_b})/2$. A graph of p against particle size will not be unlike that of the histogram of f_i vs. d_p in general appearance, but the p vs. d_p plot may be drawn validly as a smooth curve. The area under this curve between two particle sizes gives the value of f for that size interval. Figure 3 includes a curve for p as calculated from Fig. 4. Example 1 gives all the data and calculation for Figs. 2, 3 and 4.

Example 1

Grade Designation=i	Range of Sizes-μm		Number of Particles=n_i
1	0-1.5		80
2	1.5-2.3		140
3	2.3-3.2		180
4	3.2-4.5		220
5	4.5-6.0		190
6	6.0-8.0		60
	> 8.0		0
		Total	870

The calculated values for the frequency plot Fig. 3 and the cumulative plot Fig. 4 are obtained as follows:

Grade Designa-tion=i	Midpoint of Size Range = d_p - μm	Number Fraction=f_i (Eqn.2.1)	Upper Bound size of range-μm	Cumulative Fraction-less-than = F_j (Eqn.2.2)
1	0.75	80/870=0.092	1.5	80/870=0.092
2	1.90	140/870=0.161	2.3	0.092+0.161=0.253
3	2.75	180/870=0.207	3.2	0.253+0.207=0.460
4	3.85	220/870=0.253	4.5	0.460+0.253=0.713
5	5.25	190/870=0.218	6.0	0.713+0.218=0.931
6	7.00	60/870=0.069	8.0	0.93+0.07=1.000
		Total 1.00		

The calculated values for the population density plot, also shown on Fig. 3 are as follows. The p-values are plotted against d_p for the midpoint of each interval.

Grade Designation=i	F interval ΔF	d_p range Δd_p - µm	$\frac{\Delta F}{\Delta d_p} \approx p$ µm^{-1} (Eqn. 2.4)
6	1-0.931=0.069	8.0-6.0=2.0	0.035
5	0.931-0.713=0.218	6.0-4.5=1.5	0.145
4	0.713-0.460=0.253	4.5-3.2=1.3	0.195
3	0.460-0.253=0.207	3.2-2.3=0.9	0.230
2	0.253-0.092=0.161	2.3-1.5=0.8	0.201
1	0.092-0 =0.092	1.5-0 =1.5	0.061

Note that p was estimated by the approximate method of calculating the slope as $\Delta F/\Delta d_p$. For convenience, the same intervals and F values were used as above, but, of course, other intervals and F values read from the curve (Fig. 4) could have been used just as well. The smoothness of the p-curve which can be drawn with such points is one test of the satisfactory use of the approximate slope calculation.

As an example of the use of the p-plot, determine the frequency of occurrence in the particle size-interval between 2.0 and 2.4 microns. This involves evaluation of the area under the p-curve using Eqn. (2.3)

$$f_{2.0-2.4} = \int_{2.0}^{2.4} p.dd_p \approx \frac{0.205 + 0.225}{2} (2.4-2.0) = 0.043$$

The value of f is shown as the dotted bar on the histogram of Fig. 3. The integral is estimated by using the area under the p-curve on Fig. 3, using the average value of p (0.215) over the interval as ordinate.

It should now be clear that the question of what is the frequency of occurrence of particles of a precise size cannot be answered from raw data of the kind we are discussing. We may take the interval ($d_{p_a} - d_{p_b}$) to be as small as we like (e.g. $df = p dd_p = dF$) and calculate an f for it, but to take it as zero will render the calculation impossible. Unless the raw data have consisted of an actual count of individual particles by size, we cannot say what such an f-value is. It will be seen in what follows that the data are almost always handled best in terms of the F and p function, or their counterparts in terms of particle mass instead of numbers.

The real importance of the p-function lies in the fact that it has been observed to follow consistently certain rather simple and well-defined forms which are characteristic of types of aerosols. The example above is typical in that the p-curve looks something like a skewed normal probability function. A few basic types of equations have been found to give accurate representations of p, hence also of F, or vice versa. These are discussed below. With the aid of such equations various kinds of mean and median particle sizes may be defined and readily calculated. The calculation of collector performance upon an aerosol described by a p-function is also facilitated.

2. *Weight density* Grade distribution data given in terms of numbers of particles may be converted into terms of weight of particles (or vice versa) by assuming that the weight of a particle is proportional to the cube of its size. Then either all particles in a sample must be taken to have the same density or the actual density of the different kinds of particles present must be determined and used to convert volume to weight. Plots of frequency, cumulative, or weight-density (analogous to population-density) values may then be defined and constructed just as was done for the number data. The definitions are, assuming all particles to have the same density:

$$g_i = \frac{n_i d_{p_i}^3}{\sum_{i=1}^{n} n_i d_{p_i}^3} = \text{weight fraction of particle in } i^{th} \text{ grade} \qquad (2.7)$$

$$G_j = \sum_{i=1}^{j} g_i = \text{cumulative-less-than fraction by weight, up to and including } j^{th} \text{ grade.} \quad (2.8)$$

$$q = \frac{dG}{dd_p} = \text{weight-density function} \quad (2.9)$$

The calculations are illustrated in Example 2.

Example 2. Using the raw data of Example 1, convert it into size distributions in terms of weight of particles and display the corresponding graphs of G_j and q. Assume all particles have the same density.

The calculations of g_i and G_j are performed as tabulated:

Grade Designa-tion=i	Midpoint size of range d_{p_i} - μm	$(d_{p_i})^3$ -(μm)^3	$n_i \times d_{p_i}^3$	g_i (Eqn. 2.7)	G_j (Eqn. 2.8)
1	0.75	0.422	34	5×10^{-4}	5.2×10^{-4}
2	1.90	6.86	960	0.015	0.016
3	2.75	20.79	3742	0.057	0.073
4	3.85	57.07	12555	0.192	0.265
5	5.25	144.7	27493	0.421	0.686
6	7.00	343	20580	0.315	1.000
		Totals	65364	1.000	

Following the calculations of G_j, the q-values are approximated by $\Delta G/\Delta d_p$ in the same manner as was done for p in the preceding example.

Interval=i	G-interval ΔG	d_p-range Δd_p -μm	$\Delta G/\Delta d_p \approx q$ (μm)$^{-1}$ (Eq. 2.9)
6	1 - 0.686 = 0.314	2.0	0.157
5	0.686 - 0265 = 0.421	1.5	0.281
4	0.265 - 0.073 = 0.192	1.3	0.148
3	0.073 - 0.016 = 0.057	0.9	0.063
2	0.016 - 0.00052= 0.015	0.8	0.019
1	0.0052- 0 = 5.2×10^{-4}	1.5	3.5×10^{-4}

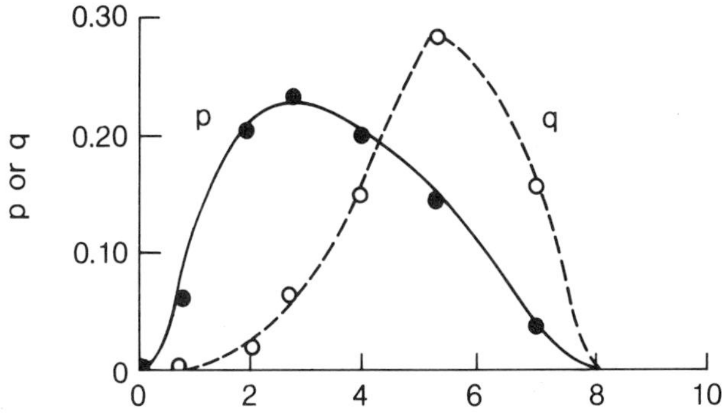

FIG. 5. *Density Functions - Examples 1 and 2, p = population density, q = weight (mass) density.*

Figs. 5 and 6 repeat the p and F values and include the q and G values for comparison. It is typical that the q and G (mass) curves are displaced toward the right with respect to the p and F (number) curves.

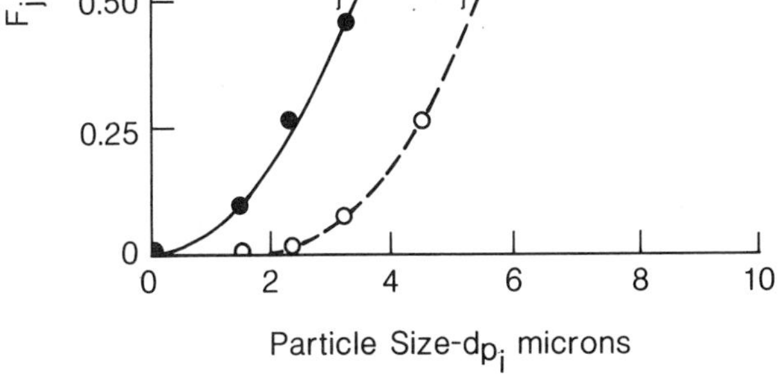

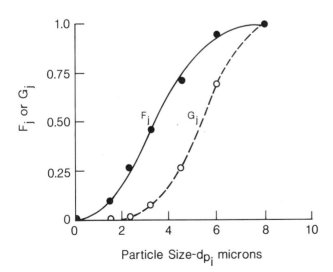

FIG. 6. *Relation between number and mass cumulative fraction less-than size. F_j number fraction less-than, G_j = weight fraction less-than.*

D. *MODE, MEDIAN AND MEAN VALUES*

It is useful to describe an aerosol by means of a single particle-size which represents in some fashion a typical size of the range of particles included in it. This may be done in terms of either a *modal, median* or *mean* value.

Where the frequency distribution (f) possesses a single well-defined peak value such as on Fig. 3, this most frequently occurring size is called the mode. For the sample data given the mode is about 2.8 μm. Occasionally there will be two peak values in the frequency distribution, a so-called bimodal pattern. When this occurs, it is significant enough to be emphasized by quoting both modal particle sizes, and to investigate the reasons for their existence. This is discussed further at the end of this chapter.

A *median size* which divides the sample into two equal portions may be defined in terms of cumulative distributions either by number of particles (F) or by mass of particles (g). Thus: (a) the number median (NMD) particle size is that for which F = 0.50, (b) the mass median (MMD) particle size is that for which G = 0.50. For the data of Example 1 the number median size is about 3.3 μm, and the mass median size is about 5.3 μm. These values may easily be read from a plot like Fig. 4.

In addition to the mode and median values, a particle size may be used to represent an aerosol. There are a number of different definitions of mean size in common use. The appropriate definition to employ in a given instance will depend usually upon the particular aspect of aerosol behavior which is under consideration. Whitby [10] has set forth a scheme by which any of the definitions of a mean size may be generated as the need arises. This scheme, somewhat modified, is as follows:

(a) Define a moment distribution factor, as the property of an individual particle which is being measured, which is proportional to a certain power of particle size d_p^m:

Size	d_p	1st moment	m=1
Surface	$\alpha\ d_p^2$	2nd moment	m=2
Volume, or Mass	$\alpha\ d_p^3$	3rd moment	m=3

(b) Define a weighting factor, as the total property measured, in all particles of the ith size, which is proportional to $(n_i d_{p_i}^w)$.

Total number	n_i	w=0
Total size	$n_i d_{p_i}$	w=1
Total surface	$n_i d_{p_i}^2$	w=2
Total volume, or mass	$n_i d_{p_i}^3$	w=3

where n_i = number of particles of ith size.

(c) Define mean size by selecting a combination of the above factors according to:

$$\text{Mean Size} = \left[\frac{\sum_i (\text{Moment Distribution x Weighting})}{\sum_i (\text{Weighting})}\right]^{1/m}$$

$$\bar{D}_{m,w} = \left[\frac{\sum_i (n_i d_{p_i}^{\ m + w})}{\sum_i (n_i d_{p_i}^{\ w})}\right]^{1/m}$$

and using (2.1):

$$\bar{D}_{m,w} = \left[\frac{\sum_i (f_i d_{p_i}^{\ m + w})}{\sum_i (f_i d_{p_i}^{\ w})}\right]^{1/m} \qquad m \neq 0 \qquad (2.10)$$

and using (2.7) also:

$$\bar{D}_{m,w} = \left[\frac{\sum g_i d_{p_i}^{\,m+w-3}}{\sum g_i d_{p_i}^{\,w-3}}\right]^{1/m}$$

Some examples of commonly used means, obtained by specifying m and w are listed below:

"average particle size" $w = 0,\ m = 1$

$$\bar{D}_{1,0} = \frac{\sum n_i d_{p_i}}{\sum n_i} = \sum f_i\, d_{p_i} \tag{2.11}$$

"surface mean" $w = 0,\ m = 2$

$$\bar{D}_{2,0} = \left[\frac{\sum n_i d_{p_i}^{\,2}}{\sum n_i}\right]^{1/2} = [\sum f_i d_{p_i}^{2}]^{1/2} \tag{2.12}$$

"volume mean" $w = 0,\ m = 3$

$$\bar{D}_{3,0} = \left[\frac{\sum n_i d_{p_i}^{\,3}}{\sum n_i}\right]^{1/3} = \left[\sum f_i d_{p_i}\right]^{1/3} \tag{2.13}$$

"surface to diameter mean" $w = 1,\ m = 1$

$$\bar{D}_{1,1} = \frac{\sum n_i d_{p_i}^{2}}{\sum n_i d_{p_i}} = \frac{\sum f_i d_{p_i}^{2}}{\sum f_i d_{p_i}} \tag{2.14}$$

Sauter mean diameter: $w = 2,\ m = 1$

"volume/surface"

$$\bar{D}_{1,2} = \frac{\sum n_i d_{p_i}^3}{\sum n_i d_{p_i}^2} = \frac{\sum f_i d_{p_i}^3}{\sum f_i d_{p_i}^2} = \frac{1}{\sum g_i / d_{p_i}} \qquad (2.15)$$

DeBrouckere diameter: $w = 3,\ m = 1$

"mass mean"

$$\bar{D}_{1,3} = \frac{\sum n_i d_{p_i}^4}{\sum n_i d_{p_i}^3} = \frac{\sum f_i d_{p_i}^4}{\sum f_i d_{p_i}^3} \qquad (2.16)$$

Note that in the manipulation of these expressions, it is implicitly assumed that the coefficient α_s and α_v are, respectively, independent of particle size.

Mugele and Evans [11] have also stated similar schemes for defining mean sizes, but in terms of continuous rather than discrete size distributions. The definitions are, therefore, expressed in the form of integration terms rather than summations.

Thus:

$$\bar{D}_{m,w} = \left[\frac{\int_0^1 d_p^{\,m + w}\, dF}{\int_0^1 d_p^{\,w}\, dF}\right]^{1/m} \qquad (m \neq 0) \qquad (2.17)$$

There are, of course, interrelationships which exist between median values and mean values for a given aerosol. The exact relation depends, however, upon the size distribution which is present. This is discussed below under the topic of size distribution functions.

Example 3. Using the data from Examples 1 and 2, calculate NMD, MMD, the "average particle size" $\bar{D}_{1,0}$, the "volume mean" $\bar{D}_{3,0}$, the Sauter mean $\bar{D}_{1,2}$ and estimate the specific surface Ω of the material.

The median diameters may be read directly from Fig. 6, as follows: NMD = 3.4 μm (at F = 0.50); MMD = 5.3 μm (at G = 0.50). The calculation of "average particle size", according to (2.11)

i	d_{p_i}	n_i	f_i	$n_i d_{p_i}$	$f_i d_{p_i}$
1	0.75	80	0.092	60	0.069
2	1.90	140	0.161	266	0.306
3	2.75	180	0.207	495	0.569
4	3.85	220	0.253	847	0.974
5	5.25	190	0.218	997	1.144
6	7.00	60	0.069	420	0.483
		870	1.00	3085	3.545

$$\bar{D}_{1,0} = \frac{\sum n_i d_{p_i}}{\sum n_i} = \frac{3085}{870} = 3.55 \ \mu\text{m}$$

$$\bar{D}_{1,0} = \sum f_i d_{p_i} = 3.55 \ \mu\text{m}$$

The calculation of "volume mean size" according to (2.13), and the calculation of $\bar{D}_{1,2}$ according to (2.15), require in addition:

$f_i d_{p_i}^3$	$g_i/d_{p_i}^3$	$f_i d_{p_i}^2$	g_i/d_{p_i}
0.04	0.00118	0.052	0.00067
1.10	0.00219	0.581	0.00789
4.31	0.00274	1.57	0.0207
14.44	0.00336	3.75	0.0499
31.54	0.00291	6.01	0.0802
23.67	0.00092	3.38	0.0450
75.10	0.01330	15.34	0.2044

$$\bar{D}_{3,0} = \left[\sum f_i d_{p_i}^3\right]^{1/3} = (75.10)^{1/3} = 4.22\ \mu\text{m}$$

$$\bar{D}_{3,0} = \left[\frac{1}{\sum g_i/d_{p_i}^3}\right]^{1/3} = \left(\frac{1}{0.0133}\right)^{1/3} = 4.22\ \mu\text{m}$$

$$\bar{D}_{1,2} = \frac{\sum f_i d_{p_i}^3}{\sum f_i d_{p_i}^2} = \frac{75.10}{15.34} = 4.89\ \mu\text{m}$$

or

$$\bar{D}_{1,2} = \frac{1}{\sum g_i/d_{p_i}} = \frac{1}{0.2044} = 4.89\ \mu\text{m}$$

The specific surface may be estimated from $\bar{D}_{1,2}$ as $\Omega = 6/\bar{D}_{1,2} = 6/4.89 = 1.23\ \mu\text{m}^{-1}$ or $1.23 \times 10^4\ \text{cm}^2/\text{cm}^3$.

III. SIZE DISTRIBUTION FUNCTIONS

A. *RECURRING PATTERNS*

When many examples of size distribution data are examined, it becomes evident that typical patterns in the frequency (f or g) or cumulative (F or G) plots recur rather regularly. A particular pattern of size distribution seems to be associated with the nature of an aerosol. Aerosols naturally occurring in the ambient air seem to follow one pattern, those generated by mechanical disintegration of solid bodies another, those generated by atomization of liquids another, those generated by chemical reactions another, and so on. It is very useful to have a technique for identifying such patterns and expressing them in mathematical form, particularly when that form can be a rather simple one.

It is most fruitful to examine these patterns in terms of the "density functions" p, or q, or the cumulative functions F or G. The sketches of size distribution shown in Figs. 3, 4, 5 and 6 represent typical characteristics of such plots. The distribution (p vs. d_p, Fig. 3) resembles in a general way the bell-shaped plot corresponding to the so-called Gaussian or "normal" distribution. However, instead of being symmetrical about a central particle size, it is usually skewed toward the larger particle sizes. Also, sometimes it may be necessary to take into account that there is a limiting small particle size below which p is zero, and/or that there is a maximum particle size present. The cumulative plots (F vs. d_p, Fig. 4) are typically S-shaped and must be related to the density plots by Eqns. (2.3) and (2.4).

Whenever possible, it is very convenient to have a simple equation to represent a given distribution pattern. This equation may give either p or F (alternatively either q or G) as a function of d_p and two or more constants or coefficients. Ideally the simplest kind of function would involve only two constants. One of these should represent the general size-magnitude of the aerosol, i.e., some kind of a defined mean particle size. The other constant should represent the spread of the size range about this mean. A third constant will, however, usually be needed whenever it is necessary to represent explicitly a limit particle-size (either maximum or minimum) in the aerosol.

Calculations of collector performance based upon grade efficiency curves, as will be explained in Chapter 3, are greatly facilitated by the use of size distribution equations. When the particle-size of the incoming aerosol and the grade efficiency of the collector may each be expressed by simple functions, computer programs may easily be written to calculate the overall efficiency of the collector and to predict the size distribution in the product aerosol emitted.

A number of semi-empirical formulas have been found to possess the features required for an adequate representation of certain types of aerosols. Some of these are listed as follows, with an example of their applicability:

(a) Log-normal probability distribution. Two constants. Ambient aerosols; many process dusts. The most commonly used and widely applicable function.

(b) Modified "upper limit" log-normal. Three constants. Used when it is necessary to specifically represent maximum particle size, e.g. in sprays.

(c) Weibull probability distribution. Three constants. Process dusts, especially those having a limiting small-sized particle.

(d) Rosin-Rammler formula. Two constants. Comparatively coarsely dispersed dusts and mists. Mathematically a special case of (c) above.

(e) Roller's formula. Two constants. Powdered industrial materials.

(f) Nukiyama and Tanasawa's formula. Three constants. Mechanically dispersed mists (atomization).

In order to select one of these to use for representing a given set of data, it is necessary first to have a knowledge of the general features and distinctive characteristics of each one. Then there must be a specific method for testing the goodness of fit and for determining the best values of the constants in each one. This information is presented in the next section.

It should also be pointed out that one of these equations may be useful in representing other kinds of relationships. For example, in Chapter 5 there are certain graphs relating to the aerodynamic capture of particles which have the typical S-shaped appearance of the cumulative plots. These can be represented usefully by one of the formulas just listed.

B. *SPECIAL FUNCTIONS*

A size-distribution function may be given initially in one of four ways: population-density, fraction by number of particles: p; weight-density, fraction by mass (or volume) of particles: q;

cumulative less-than-fraction by number of particles: F; cumulative less-than-fraction by mass (or volume) of particles: G. Alternatively the last two may easily be stated as cumulative-greater-than just as well. If a distribution formula is given in any one of these four forms, in principle the other three can be derived from it by the use of these relationships.

$$p = \frac{dF}{dd_p} \qquad q = \frac{dG}{dd_p} \tag{2.4) and (2.9}$$

$$F = \int_0^{d_p} p\, dd_p \qquad G = \int_0^{d_p} q \cdot dd_p \tag{2.18}$$

$$p = \frac{q}{d_p^3 \int_0^\infty \frac{q}{d_p^3}\, dd_p} \qquad q = \frac{d_p^3 p \cdot p}{\int_0^\infty d_p^3 p\, dd_p} \tag{2.19}$$

$$F = \frac{1}{\int_0^\infty \frac{q}{d_p^3}\, dd_p} \int_0^{d_p} \frac{q}{d_p^3}\, dd_p \qquad G = \frac{1}{\int_0^\infty d_p^3 \cdot p\, dd_p} \int_0^{d_p} d_p^3 p \cdot dd_p \tag{2.20}$$

However, in some cases these conversions are not easy to perform in practice.

In order for a function to represent real data satisfactorily, it must usually possess the following properties:

As $d_p \to 0$: $p = F = q = G \to 0$

$$\frac{dp}{dd_p} = \frac{dF}{dd_p} = \frac{dq}{dd_p} = \frac{dG}{dd_p} = 0$$

As $d_p \to \infty$: $p = q \to 0;\ F = G \to 1$

$$\frac{dp}{dd_p} = \frac{dF}{dd_p} = \frac{dq}{dd_p}\ \frac{dG}{dd_p} \to 0$$

Although, of course, there are no particles either of zero or of infinite size, there is usually a negligible error introduced by using a function which includes these limits, and such a function is usually of the simpler two-constant variety. In case it is important to represent either a minimum particle size significantly greater than zero, and/or to represent a known upper limit to the particle size, one of the more elaborate functions involving three (or four) constants is required. The limiting conditions are modified accordingly. In addition, p and q must each possess a maximum point corresponding to a point of inflection in F and G respectively. Functions possessing these characteristics will only be capable of describing unimodal distributions.

These general statements, with a few exceptions as noted, may be applied to each of the functions given below. Each of these has been discussed in detail in standard references which are cited. The following treatment is devoted to a practical working summary of the characteristics of each function and the method of testing its fit to actual data.

1. *Log-normal probability* [*1,3,4,12*] This is considered to be the most generally applicable function to describe the aerosols encountered in the ambient air, and to describe many process dusts encountered in emission control problems. It is also relatively convenient to use, and for this reason may be used as a satisfactory approximation in some cases.

$$\frac{dF}{ddp} = p(d_p) = \frac{1}{dp\,\sqrt{2\pi}\,\ln\sigma_g}\exp - \left[\frac{\ln d_p/\bar{d}_p}{\sqrt{2}\,(\ln\sigma_g)}\right]^2 \tag{2.21}$$

$$\frac{dF}{d\ln d_p} = \frac{1}{\sqrt{2\pi}\,\ln\sigma_g}\exp - \frac{\ln d_p/\bar{d}_p}{\sqrt{2}\,(\ln\sigma_g)}^{2}$$

$$F(d_p) = \frac{1}{\sqrt{2\pi}\ \ln\sigma_g} \int_{-\infty}^{\frac{\ln(d_p/\bar{d}_p)}{\sqrt{2}\ \ln\sigma_g}} \exp\left[-\frac{(\ln d_p/\bar{d}_p)^2}{\sqrt{2}\ (\ln\sigma_g)^2}\right] d\ln d_p \qquad (2.22)$$

These expressions result from applying the Gaussian or normal law of distribution to $\ln d_p$ instead of d_p itself. Irani and Callis [4], and Mugele and Evans [11] have given derivations to indicate how this would be consistent with certain physical facts about aerosol and particle properties. The two constants are: $\bar{d}_p$ = number median particle size (NMD) for which F = 0.50, and σ_g = geometric standard deviation (dispersion indicator) $\geqq 1$. When $\sigma_g = 1$, the aerosol is said to be monodisperse, i.e. all particles are the same size. The maximum value of p, corresponding to the inflection in F, occurs at

$$d_{p_{max}} = \bar{d}_p / \sigma_g^{\ln\sigma_g} \qquad (2.23)$$

for which

$$p_{max} = \frac{\sigma_g^{\ln\sigma_g}}{d_p\sqrt{2\pi}\ \ln\sigma_g} \exp\left(-\frac{\ln^2\sigma_g}{2}\right) \qquad (2.24)$$

For ease in evaluation, F may readily be transformed into terms of the "error function" erf. Thus

$$f_{a-b} = F(d_{p_a}) - F(d_{p_b})$$

$$= \frac{1}{2}\left[\operatorname{erf}\left(\frac{\ln d_{p_a}/\bar{d}_p}{\sqrt{2}\ \ln\sigma_g}\right) - \operatorname{erf}\left(\frac{\ln d_{p_b}/\bar{d}_p}{\sqrt{2}\ \ln\sigma_g}\right)\right] \qquad (2.25)$$

$$F(d_p) = \frac{1}{2}\left[1 + \text{erf}\left(\frac{\ln d_p/\bar{d}_p}{\sqrt{2}\,\ln\sigma_g}\right)\right] \tag{2.26}$$

Tables of erf are available in standard handbooks. Smith [13] gives a very good approximation formula which is well-suited for use with a pocket calculator.

It may be shown that the mass (or volume) distribution functions q ang G are of the same forms as those for p and F and have the same value of σ_g. The only difference is that $\bar{d}_p$ (NMD) is replaced by $\bar{d}_{p_m}$, the mass median size (MMD), for which G = 0.50. The interrelationship is:

$$\ln\bar{d}_{p_m} = \ln\bar{d}_p + 3\,\ln^2\sigma_g \tag{2.27}$$

or

$$\ln \text{MMD} = \ln \text{NMD} + 3\,\ln^2\sigma_g$$

It is also possible to consider the distribution function in terms of the diameter, and of the surface, of the particles in addition to simply their number count of their volume (mass). In each case, this distribution law applies with the same value of σ_g, but a different value of the median size:

$$\ln\text{DMD} = \ln\text{NMD} + \ln^2\sigma_g \qquad \text{(diameter median)} \tag{2.28}$$

$$\ln\text{SMD} = \ln\text{NMD} + 2\,\ln^2\sigma_g \qquad \text{(surface median)} \tag{2.29}$$

These median values are always in the order MMD > SMD > DMD > NMD.

A general relationship exists among any of these medians and among the several mean values as defined in section II.D:

$$\ln\bar{D}_{m,w} = \ln\bar{d}_{p_x} + \frac{m + 2w - 2x}{2} \ln^2\sigma_g \tag{2.30}$$

Here $\bar{D}_{m,w}$ = mean value, as defined by m and w in Eqn. (2)

$\bar{d}_{p_x}$ = median value, based upon

x = 0 for number count $\quad \bar{d}_{p_o}$ = NMD

x = 1 for diameter $\quad \bar{d}_{p_1}$ = DMD

x = 2 for surface $\quad \bar{d}_{p_2}$ = SMD

x = 3 for volume (mass) $\quad \bar{d}_{p_3}$ = MMD

The easiest way to test data for conformity with this function, and at the same time to obtain values for σ_g and the appropriate median, is to use a special graph paper commercially available, which is known as log-probability paper. One axis is a standard log scale, usually of 2 cycles, used for plotting log d_p. The other axis is a specially constructed scale corresponding to the error function integral as in (2.26), and used for plotting cumulative-percent-less-than on whichever basis is desired, F or G. An illustration is shown in Fig. 7.

Whenever experimental data points are plotted on this paper, a straight line will be obtained if the data agree with this distribution function. The "slope" of this line is determined by the value of σ_g. This is easily calculated by reading the values of d_p corresponding to $(+\ \sigma_g\bar{d}_p)$ and $(-\ \sigma_g\bar{d}_p)$. From (2.24) it may be shown that $(+\ \sigma_g\bar{d}_p)$ is located at 84.13%-less-than, and $(-\ \sigma_g\bar{d}_p)$ at 15.87%-less-than. These values, designated as $d_{p_{84}}$ and $d_{p_{16}}$ respectively, are used together with the median size $d_{p_{50}}$ to calculate σ_g

$$\sigma_g = \sigma\bar{d}_p/\bar{d}_p = d_{p_{84}}/d_{p_{50}} = d_{p_{50}}/d_{p_{16}} = \sqrt{d_{p_{84}}/d_{p_{16}}} \tag{2.31}$$

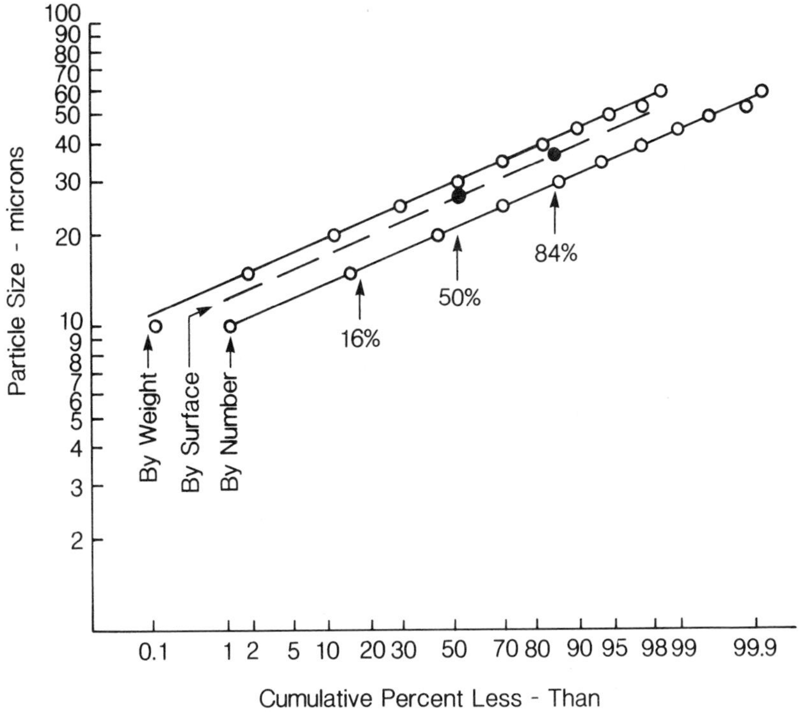

FIG. 7. *Log-probability size distribution, Example 3: sample of glass beads* [4]; *MMD = 30 microns; NMD = 21.5 microns* $\sigma_g = 1.38$.

The straight lines on the log-probability plot corresponding to each of the four kinds of distribution (count, diameter, surface, mass) will be parallel to each other and each will have the same value of σ_g. The spacing between them will be determined by the median values $\bar{d}_{p_x}$, as given above in (2.27), (2.28) and (2.29).

Another method of plotting which is sometimes useful is to let $y = dF/d\ln d_p$, and $x = \ln d_p$, then plot y vs x. Values of y may be approximated by taking $\Delta F/\Delta \ln d_p$. If the data obey this distribution law the symmetrical bell-shaped curve of the Gaussian distribution will be obtained. For according to the second form of Eqn. (2.21)

$$y = \frac{1}{\sqrt{2\pi}\ \ln\sigma_g} \exp - \left[\frac{x - \bar{x}}{\sqrt{2}\ \ln\sigma_g} \right]^2 \qquad (2.32)$$

On such a plot $\bar{x}$ will be found at the peak of the curve (for which $dy/dx = 0$), and the peak value of y_{max} will give σ_g:

$$\sigma_g = \exp\ (1/\sqrt{2\pi}\ y_{max}) \qquad (2.33)$$

Example 4. The table gives data on the cumulative size distribution of a sample of glass beads (all having the same density) as quoted by Irani and Callis [4]. Test whether these data follow the log-normal probability function and if so, determine σ_g, NMD, MMD and the Sauter mean diameter ($\bar{D}_{1,2}$). Show how to use the constants found to calculate the frequency distribution in 5-micron intervals.

Size Distribution of a Sample of Glass Beads
Cumulative Distribution

Weight Distribution			Number Distribution	
Size (microns)	% Cumulative (Greater-than)	G (% Finer-than)	% Cumulative (Greater-than)	F %Finer-than
5	100.0	0.0	100.00	0.0
10	99.9	0.1	99.0	1.0
15	98.4	1.6	86.2	13.8
20	89.5	10.5	58.0	42.0
25	71.5	28.5	32.0	68.0
30	50.0	50.0	15.0	85.0
35	32.1	67.9	7.0	93.0
40	19.2	80.8	2.8	97.2
45	10.8	89.2	1.2	98.8
50	6.0	94.0	0.5	99.5
55	3.0	97.0	0.15	99.85
60	1.9	98.1	0.09	99.91
100	0.0	100.0	0.0	100.0

Source: Ref. [4].

If the raw data points are plotted on regular graph paper, the curves will be similar to those of Fig. 6. However, it is better to plot them on the special log-probability paper as shown in Fig. 7. A straight line is clearly defined and drawn through each set of points. The following values are read from these lines;

	F Number Distribution	G Weight Distribution
$d_{p_{84}}$ - μm	29.7	41.7
$d_{p_{50}}$ - μm	21.5 (NMD)	30 (MMD)
$d_{p_{16}}$ - μm	15.5	21.7

Number Distribution (2.31), F line: $\sigma_g = 29.7/21.5 = 1.38$ or $\sigma_g = 21.5/15.5 = 1.39$, or $\sigma_g = \sqrt{29.7/15.5} = 1.38$. Since the values of σ_g are the same for each line, the lines are proven to be parallel, as indeed they appear to be to the eye.

Relationship between medians (2.27): solving for

$$\ln\sigma_g = \sqrt{\frac{\ln \mathrm{MMD/NMD}}{3}} = \sqrt{\frac{\ln 30/21.5}{3}} = 0.3332;\ \sigma_g = \underline{\underline{1.395}}.$$

This value checks very well with those obtained from the slope of each line. Taking $\sigma_g = 1.39$ the Sauter mean diameter is calculated from (2.30), using m = 1, w = 2, x = 3, as

$$\ln \bar{D}_{1,2} = \ln 30 + \frac{1 + 2(2)-2(3)}{2} \ln^2 1.39;\ \bar{D}_{1,2} = \underline{28.4}\ \mu\mathrm{m}.$$

If desired, lines could also be constructed on Fig. 7 to give the distribtuion by diameter, or by surface. They would be drawn parallel to the count and mass lines, i.e. with $\sigma_g = 1.39$, and pass through a median size as calculated from (2.28) and (2.29). For example, the surface median size would be: ln SMD = ln 21.5 + $2\ell n^2 1.39$, SMD = 26.7 μm. $d_{p_{84}} = 26.7 \times 1.39 = 37.1$ μm. This line is shown dashed on Fig. 7.

To construct the number frequency distribution in 5-μm intervals, (2.3) or (2.23) is used. The integrals are approximated by using the value of p at the center of the interval:

$$f_{0-5} = \int_0^5 p \cdot dd_p \approx p_{2.5}(5-0) \text{ or } g_{0-5} = \int_0^5 q \cdot dd_p \approx q_{2.5}(5-0).$$

The values of p, or q, are calculated from (2.21). To cover the range of sizes from 0 to 60 µm, twelve intervals of 5 µm each are used, and a final interval is taken from 60 to 100 µm because this represents a small portion of the total. The results are tabulated for both number size-distribution f_i and mass size-distribution g_i. An alternate approach is to calculate f from the erf relationship given by (2.23). This involves no approximation. The results of both methods are tabulated for the number and mass frequency. They are seen to agree well with each other and with the original data cited by Irani and Callis, except for the upper end of the range where the original data show an unexpected increase in occurrence of particles. Figure 8 is a plot of these results.

The method of calculation for each column in Table 3 is shown below and keyed to the table by the letters (A), (B), etc., appearing at the head of the column.

$$\text{(A)}\ p_{1/2} = \frac{1}{d_{p_{1/2}} \sqrt{2\pi}\ \ln 1.39} \exp - \left[\left(\frac{\ln d_{p_{1/2}}/21.5}{\sqrt{2}\ \ln 1.39} \right) \right]^2 \qquad (2.21)$$

$$\text{(B)}\ f_i \approx p_{1/2}\,(d_{p_a} - d_{p_b}) \qquad (2.23)$$

$$\text{(C)}\ f_i = F_a - F_b = \frac{1}{2}\left[\text{erf}\left(\frac{\ln d_{p_a}/21.5}{\sqrt{2}\ \ln 1.39} \right) - \text{erf}\left(\frac{\ln d_{p_b}/21.5}{\sqrt{2}\ \ln 1.39} \right) \right] \qquad (2.25)$$

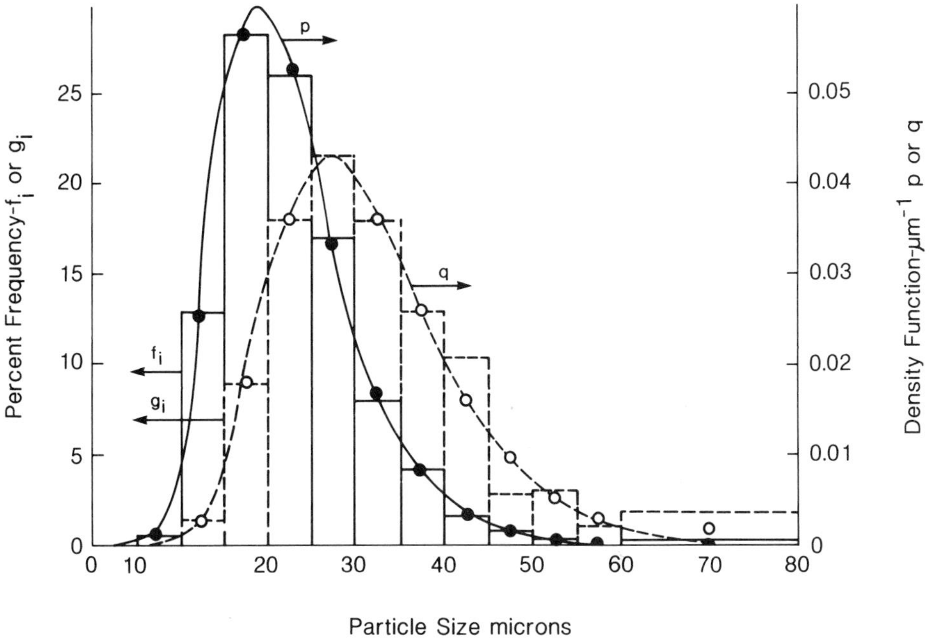

FIG. 8. *Density and frequency distribution, Example 3: sample of glass beads [4]. Solid lines by number, dashed lines by mass.*

$$\text{(D)}\quad q_{1/2} = \frac{1}{d_{p_{1/2}}\sqrt{2\pi}\,\ln 1.39}\exp - \left[\left(\frac{\ln d_{p_{1/2}}/30}{\sqrt{2}\,\ln 1.39}\right)\right]^2 \qquad (2.21)$$

$$\text{(E)}\quad g_i \approx q_{1/2}\,(d_{p_a} - d_{p_b}) \qquad (2.3)$$

$$\text{(F)}\quad g_i = G_a - G_b = \frac{1}{2}\left[\operatorname{erf}\left(\frac{\ln d_{p_a}/30.0}{\sqrt{2}\,\ln 1.39}\right) - \operatorname{erf}\left(\frac{\ln d_{p_b}/30.0}{\sqrt{2}\,\ln 1.39}\right)\right] \qquad (2.25)$$

TABLE 3 *Results of example 4*

Size Interval $(d_{p_b}-d_{p_a})$ –µm	Midpoint of Interval $-d_{p_{1/2}}$	(A) $p_{1/2}$	(B) f_i	(C) f_i	Exptl f_i	(D) $q_{1/2}$	(E) g_i	(F) g_i	Exptl g_i
0–5	2.5	2×10^{-10}	10^{-9}	0	0	2×10^{-13}	1×10^{-12}	0	0
5–10	7.5	1×10^{-3}	0.005	0.010	0.010	2×10^{-5}	1×10^{-4}	0	0.001
10–15	12.5	0.0249	0.125	0.127	0.128	0.0028	0.014	0.017	0.015
15–20	17.5	0.0569	0.285	0.279	0.282	0.0181	0.091	0.092	0.089
	(19.3) Max	(0.0595)							
20–25	22.5	0.0533	0.267	0.260	0.260	0.0368	0.184	0.181	0.180
	(26.9) Max					(0.0426)			
25–30	27.5	0.0333	0.167	0.168	0.170	0.0425	0.213	0.210	0.215
30–35	32.5	0.0170	0.085	0.086	0.080	0.0362	0.181	0.180	0.179
35–40	37.5	0.0078	0.039	0.040	0.042	0.0257	0.129	0.129	0.129
40–45	42.5	0.0033	0.017	0.017	0.016	0.0163	0.082	0.082	0.104
45–50	47.5	0.0014	0.007	0.007	0.007	0.0096	0.048	0.049	0.028
50–55	52.5	0.0006	0.003	0.003	0.0035	0.0054	0.027	0.028	0.030
55–60	57.5	0.0002	0.001	0.001	0.0006	0.0030	0.015	0.015	0.011
60–100	80	5×10^{-6}	2.5×10^{-5}	0.0009	0.0009	0.00018	0.0072	0.016	0.019
	TOTALS		1.001	0.999	1.000		0.991	0.999	1.000

COMMENTS ON RESULTS OF EXAMPLE 4

1. The numbers listed in columns headed (A), (B), (C), (D), (E), (F) were calculated by the corresponding equations as indicated opposite each calculation.

2. The values in column (A) are plotted on Fig. 8, solid curve, those of column (D) on Fig. 8, dashed curve.

3. Note that the smooth curves for p and q are not based upon the histograms for f_i and g_i, but rather upon the calculations for F and G.

4. The calculated value of f_i in columns (B) and (C) are to be compared with each other and with the experimental values. The agreement is generally quite good.

5. The calculated values of g_i in columns (E) and (F) are to be compared with each other and with the experimental value. This agreement is also quite good.

6. The raw data show an increase in the number and the mass of particles which would be expected in the upper end of the range (above 60 microns) according to the log-probability distribution.

Calculation for maximum points

$$p_{max} = \frac{1.39^{\ln 1.39}}{21.5\sqrt{2\pi}\ \ln 1.39} \exp\left(-\frac{\ln^2 1.39}{2}\right) = 0.0595 \qquad (2.24)$$

$$d_{p_m} = 21.5\ {}_{1.39}{}^{\ln 1.39} = 19.3\ \mu m \qquad (2.23)$$

$$q_{max} = \frac{1.39^{\ \ln 1.39}}{30\sqrt{2\pi}\ \ln 1.39} \exp\left(-\frac{\ln^2 1.39}{2}\right) = 0.0426 \qquad (2.24)$$

$$d_{p_m} = 30\ {}_{1.39}\ \ln 1.39 = 26.9\ \mu m \qquad (2.23)$$

2. *Modified "upper limit" log-normal [11]* This function was devised by Mugele and Evans in order to represent the size distribution in sprays and other aerosols having a similar mechanism of formation. It incorporates a parameter to represent the maximum stable drop-size $d_{p_{max}}$ as the upper-limit to the particle distribution. It assumes then that the ratio $u = d_p/(d_{p_{max}} - d_p)$ should be distributed log-normally, as d_p is in Section 1 above. This results in a three-constant equation: $d_{p_{max}}$, σ_g, and u_{50} (median).

To test data for a fit to this function the special log-probability graph paper is used to determine whether a value of $d_{p_{max}}$ can be found such that a plot of log u vs. weight (or volume)-cumulative-percent-less-than, G, on a probability scale will yield a straight line. The value of $d_{p_{max}}$ may be determined by trial-and-error or it may be estimated, at least for a first trial, from:

$$\frac{d_{p_{max}}}{d_{p_{50}}} = \frac{d_{p_{50}}\ (d_{p_{90}} + d_{p_{10}}) - 2\ d_{p_{90}}\ d_{p_{10}}}{d_{p_{50}}^2 - d_{p_{90}}\ d_{p_{10}}} \qquad (2.34)$$

Here $d_{p_{10}}$, $d_{p_{50}}$, and $d_{p_{90}}$ represent values of particle size at 10%-, 50%- (median), and 90%-finer-than as read from an initial plot of d_p vs. G.

Having found a straight line on this plot, the mass median value of $u = u_{50}$ at G = 50%, is read. Then from the values of u_{50} and u_{84} (also read from the straight line), the value of σ_g is found:

$$\sigma_g = u_{84}/u_{50} \tag{2.35}$$

Finally, the distribution functions are:

$$q = \frac{u_{50}}{\sqrt{2\pi}\ \ln \sigma_g} \exp - \left[\left(\frac{\ln u/u_{50}}{\sqrt{2}\ \ln \sigma_g} \right) \right]^2 \tag{2.36}$$

and

$$G = \frac{1}{\sqrt{2\pi}\ \ln \sigma_g} \int_{-\infty}^{\frac{\ln u/u_{50}}{\sqrt{2}\ \ln \sigma_g}} \exp \left[- \left(\frac{\ln u/u_{50}}{\sqrt{2}\ \ln \sigma_g} \right) \right]^2 d(\ln u) \tag{2.37}$$

or

$$G = \frac{1}{2} \left[1 + \text{erf} \frac{\ln u/u_{50}}{\sqrt{2}\ \ln \sigma_g} \right] \tag{2.37}$$

It would seem that there is no reason why similar functions could not be found in terms of number distribution (p and F) as well, noting, however, that the upper limit of the integrals in (2.18), (2.19), and (2.20) must be $d_{p_{max}}$.

Mugele and Evans tested this "upper-limit distribution" against a variety of data on sprays and found it to give a better fit in each case than the log-probability, the Rosin-Rammler, and the Nukiyama-Tanasawa distributions. The last two are described below. An additional test is given in Example 5.

Example 5. Kim and Marshall [14], in a study of the production of sprays from a pneumatic atomizer, determined a generalized drop-size distribution by weight as given in the first two columns of the table below. Here, the drop-size d_p^* is a normalized value obtained by dividing the actual size in each case by the mass-median diameter. Test these data against the upper-limit function, and establish the maximum drop size.

Since $d_{p_{10}}^*$, $d_{p_{50}}^*$ and $d_{p_{90}}^*$ are all actual entries in the table, the value of $d_{p_{max}}$ may be estimated at once: $d_{p_{10}}^* = 0.33$; $d_{p_{50}}^* = 1.0$; $d_{p_{90}}^* = 2.0$.

$$\frac{d_{p_{max}}^*}{1.0} = \frac{1.0(2.0 + 0.33) - 2(2.0 \times 0.33)}{(1.0)^2 - (2.0 \times 0.33)} = 2.97 \qquad (2.34)$$

Using this value, the entries for u in the third column of the table are calculated. These points are then plotted against G on the log-probability graph Fig. 9. A good straight line is obtained, confirming the value of $d_{p_{max}}^*$ and a fit to the upper-limit equation.

From the straight line $u_{50} = 0.52$, $u_{84} = 1.53$ and $\sigma_g = 1.53/0.52 = 2.94$. Then q (if desired) and G may be calculated from

$$q = \frac{0.52}{\sqrt{2\pi}\ \ln 2.94} \exp - \left[\left(\frac{\ln u/0.52}{\sqrt{2}\ \ln 2.94}\right)\right]^2 \qquad (2.36)$$

$$G = \frac{1}{2}\left[1 + \text{erf}\ \frac{\ln u/0.52}{\sqrt{2}\ \ln 2.94}\right] \qquad (2.37)$$

The calculated values of G are listed in fourth column and are seen to reproduce the original values very well.

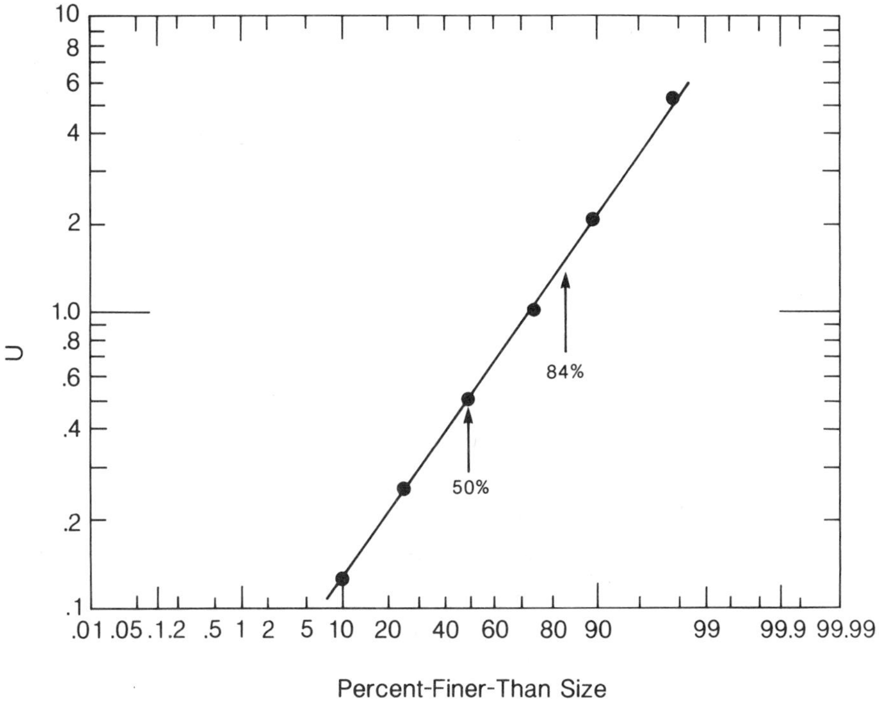

FIG. 9. *Test of upper-limit function. Example 5: data on sprays [14]; $u = d_p^*/(d_{p_{max}}^* - d_p^*)$, $d_{p_{max}}^* = 29.7$ microns; $u_{50} = 0.508$, $\sigma_g = 2.94$. Reproduced by permission from American Institute of Chemical Engineers Journal. Ref. [15]*

G Exp't'l	d_p^*	$u = d_p^*/(2.97 - d_p^*)$	G Calculated
0.10	0.33	0.125	0.093
0.25	0.60	0.253	0.252
0.50	1.0	0.508	0.491
0.75	1.5	1.02	0.734
0.90	2.0	2.06	0.899
0.98	2.5	5.32	0.984

3. *Weibull distribution* [16,17] A distinctly different and wholly empirical equation, so far as particle size distribution is concerned, is the Weibull function. Originally derived and widely used for other purposes [16], its applicability has been demonstrated by Steiger [17] for describing size distribution in various kinds of dusts.

It is a three-constant equation basically stated in the cumulative form:

$$F(d_p) = 1 - \exp - \frac{(d_p - \gamma)^\beta}{\alpha} \tag{2.38}$$

The utility of this form lies in the constant γ, which may be taken to represent the size of the smallest particle present, in those cases when it is imperative or simply more accurate to describe a distribution without letting $d_p \to 0$. Furthermore, α may be expressed as a "special" particle size $\bar{d}_p$, raised to the β power. Rewriting (2.38),

$$F(d_p) = 1 - \exp \left[\frac{d_p - d_{p_{min}}}{\bar{d}_p} \right]^\beta = 1 - \exp - X^\beta \tag{2.39}$$

it is seen how the ratio $X = (d_p - d_{p_{min}})/\bar{d}_p$ plays a role somewhat analogous to that of u in the "upper-limit" form, as in Section 2 above. Notice, however, that $\bar{d}_p$ may not be a particle size actually present, for whenever $\alpha^{1/\beta} < \gamma$, $\bar{d}_p$ will be smaller than $d_{p_{min}}$. The third constant β is a measure of the degree of dispersion of the sizes present.

The equation may be discussed and its properties demonstrated either in the form of (2.38) or the generalized form of (2.39). The population density is derived using (2.4), as

$$p(d_p) = \frac{\beta}{\alpha} (d_p - \gamma)^{\beta-1} \exp - \frac{(d_p - \gamma)^\beta}{\alpha} \tag{2.40}$$

or

$$p(x) = \frac{\beta X^{\beta-1}}{\bar{d}_p} \exp - X^{\beta} \qquad (2.41)$$

Examples of the graphs which the Weibull equations may describe are best shown in generalized form, as in Fig. 10 (cumulative) and Fig. 11 (density). It is seen that provided $\beta > 1$ a variety of S-shaped curves may be represented by the form of (2.38), and that the forms given by (2.40) are not unlike the skewed type of probability curves represented by the log-normal function (2.21). Steiger [17] states that when β = 3.25, most of the Weibull curve is identical with the normal (not log-normal) distribution function. Usually β will lie between 1.0 and 3.0.

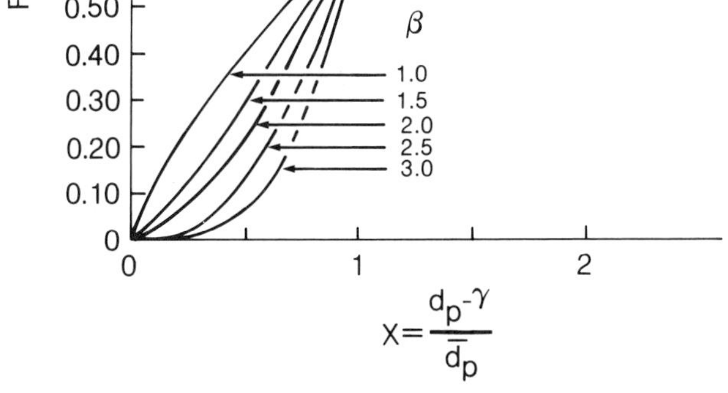

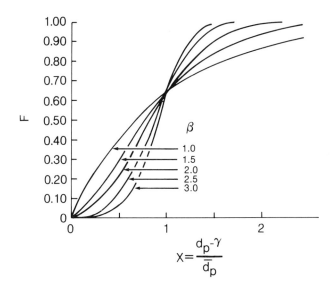

FIG. 10. *Examples of Weibull cumulative plots in generalized form. See equation (2.39).*

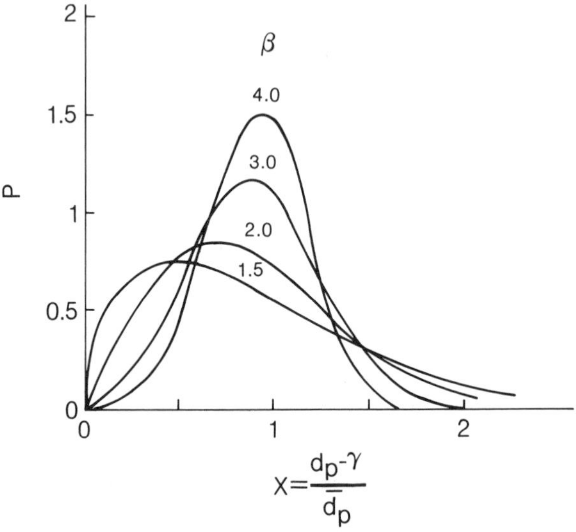

FIG. 11 *Examples of Weibull density plots in generalized form. See equation (2.41).*

The values of α, β, and γ may be found from a set of data by a graphical technique utilizing a linearized plot. For this purpose, plot log ln 1/(1-F) vs. log $(d_p - \gamma)$, for from (2.38)

$$\log \ln 1/(1 - F) = \beta \log (d_p - \gamma) - \log\alpha \qquad (2.42)$$

That is, plot ln 1/(1-F) vs. $(d_p - \gamma)$ on log-log coordinate scales. There is also a special graph paper available for this purpose [18]. If the value of γ cannot be seen at once from the data, i.e. $\gamma = d_{p_{min}}$ for which f = F = 0, then several values of γ will have to be tried until one is found that does give a straight line on such a plot. If none can be found, the Weibull function does not fit the data.

4. *Rosin-Rammler distribution* [*19*] Of lesser importance for aerosol work is this empirical relationship, originally proposed

to represent size distribution of relatively coarse dusts produced by grinding of solids. It is basically of the two-constant form.

$$G = 1 - \exp - ad_p^s \tag{2.43}$$

which is seen to be a special case of the Weibull (2.38) in which $\alpha = 1/a$, $\beta = s$ and $\gamma = 0$. A mean size may be defined as $\bar{d}_p = 1/a^{1/s}$, and the function written in the alternate form

$$G = 1 - \exp - (d_p/\bar{d}_p)^s \tag{2.44}$$

Data may be tested for a fit to this form in a manner similar to that used with the Weibull. Thus (2.42) becomes:

$$\log \ln 1/(1-G) = s \log d_p + \log a \tag{2.45}$$

The same kind of special paper may be used.

5. *Roller distribution* [20] Another two-constant empirical formula, said to be applicable over a wide range of particle sizes of powdered industrial materials (but probably of limited value in describing aerosols) is given as:

$$G = ad_p^{1/2} \exp(- s/d_p) \tag{2.46}$$

A severe limitiation of the use of this function is that it does not fulfill one of the basic conditions, namely that as $d_p \to \infty$, $G \to 1$. Instead, as $d_p \to \infty$, $G \to \infty$, and $G\sqrt{d_p} \to a$. Where G = 1, d_p has a value $d_{p_{100}}$ in accord with

$$\frac{1}{a\sqrt{d_{p_{100}}}} = \exp - \left(\frac{s}{d_{p_{100}}}\right) \tag{2.47}$$

The weight-density function is

$$q = a\left[\sqrt{d_p}\ \exp\left(-\frac{s}{d_p}\right)\right]\left[\frac{1}{2d_p} + \frac{s}{d_p^2}\right] = G\left[\frac{1}{2d_p} + \frac{s}{d_p^2}\right] \quad (2.48)$$

which likewise does not approach zero as $G \to 1$. Great care must, therefore, be taken in using this function, not to apply it in the region of sizes near that for which $G = 1$. For this reason, it does not seem to warrant more detailed study.

Data may be tested for accord with it by plotting $\log (G/\sqrt{d_p})$ vs. $1/d_p$, for

$$\log G/\sqrt{d_p} = \log a - 0.4343\ s/d_p \quad (2.49)$$

If a straight line is obtained its slope will be equal to $-0.4343s$, and its intercept is $\log a$.

6. *Nukiyama-Tanasawa distribution* [21] This expression was offered in the population-density (p) form to express the size distribution in mechanically dispersed mists, formed by atomization:

$$p = a\ d_p^2 \exp\ (-\ bd_p)^s \quad (2.50)$$

This appears to have three constants a, b, s, but they are not independent. For, in order to fulfill the requirement of (2.4), an interrelationship

$$\frac{a\ b^{-3/s}\Gamma(3/s)}{s} = 1 \quad (2.51)$$

must be fulfilled, where Γ indicates the gamma function or generalized factorial. Therefore, only two of the constants are independent.

Through application of (2.19), the corresponding weight-density function is found to be

$$q = \frac{sb^{6/s}}{\Gamma(b/s)} d_p^s \exp - bd_p^s \qquad (2.52)$$

The cumulative forms for F and G may be found by application of (2.20).

If data are available for p, a test of fit to (2.50) may be made by plotting $\ln p/d_p^2$ vs. d_p^s to see if a straight line may be determined. For from (2.50) and (2.51) combined

$$\ln p/d_p^2 = \ln sb^{3/s}/\Gamma(3/s) - b\, d_p^s \qquad (2.53)$$

Values of s must be tried e.g. as 1/2, 1/3, 1/4 to see whether one can be found to produce a straight line. If so, then -b is the slope of that line. Then the values of b and s so determined may be checked to see whether $\ln sb^{3/s}/\Gamma(3/s)$ agrees with the measured intercept of the line. An analogous procedure could be carried out with data for q using the linearized form of (2.52).

The value of this function has been shown by Mugele and Evans [11] to be somewhat limited. They found that the upper-limit form of the log-normal function fitted data for atomized sprays better. Furthermore, the information on pneumatic atomizers has been greatly extended and improved by Kim and Marshall [14] whose generalized drop size distribution function has been cited in Example 5. This may be considered to supercede the Nukiyama-Tanasawa function.

7. *Multi-modal distributions* If the plot of F (or G) vs. d_p on log-probability paper is not straight, the reason may be that the distribution has more than one mode. Each mode is identified by a peak value of p (or q) and corresponds to a point of inflection on the F (or G) plot. It may be that there are two or more "parent"

distributions, each of which is log-normally distributed, mixed together in such proportions as to produce a multi-modal composite. Examples of this are given by Allen [1], one of which is shown in Fig. 12.

The data for a mixture of this sort may be tested to determine whether log-normal parent distributions may exist, by using the relationships of Equations (2.32) and (2.33). The experimental distribution data for the mixture are replotted as $\Delta G/\Delta \ln d_p$ vs $\ln d_p$, i.e. as y vs x of (2.32)., as shown in Fig. 13. If two peaks are found, around each of which a symmetrical bell-curve may be discerned , each parent distribution will

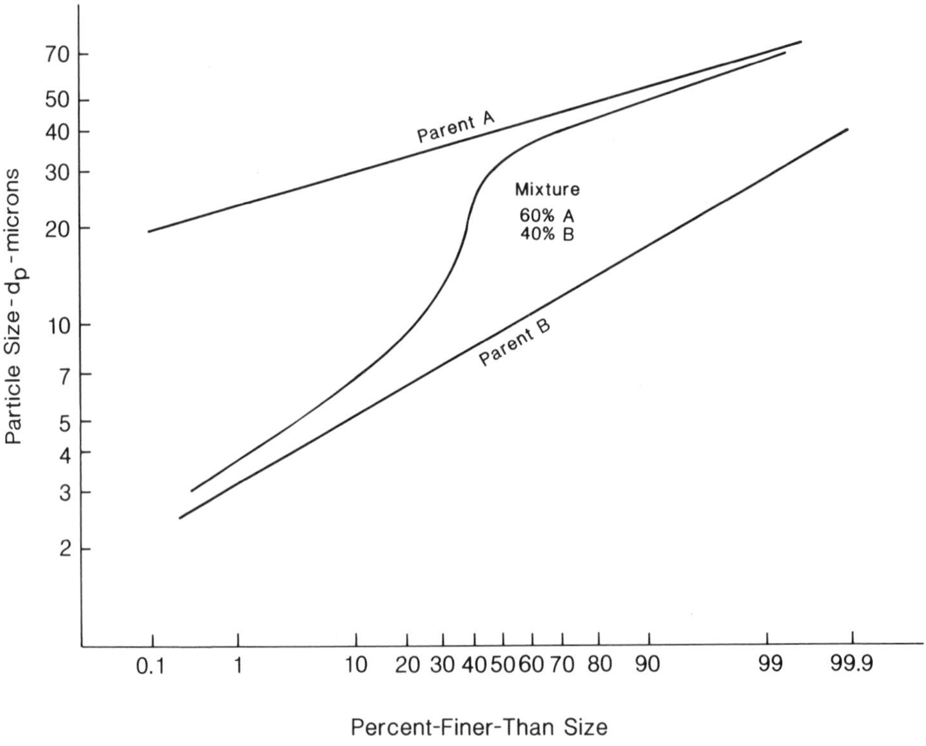

FIG. 12. *A bimodal distribution which is a mixture of log-probability parent distributions.*

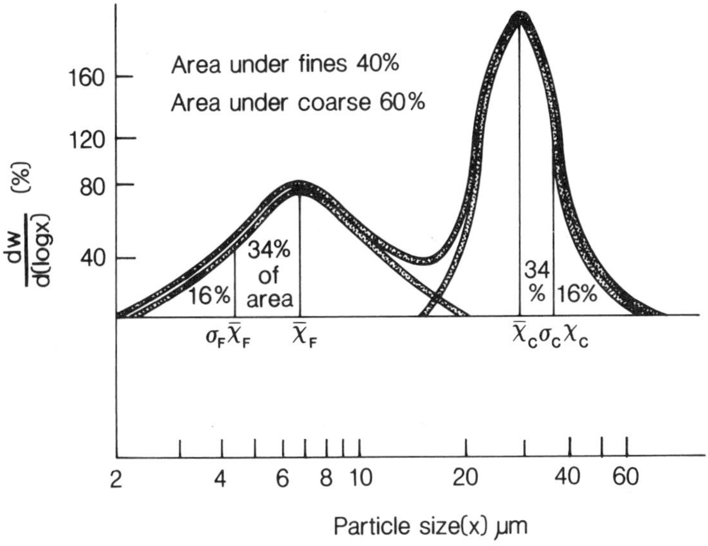

FIG. 13. *Bimodal distribution of Fig. 12 replotted.*

correspond to one of these. The values of d_p at which the peaks are found will be the values of $\bar{d}_p$ = MMD for each parent. From the two values of y_{max} observed, the respective values of σ_g may be calculated by using Eqn. 2.33. Thus the straight lines representing each parent distribution may be constructed as shown in Fig. 12. Finally the relative proportion of each parent in the mixture may be estimated from the relative areas under each of the bell-curves. As a check, the curve on Fig. 12 for the mixture may be reconstructed by calculation, using the information just obtained about the parents. This curve may then be compared with that for the original data.

One of the most important examples of a multi-modal distribution is that of the particles in the usual ambient urban atmosphere. In terms of mass measurements (g) two peaks and often three are observed. These have been distinguished according to characteristic sources of origins of particles in each mode. The table below from Willeke and Whitby [22] summarizes these facts.

Names	Transient nuclei range	Accumulation range	Mechanical aerosol range
Size 0.001	0.01	0.1 1	10 100
microns	1 2	3	4 5
Sources	Combustion Heterogeneous nucleation	Coagulation from transient nuclei Condensation Combustion	Windblown dust Large particle emissions Sea spray
Lifetime	< 1 hour	Days	Hours Minutes Days Hours

C. *Note of Caution*

In performing the calculations explained above, one must not lose sight of the accuracy and precision of the raw data upon which they are based. Particle size measurements, counting and weighing, are difficult techniques at best. It requires painstaking care, and close attention to the principles of the method of measurement, in order to obtain reliable, meaningful, and reproducible values.

Elaborate calculations of density functions and mean particle sizes are not worthwhile unless the raw data are reliable and well-determined. Consequently, the manipulator of such data must have a good understanding of the problems and techniques of the measurer who obtained the data. In deciding what calculations to undertake, one must bear in mind the purpose for which the data are to be used. In fact this should be the primary consideration when the method of measurement is selected at the outset.

For example, when a cumulative distribution by mass is considered, little attention need be given to the particle sizes corresponding to either the upper two- or three-percent or the lower two- or three-percent of the aerosol as those will represent a relatively small proportion of the total mass. A size-distribution function which fits quite well through the mid-range of sizes may be quite acceptable in describing the aerosol, say for purposes of calculation of a dust collector performance on the mass basis. On the other hand, the principle concern may be with the number of particles in the finest size-range because of the health hazard they represent. This might call for emphasis upon a different kind of distribution function in order to represent this range most accurately.

REFERENCES

1. T. Allen, Particle Size Measurement, Chapman and Hall, London, 3rd Edition (1981).
2. J.K. Beddow, Particle Characterization in Technology: Vol. I, Applications and Microanalysis, CRC Press Inc., Boca Raton, (1984).
3. J.D. Stockham and E.G. Fochtman, Particle Size Analysis, Ann Arbor Science, Ann Arbor, (1977).
4. R.R. Irani and C.F. Callis, Particle Size: Measurement, Interpretation and Application, Wiley, New York, (1963).
5. W.C. McCrone and J.G. Deely, The Particle Atlas, Ann Arbor Science, Ann Arbor, 3rd Edition (1977).
6. J.K. Beddow, Particle Characterization in Technology: Vol II, Morphological Analysis, CRC Press Inc., Boca Raton (1984).
7. H. Waddell, Physics, 5: 281 (1984).
8. D. Kunii and O. Levenspiel, Fluidization Engineering, Wiley, New York, p.65 (1969).
9. L.E. Sparks, Handbook of Air Pollution Technology (S. Calvert, and H.M. Englund, eds) Wiley Interscience, New York, Chap. 31 (1984).
10. K.F. Whitbey and B.H.Y. Liu, Encyclopedia of Chemical Technology (H.F. Mark, J.J. McKetta, Jr., and D.F. Othmer, eds), Wiley Interscience, New York, 2nd Edition, Vol. 7, p.430 (1965).
11. R.A. Mugele and H.D. Evans, Ind. Eng. Chem. 43: 1317 (1951).
12. L. Silverman, C.E. Billings, and M.W. First, Particle Size Analysis in Industrial Hygiene, Academic Press, New York (1971).
13. J.M. Smith, Scientific Analysis on the Pocket Calculator, Wiley Interscience, New York, p. 125 (1975).
14. K.Y. Kim and W.R. Marshall, AIChE Journal, 17: 581 (1971).

15. W. Licht, AIChE Journal, 20: 595 (1974).
16. W. Weibull, J. Appl. Mech., 18: 293 (1951).
17. F.H. Steiger, Chem. Tech., 1: 225 (1971).
18. C.C. Harris, Powder Tech., 5: 40 (1971/72).
19. P. Rosin and E. Rammler, Koll. Zeit., 67: 16 (1934).
20. N. Roller, J. Franklin Inst., 223: 609 (1937).
21. S. Nukiyama and Y. Tanasawa, Trans. Soc. Mech. Engrs. (Japan); 5: 63 (1939).
22. K. Willeke and K.T. Whitby, Journ. Air Poll. Control Assoc. 25: 529 (1975).

PROBLEMS

1. The range of droplet sizes in a cloud was determined to be as follows:

Range of drop diameter -microns	Number of drops
5–8	4
8–11	6
11–14	15
14–17	24
17–20	24
20–23	12
23–26	4
26–29	4
29–32	4
32–38	3

a. Determine the number median diameter.
b. Determine the mass median diameter.
c. Determine the Sauter mean diameter.
d. What weight fraction of the sample is represented by drops greater than 20 μm in diameter?
e. What is the population density of the 20–23 μm grade?

f. Can the distribution be reasonably well described as log-normal? (Hint: try plotting $\Delta n/\Delta \ln d_p$ vs $\ln d_p$; also try the upper-limit function). If so, find the two constants for the distribution.

2. Aerosol A has an MMD of 10 microns and $\sigma_g = 3.0$, while aerosol B has an MMD of 20 microns and $\sigma_g = 1.5$, both aerosols being log-normally distributed. Which aerosol has the larger percent by weight of particles finer than 40 microns? Which has the larger De Brouckere diameter?

3. An aerosol sample has been determined to follow the log-normal particle size distribution with a standard geometric dispersion of 1.83. If the Sauter mean diameter of the sample is 50 microns, what fraction of the weight of the sample will be made up of particles between 50 microns and 60 microns in size?

4. Prove that the Sauter mean diameter, as defined in Eqn. (2.15) of the text, may be computed by $1/\sum g_i/d_{p_i}$ i.e. prove that the first form and the last form of Eqn. (2.15) are equal to each other. Do any assumptions need to be made?

5. The spray from a certain nozzle gave a drop-size distribution which was log-normal, with an NMD of 240 microns and a standard geometric dispersion of 2.00. For this spray:

a. What fraction of the total surface would be on drops between 100 and 200 microns in diameter?

b. What is the value of the "surface to diameter" mean $\bar{D}_{1,1}$?

c. What is the value of the maximum population density?

d. At what size does this value occur?

6. A sample of glass beads (density = 2.60 gm/cm^3) has the size distribution given by the following analysis:

Cumulative percent by weight finer-than	Size - microns
95	41.5
91.5	38.5
85	34.5
80	31.5
60	26
50	24
40	22
29	20
20	18.3
3.5	12.5

a. Find the MMD and the NMD.

b. What is the value of the mode size?

c. What is the value of the q-function (Eqn. 2.9) at 30 microns?

d. Find a size distribution function to represent this sample. Try log-normal, upper-limit log-normal, and Weibull.

7. The following results were obtained from a Bahco size-analysis of a sample of fly-ash (sp. gr. = 2.63).

Weight percent finer-than	Size-microns
99.54	250
96.72	149
78.66	61
66.55	25.4
66.44	23.7
66.10	19.7
62.56	11.1
54.31	7.7
33.94	4.4
8.94	2.1
2.84	1.3

a. In 750 kg of fly ash, how much will lie between 10 microns and 50 microns?

b. What is the mass median size?

c. What is the Sauter mean size?

d. Does this distribution fit any regular pattern? (i.e. one of the basic equations?)

3
General Concepts of Particulate Collection

I. MECHANISMS FOR COLLECTION

In order for suspended aerosol particles to be removed from a gas (i.e. collected) the gas must be passed through a zone in which the particles come under the influence of some kind of force (or forces) which causes them to be diverted from the flow direction of the stream. They must remain under the influence of the collecting force(s) a sufficient length of time to be diverted to contact some collecting surface where they are removed from the stream. In some cases, the initial path of the particle is such that it collides directly with the collector surface unaided by any force. This is referred to as direct interception.

The hopeful assumption is made that once a particle collides with the collecting surface, it adheres to it and is not subsequently reentrained into the gas stream. Provision must be made to maximize the validity of this assumption, as well as to remove the collected particles from the collecting surface in order that it may continue to function properly.

The basic situation may be illustrated in one of the sketches of Fig. 1. The nature of the collecting force may be illustrated

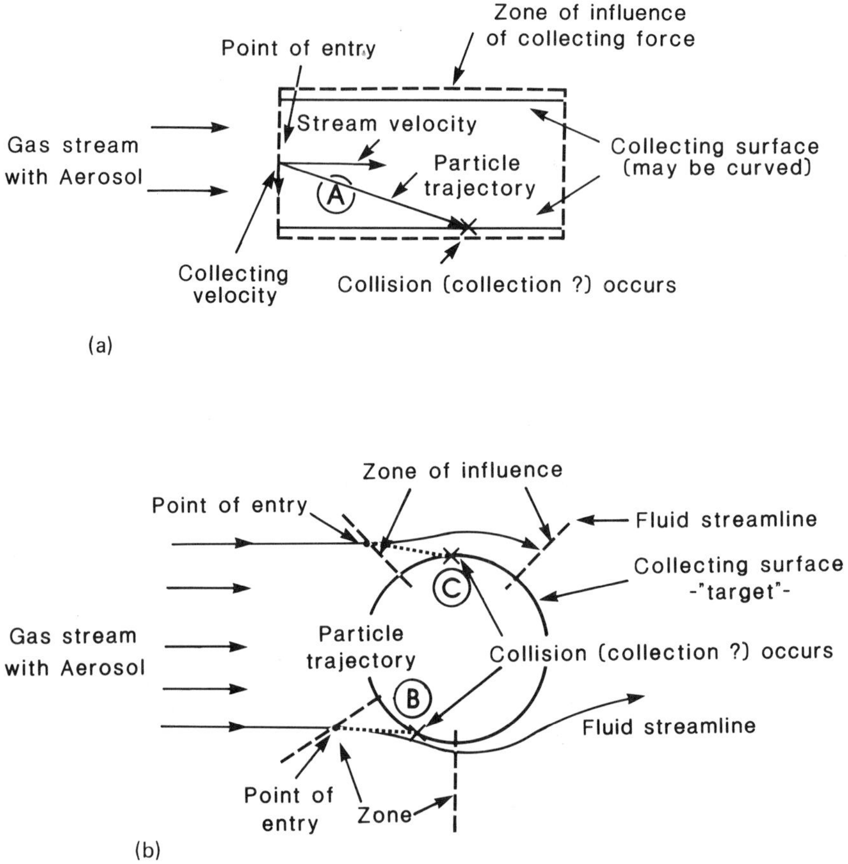

FIG. 1 *Operation of collection forces within a zone of influence. (a) Boundary surface of a containment zone, (b) outside surface of a target zone.*

by the designated locations: gravitational, at A or C; centrifugal, at A; inertial, at B; direct interception, at D; diffusional, at C (Brownian, eddy); electrostatic, at A or C; thermal, at A or C; and possibly others such as sonic, diffusiophoretic, photophoretic, etc. The agglomeration of small particles into large ones by the action of some of these forces (diffusional, electrostatic, sonic, etc.) may also aid the collection process. Several modes of collection may occur simultaneously. In fact collectors deliberately involving combinations of forces have been developed in recent years in order to enhance the collection of the finer particles.

There are two general categories of collecting surfaces as shown in Fig. 1. One is the boundary surface of a containment zone through which the air is made to flow. This type of surface may be a plane wall or the inside of a hollow cylinder, as in A. The other is the outside of a target such as a cylinder (fiber) or sphere (drop of liquid) as in B and C, or some irregular shape such as that presented by the particles in a granular packing, or the fibers of a media filter. A target type of collector consists of an array or assembly of many target elements, as in a filter made up of a mat of fibers, or in a scrubber consisting of a stream of suspended droplets. Direct interception upon previously deposited particles may be principal mode of collection by fabric filters.

The several types of collectors in common use may be classified according to the nature of the collecting force and collecting surface as shown in the following table:

Collector	Primary Collecting Forces	Collecting Surface
	Surface Collectors	
Settling chamber	Gravitational	Plane
Momentum	Gravitational Inertial	Plane or cylindrical
Cyclone	Centrifugal	Cylindrical
Electrostatic Precipitator	Electrostatic	Plane or cylindrical
Thermal collector[a]	Thermophoresis	Plane
Impinger[a]	Inertial	Plane
	Target Collectors	

Collector	Primary Collecting Forces	Collecting Surface
	Surface Collectors	
Filter (media)	Inertial Diffusional Direct interception Electrostatic	Cylindrical fiber or granular
Filter (fabric)	Direct interception	Layer of particles irregular
	Target Collectors	
Scrubber	Inertial Diffusional Direct interception	Spherical or irregular

[a]Usually used only for collection of small samples of aerosol.

Combinations of collecting forces e.g. charged droplet scrubber, or charged media filters, are also employed in advanced devices.

In order to analyze or predict the performance of a collector, it is necessary to know how a given type of force acts upon a particle. The magnitude and direction of the force on a particle of given density and size must be predictable, so that a particle trajectory can be determined. This trajectory is a resultant of the collecting velocity set up by the force or forces, and the gas stream velocity with which the particle enters the zone of influence. The motion of the particle in the direction of collection is always opposed by the frictional drag exerted by the gas on the particle. This must also be calculated and reckoned into the determination of the particle trajectory.

The interaction of these forces and velocities determine the differential equation which represents the net motion of the particle. The extent of our knowledge and ability to set up and solve such differential equations is treated in Chapters 4 and 5. The application of these results to the mathematical modelling of the collection performance of various kinds of collectors is the subject matter of Chapters 6, 7, 8, and 9.

II. EXPRESSING THE EFFECTIVENESS OF COLLECTION

The success of a particulate collection operation may be expressed either in terms of the amount of aerosol removed from the air stream, or in terms of the amount permitted to remain in it. Several different kinds of measures of collection effectiveness may be defined, with reference to the sketch.

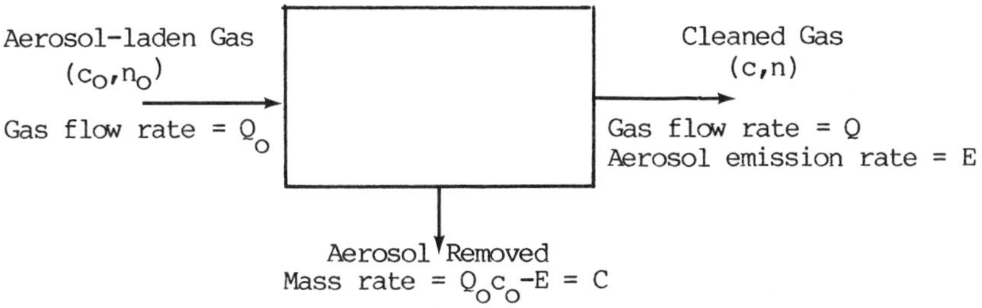

Let Q = volume rate of flow of gas stream — e.g., m^3/min; cfm

c = mass of aerosol per unit volume of gas, called "grain-loading" — e.g., gm/m^3;gr/ft^3 [a]

n = number of aerosol particles per unit volume of gas — e.g., m^{-3}; ft^{-3}

E = mass emission rate of uncollected aerosol — e.g., gm/min; lbm/hr

E = Qc

C = mass rate of aerosol collection — e.g., gm/min; lbm/hr

Subscript o refers to entering conditions.

A. *OVERALL PERFORMANCE*

The following "overall collection efficiencies" have been defined and are in use for different purposes. These definitions, as written, assume that Q_o = Q.

---

[a] A grain (gr.) is defined as 1/7000 of a pound mass.

Collector Efficiency, by Mass of Particles:

$$\eta_M = \frac{c_o - c}{c_o} = \frac{Q_o c_o - E}{Q_o c_o} = \frac{C}{Q_o c_o} = 1 - \frac{E}{Q_o c_o} = \frac{Q_c}{Q_o c_o} \tag{3.1}$$

Collector Efficiency, by Surface Area of Particles:

$$\eta_A = \frac{\text{Surface area of particles collected}}{\text{Total surface area of particles entering}} \tag{3.2}$$

Collector Efficiency, by Number of Particles:

$$\eta_N = \frac{n_o - n}{n_o} = \frac{Q_o n_o - Qn}{Q_o n_o} \tag{3.3}$$

These three efficiencies depend upon d_p^3, d_p^2 and d_p respectively, where d_p represents particle size (see Chapter 2.I.B). They are, therefore, quite different numerically when applied to the same collection. The experimental determination of any of these quantities depends greatly upon the techniques available for measuring either number of particles, surface area, or mass.

The overall collector efficiency by mass of particles is usually the easiest to measure experimentally. The inlet and outlet gas streams in steady flow may be sampled by a collecting device, such as an absolute filter, which collects virtually all of the dust. By weighing the dust collected from these samples, and by measuring the volume of gas which simultaneously passed through the samples, the values of c_o and c may be calculated. Alternatively, only the inlet (or outlet) stream need be sampled, and the dust collection rate C measured. In these calculations, if the temperature and/or pressure of the outlet stream differs much from that of the inlet, correction will have to be made for the fact that $Q_o \neq Q$.

Some on-stream monitoring devices actually function by counting particles, or by measuring their surface through an indirect effect such as light-scattering. In a case of this kind, either η_N or η_A is obtained by the test procedure. If a sample is taken for examination under the microscope, η_A will usually be based upon the projected area of particles.

Designing a collection system to minimize the overall dust emission rate, E, will usually be an important criterion of performance. This will be true especially whenever a system must meet legally specified emission limitations such as those discussed in Chapter 1. The following terms are defined to focus attention upon emission rather than collection.

Penetration, based upon mass emission:

$$P = E/Q_o c_o = c/c_o = 1 - \eta_M \tag{3.4}$$

(alternatively, may sometimes be based upon number of particles n/n_o)

Decontamination Factor, reciprocal of penetration:

$$\text{D.F.} = c_o/c = 1/P \tag{3.5}$$

(alternatively, may be taken as n_o/n)

$$\text{Decontamination Index} = \log_{10}(\text{D.F.}) = \log_{10} 1/P \tag{3.6}$$

$$\text{Number of transfer Units NTU} = \ln 1/P = 2.30 \text{ D.I.} \tag{3.7}$$

Equations (3.4), (3.5) or (3.6) are especially useful in expressing collector performance whenever there is a very small amount of uncollected dust. For example, it is more graphic to compare a performance of 99.0% mass collection efficiency with one of 99.9%, or 99.99% in terms of one of these items:

Overall mass efficiency	($100\eta_M$)	:	99.00%	99.90%	99.99%
Penetration	(100 P)	:	1.0%	0.1%	0.01%
Decontamination Factor	D.F.	:	100	1000	10,000
Decontamination Index	D.I.	:	2	3	4
Number of Transfer Units	NTU	:	4.605	6.908	9.210

An improvement in performance by one unit in the Decontamination Index, represents reducing the dust emission rate to one-tenth of its previous value from the same incoming stream. The concept of a Transfer Unit has been borrowed from the N.T.U. used by chemical engineers to express the performance of a gas absorption tower or a distillation column. It is useful because several of the mathematical models of collector performance are exponential in nature, as will be shown later.

Example 1

An air stream containing 5 grains of dust per standard cubic foot is passed through a particle collector at the rate of 10,000 std. cu. ft. per min. (commonly abbreviated scfm). Samples of the aerosol are taken before and after the collector, and the number of particles of various sizes in an equal volume of air are counted (say by a technique such as the Coulter Counter). The results of the counts are tabulated below. They are presented in an over-simplified form in comparison with actual test results, in order to reduce complexities in calculation and in order to focus attention upon the basic principles.

Particle Size -Microns	Number/Volume n_o - Input	Number/Volume n - Output
100	100	1
50	900	27
10	1100	44
6	1500	255
2	1000	280
1	500	250

The particles may be regarded as spherical with the size given being the diameter, and as all having a density of 2.5 gm/cm^3.

(a) What is the overall efficiency of this collector as expressed by mass, by surface area, and by number of particles? (b) What is the penetration? The outlet dust loading? (c) What is the emission rate in lbm/hr? (d) What is the collection rate?

Solution: it is assumed that the surface of each particle is proportional to the square of its diameter and the volume of each particle proportional to the cube of the diameter. Only relative values of surface and volume are required. It is assumed that the values of α_s and α_v (as defined in Chapter 2) are respectively the same for all particle sizes, hence cancel out. A table may be constructed as shown on page 80 in which the relative surface proportional to nd_p^2, and the relative volume proportional to nd_p^3, of all particles of each size are tabulated. Calculations are then carried out as shown on page 81, with references to the numbered equations. Note that particles of different sizes are collected with different degrees of effectiveness.

B. *GRADE OR FRACTIONAL EFFICIENCY*

It might be imagined that, for a given kind of collection mechanism operating upon particles of a given kind, there would be a critical particle size such that all particles larger than this size would be collected completely, and all smaller particles not collected at all. But this is not true. Every type of collecting force operates in a manner which depends upon and varies with particle size, as well as shape, and density. Consequently, different particle sizes are collected with different degrees of effectiveness, as indicated in the above example. The overall collection is a composite or summation of these actions.

The relationship between collection efficiency and particle size, for particles of a given kind being treated in a given collector, is called the "grade-efficiency" or "fractional efficiency." It may be expressed either in tabular, graphical, or explicit functional form, $\eta_i = f(d_{p_i})$. Here d_{p_i} represents the size of an individual particle of a certain "kind" as indicated by the subscript i. Kind refers to particles having a certain density, shape and composition. The grade-efficiency of i is the efficiency with which particles of a certain kind are collected, as a function of their size.

Example 2.1

Particle Size d_p -microns	No./Vol. Input No.	No./Vol. Output No.	d_p^2	d_p^3	$n_o d_p^2$	nd_p^2	$n_o d_p^3$	$n \times d_p^3$
100	100	1	1×10^4	1×10^6	1×10^6	1×10^4	1×10^8	1×10^6
50	900	27	2.5×10^3	1.25×10^5	2.25×10^6	6.75×10^4	1.125×10^8	3.375×10^6
10	1100	44	1×10^2	1×10^3	1.10×10^5	4.4×10^3	1.10×10^6	4.4×10^4
6	1500	255	36	216	5.40×10^4	9180	3.24×10^5	55080
2	1000	280	4	8	4×10^3	1120	8×10^3	2240
1	500	250	1	1	500	250	500	250
Totals	5100	857	------	-----	3.419×10^6	9.245×10^4	2.139×10^8	4.477×10^6

(a) Mass Efficiency

$$= \eta_M = \frac{\Sigma n_o d_p^3 - \Sigma n d_p^3}{\Sigma n_o d_p^3} \times 100 \qquad (3.1)$$

$$= \frac{2.139 \times 10^8 - 4.477 \times 10^6}{2.139 \times 10^8} = 97.9\%$$

Surface Efficiency

$$= \eta_A = \frac{\Sigma n_o d_p^2 - \Sigma n d_p^2}{\Sigma n_o d_p^2} \times 100 \qquad (3.2)$$

$$= \frac{3.419\text{x}10^6 - 9.245\text{x}10^4}{3.419 \times 10^6} \times 100 = 97.3\%$$

Number Efficiency

$$= \eta_N = \frac{\Sigma n_o - \Sigma n}{\Sigma n_o} \times 100$$

(3.3)

$$= \frac{5100 - 857}{5100} \times 100 = 83.2\%$$

(b) Penetration $= P = 1 - \eta_M = (1 - 0.979) \times 100 = 2.1\%$

(3.4)

Outlet dust grain loading $= c = P_{c_o} = 0.021 \times 5$

$= 0.105$ grains/std. cu. ft.

(c) Emisson Rate $= E = PQ_o c_o = 0.021 \times 10{,}000 \times \dfrac{5 \times 60}{7000}$

$= 9.0$ lbm/hr

(d) Collection Rate $= C = Q_o c_o - E = \dfrac{10{,}000 \times 5 \times 60}{7000} - 9.0$

$= 419.6$ lbm/hr

The grade-efficiency may be expressed in terms of any of the same measures as are used for overall efficiency. However, for particles of fixed size and all of the same "kind," the grade-efficiency by number of particles, by surface area of particles and by mass or volume of particles are all identical numbers

For most collectors, operating upon a dust composed of particles all of the same kind, there is a unique single-valued function f for the particular set of operating conditions. This function will, however, depend upon such parameters as the nature and design dimensions of the collector, rate of flow and loading of gas stream, temperature, nature of collecting forces, etc. In the case of fabric filters ("baghouses") these statements must be modified, as will be shown Chapter 8.

If the dust is a mixture of particles of more than one kind, there will be a different function of particle size for each kind. These functions will differ because of the effect of density and shape of particles upon the collection process. In what follows, it will be assumed usually that we are dealing with a single kind of particle, although whenever possible the effect of particle density will be explicitly stated.

A study of experimentally determined grade-efficiency curves for various kinds of collecting devices reveals that each kind of device tends to have a typical performance curve. The curves of Figs. 2, 3, and 6 are a selection of graphical representations of typical performance associated with certain standard types of equipment. Qualitatively they show, for example, that cyclones are generally less efficient than other kinds of equipment, that Venturi scrubbers are more effective on finer particles than other scrubbers, that there may be a certain particle size for which collection is at a minimum, and other features. Each such graph is valid only for a fixed set of operating conditions for the collector is represents, all of which must be stated in order for the graph to have quantitative meaning.

Figs. 3 and 6 in particular show the results of some experimental measurements in which special attention was directed

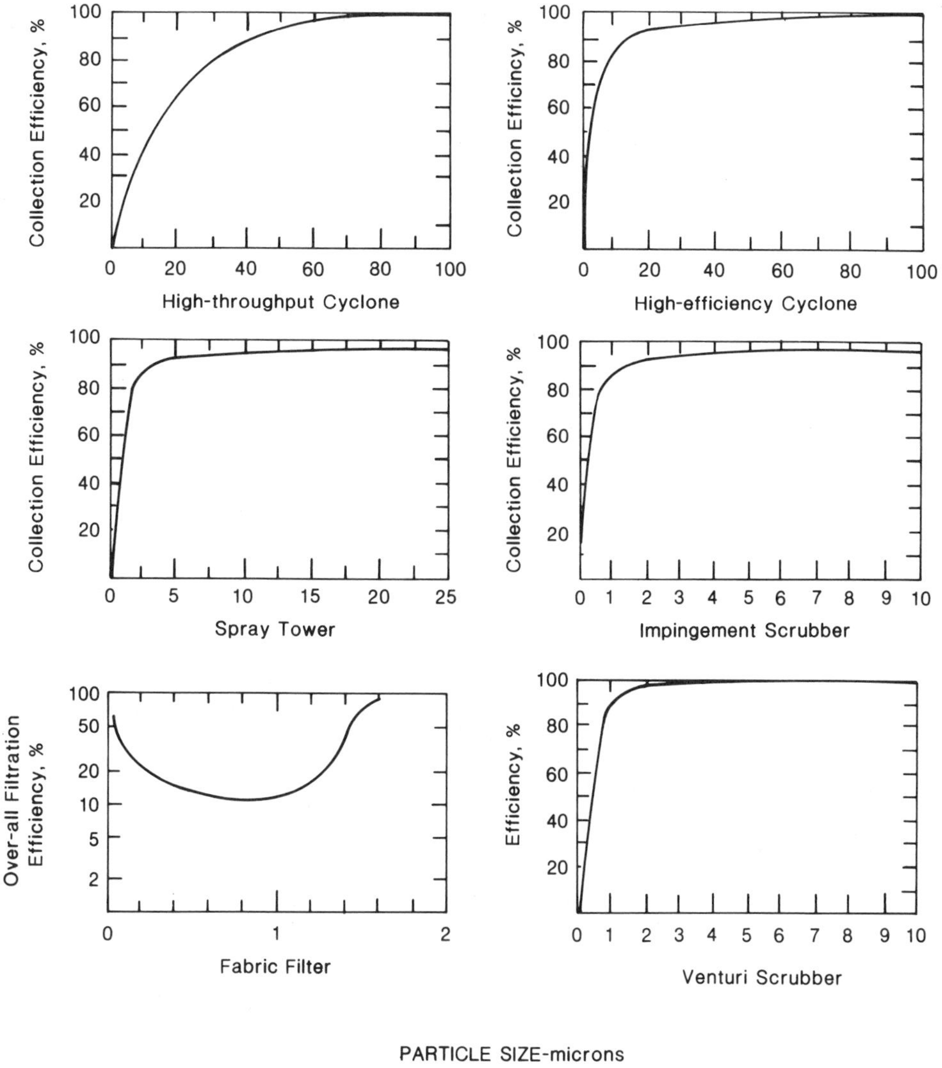

FIG. 2 *Grade efficiency curves for various types of dust collecting equipment. Excerpted by special permission from CHEMICAL ENGINEERING, Jan. 27, 1969, c. 1969 by Mc Graw-Hill, Inc., New York, N.Y. 10020. Ref.[1].*

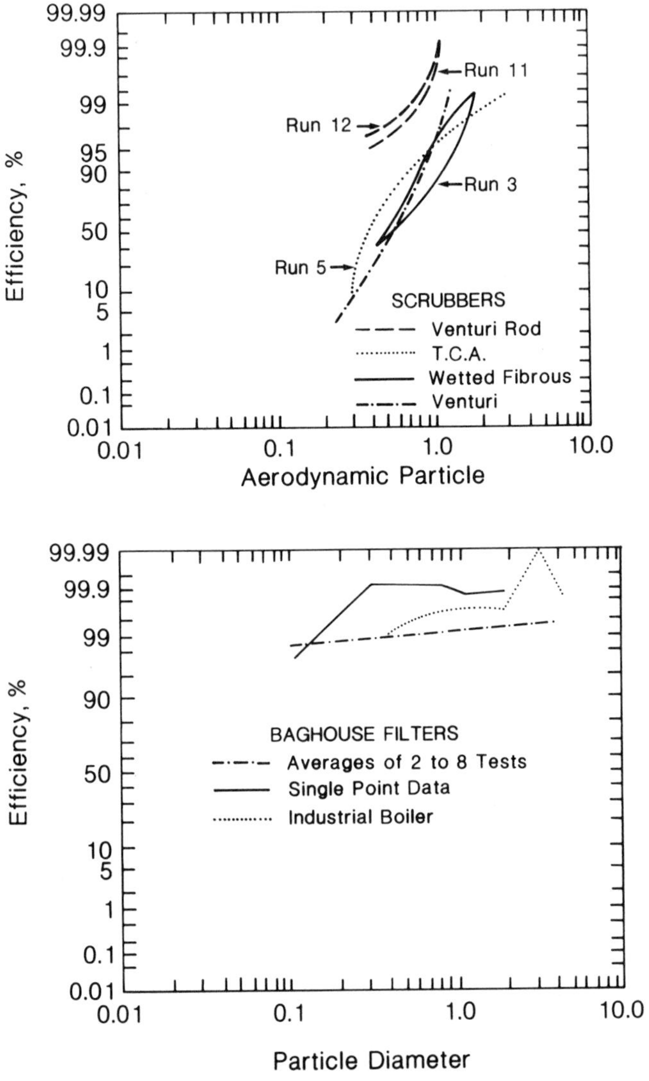

FIG. 3 *Experimental measurements of grade efficiency [2]. Reproduced by permission of the American Institute of Chemical Engineers*

toward the collection of very fine particles. They illustrate the fact that the experimental determination of grade efficiency typically will produce irregularly shaped graphs for which the smoothed curves, such as those in Fig. 2, represent a convenient approximation or "picture". This is partially due to the difficulties of the measurement techniques. Often two or more different methods of particle size distribution measurement are required to cover the range of sizes in the inlet and outlet of the same collector [3,4]. It may also be due to the inherent nature of the dust, as is mentioned in connection with Fig. 6.

For purposes of illustration, the graph shown in Fig. 4 was selected as an example of a cyclone collector operating upon a specific dust of constant density. It is one of the goals of theoretical modelling of collector action to be able to predict a grade-efficiency function relationship in the form of an equation.

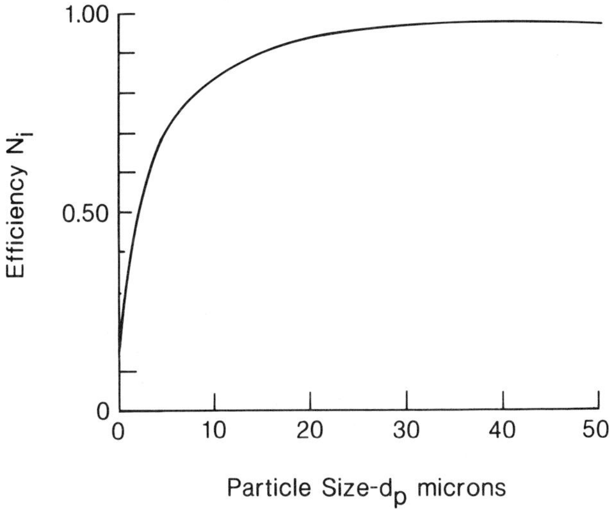

FIG. 4 *Typical performance of cyclone collector. Design of cyclone, operating conditions, and kind of dust are fixed.*

For example, the "lateral-mixing" model (see Eqn. 3.33) for a cyclone collector leads to an equation of the form:

$$\eta_i = 1 - \exp - 2(A\, d_{p_i}^2)^{\frac{1}{2n+2}} \tag{3.8}$$

where A and n are given explicitly by the model theory in terms of design and operating parameters. This is discussed in detail in Chapter 6. This kind of equation does correspond to a graph like that in Fig. 4.

An important special point on the grade-efficiency function is that at which $\eta_i = 0.50$. The particle size at this point is called the "cut size" or "cut diameter". This value will be referred to frequently in discussing collector performance.

C. *RELATION BETWEEN GRADE EFFICIENCY AND OVERALL EFFICIENCY*

When the grade-efficiency relationship is known it may be used to predict the overall collection efficiency of an aerosol of a given size distribution. There are two ways of doing so, depending upon whether the data are provided by grade-intervals, or by continuous distribution functions. In making these calculations, it is customary to assume that the dust is uniformly distributed in any inlet or outlet stream.

1. *Grade-Interval calculations* Given the grade- or size-distribution make-up of an aerosol, say in terms of mass fraction, g_i, of particles of size and kind designated by d_{p_i}: $g_i = g_i(d_{p_i})$, where $\Sigma\, g_i = 1$, as discussed in detail in Chapter 2. Given also the grade-efficiency performance of the collector which is to be used to treat an aerosol such as described above, $\eta_i = \eta_i\,(d_{p_i})$.

The incoming aerosol mass rate of flow may be calculated for each size range as

$$\text{mass rate of input in } i^{th} \text{ grade} = Q_o c_o g_i \tag{3.9}$$

and emission rate for each size range as

$$\text{mass rate of emission in } i^{th} \text{ grade} = (1 - \eta_i)\, g_i Q_o c_o \tag{3.10}$$

$$= P_i g_i Q_o c_o$$

Overall emission rate will be the summation of the grade rates, over all values of i:

$$E = Q_o c_o \sum_i (1 - \eta_i) g_i = Q_o c_o (1 - \sum_i \eta_i g_i) = Qc \tag{3.11}$$

Overall penetration will be given by (3.4):

$$P = E/Q_o c_o = 1 - \sum_i \eta_i g_i = \sum_i P_i g_i \tag{3.4}$$

Overall collector efficiency by mass will be given by (3.1):

$$\eta_M = \frac{Q_o c_o - E}{Q_o c_o} = 1 - P = \sum_i \eta_i g_i \tag{3.1}$$

The calculation procedure is illustrated by the following example.

Example 2. Consider the dust of size-distribution as given in Example 1 as being collected according to the grade-efficiency given in Fig. 4. (a) Calculate the penetration of the collector, and (b) determine the outlet dust size distribution.

Solution: (a) the procedure may be displayed conveniently in tabular form with the calculations represented by each column associated with a numbered equation above.

Particle Size d_{p_i} -microns	No/Vol Input- n_{o_i}	(A) $n_{o_i} d_{p_i}^3$	(B) g_{o_i}	(C) η_i	(D) $\eta_i g_{o_i}$
100	100	1×10^8	0.4675	1.00	0.467
50	900	1.125×10^8	0.5260	0.97	0.510
10	1100	1.10×10^6	0.0051	0.84	0.0043
6	1500	3.24×10^5	0.0015	0.73	0.0011
2	1000	8×10^3	3.7×10^{-5}	0.50	1.9×10^{-5}
1	500	500	2.3×10^{-6}	0.40	9.2×10^{-7}
TOTALS	5100	2.139×10^8	1.000	- -	0.982

The columns are determined as follows: (A) values are repeated from Example 1, (B) $g_{o_i} = n_o d_{p_i}^3 / \sum n_o d_{p_i}^3$ = mass fraction of particles of size d_{p_i}, (C) grade-efficiency read from Fig. 4, (D) product of entries in columns (B) and (C).

Using Equation (3.4):

$$\text{Penetration} = P = 1 - \sum \eta_i m_i = 1 - 0.982$$

$$= 0.0176 \text{ or } \underline{1.76\%}$$

It is clear from column (D) that the overall performance (penetration) of this collector is determined almost entirely by the grade-performance on the two coarsest size particles in the dust, for which the penetration is 1-(0.467 + 0.510) = 0.023, or 2.3%. However, if it should become necessary to reduce the total penetration say to below 0.5%, considerably improved collections of the finer sizes would be required.

(b) To determine the size distribution of the uncollected dust, it is simply necessary to calculate the uncollected portion of each size range as shown in this tabulation:

Particle Size-d_p -microns	Output $(1-\eta_i)n_i$	Input	Output
		Number fraction - f_i	
100	0	0.020	0
50	27	0.176	0.019
10	176	0.216	0.125
6	405	0.294	0.287
2	500	0.196	0.355
1	300	0.098	0.213
TOTALS	1408	1.000	1.000

A similar calculation could be done in terms of weight fractions.

For the sake of simplicity the particle size analysis has been given at exact size values in this example. Commonly these data would be expressed in terms of a range of particle size Δd_{p_i} for each grade. In this case the midpoint particle size of each grade interval would be used as the d_{p_i} value for the calculation in Columns A and B, and the grade-efficiency value in Column C would also be taken at this midpoint size.

2. *Continuous distribution calculations* Given the size distribution in terms of a cumulative fraction-less-than-size d_p, $G_o = G_o(d_p)$ or $F_o = F_o\,(d_p)$ such as one of the functions described in Chapter 2. Given also the grade efficiency performance of the collector by an equation such as (3.8), $\eta = \eta\,(d_p)$. Then the overall efficiency, in terms of mass, will be given by

$$\eta_M = \int_0^1 \eta dG_o = \int_0^\infty \eta q_o\, dd_p \tag{3.12}$$

The overall penetration will be given by

$$P = 1 - \eta_M = \int_0^1 (1 - \eta)dG_o = \int_0^\infty (1 - \eta)q_o dd_p \tag{3.13}$$

The overall emission rate will be given by (3.4). The size distribution of the emission particulate, by cumulative fraction-less-than size, $G = G(d_p)$, may be determined by a material balance taken over all particles finer than a given size d_{p_i}:

$$(Q_o c_o)G_{o_i} = G_i \cdot E + G_{c_i} \cdot C \tag{3.14}$$

and

$$Q_o c_o \int_0^{G_{o_i}} \eta_i \, dG_o = C \, G_{c_i} \tag{3.15}$$

Combining

$$G_i = \frac{1}{P}\left[G_{o_i} - \int_0^{G_{o_i}} \eta_i \, dG_o\right] \tag{3.16}$$

Evaluation of the integrals in (3.12), (3.15) and (3.16), may be done either analytically or graphically. If equations are given explicitly in terms of d_p for both the η-function and the G-function, it may be possible to work out the integral rigorously. Otherwise, graphical or numerical approximate integrations will be required.

a. Analytical Evaluations. A fairly common case will be that where G (and q) is a log-normal distribution (Eqns. (2.21), (2.22), (2.26)) and η is an exponential function like (3.8), viz

$$\eta = 1 - \exp - M d_p^N \tag{3.17}$$

This is a Weibull type function (Eqn. 2.38). Where these relationships hold (3.13) may be displayed as

$$P = \frac{1}{\sqrt{2\pi}} \int_{\infty}^{\infty} e^{\frac{-t^2}{2} - M \bar{d}_{p_m}^{N} \sigma_g^{Nt}} \, dt \tag{3.18}$$

where $t = [\ell n\, d_p/\bar{d}_{p_m}]/\ell n \sigma_g$. While this cannot be integrated analytically, charts could be prepared by numerical integration, for given sets of values of $\bar{d}_{p_m}$ (MMD), σ_g, M and N. Chapter 6 gives such a chart for cyclones.

Another case, studied by Sundberg [5], is that in which both G_o and η may be represented by log-normal functions. Sundberg shows empirically that grade-efficiency test data for some collectors may be represented approximately by straight lines on the special log-probability graph paper described in Chapter 2. Then the "cut diameter" d_{p_c} (at which $\eta = 0.50$) and a standard geometric deviation for grade-efficiency σ_{g_c} are the parameters for η, just as $\bar{d}_{p_m}$ and σ_g are those for G. Sundberg proves that (3.12) yields:

$$\eta_M = \mathrm{erf}\left[\frac{\ln\ \bar{d}_{p_m}/d_{p_c}}{\sqrt{\ln^2 \sigma_g + \ln^2 \sigma_{g_c}}}\right] \tag{3.19}$$

and gives a pair of charts by which this calculation may be made. It could also be done on a pocket calculator using the method mentioned in Chapter 2 to find the value of erf.

Vatavuk [6] assumed that particle size distribution might sometimes be represented by an exponential function also, at least over the major part ($0.15 < G < 0.85$) of the range:

$$G_o = 1 - e^{-\beta d_p}$$

This is a special case of the Rosin-Rammler distribution Eqn. (2.43) in which s = 1. Coupling this with (3.17), Eqn. (3.13) becomes

$$P = \beta \int_o^\infty \exp - (Md_p^N + \beta d_p)\ dd_p \tag{3.20}$$

In general this must be evaluated numerically, but when N = 1 (a special case of some interest) this result is simply

$$P = \beta/(M + \beta) \text{ or } \eta_M = M/(M + \beta) \tag{3.21}$$

Equation (3.16) will yield

$$G = 1 - \exp - (M + \beta)\, d_p \tag{3.22}$$

b. Graphical Methods. The integral in (3.12) may be evaluated by making a plot of η vs. G on which each point corresponds to a certain particle size. This is shown pictorially in Fig. 5. The area under the entire curve is η_M. This value will be located at a horizontal line such that the shaded area above and to the left of the curve will be equal to the shaded area below and to the right.

The integral in (3.16) is represented by the cross-hatched area under the curve up to any desired value of G_o. This will have to be determined for a number of values of G_o in order to construct the outlet size distribution. The value of this integral, subtracted from G_o at each particle size, is used in (3.16) in order to locate points on the outlet curve. This is shown as the dashed curve on Fig. 5. The Gauss method of quadrature has been found to be an effective method of numerical integration in these cases. It lends itself well to pocket calculator computations.

Numerical illustrations of these methods of calculation described in 1. and 2. above will be given later along with the discussions of individual collector performance.

D. *INTERRELATIONSHIPS AMONG PARTICLE-SIZE FUNCTIONS*

It is now evident that there are three important and interrelated functions of particle size: (i) the size-distribution of particles per unit volume in the input stream; (ii) the grade-

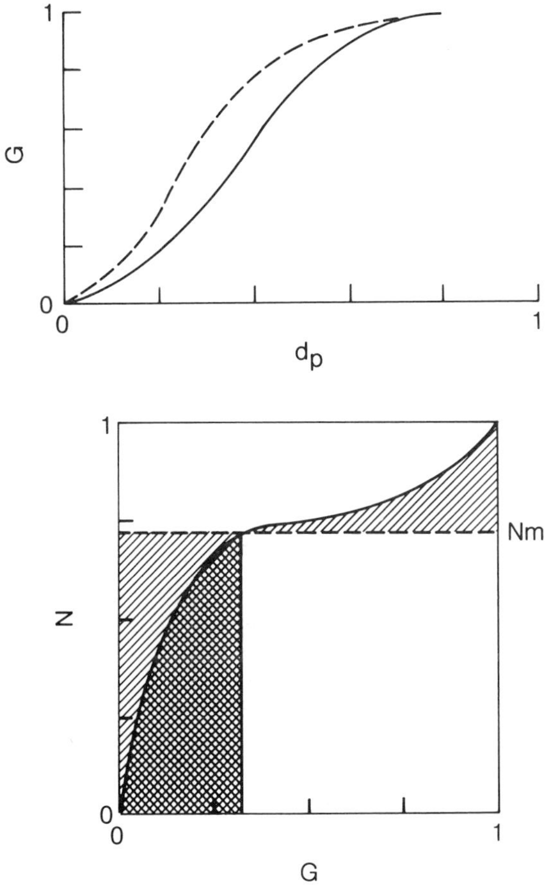

FIG. 5 *Graphical determination of overall efficiency (3.12) and outlet particle size distribution (3.16).*

efficiency relationship; (iii) the size-distribution of particles per unit volume, in the output stream. Only two of these are independent; if any two of them are given the third one is fixed and may be calculated. The three possible cases are: (1) to predict the output (iii), from a specified collection process (ii) acting upon a specified input dust (i) - a collector design procedure, as illustrated in Example 2 above. (2) To determine the grade-efficiency performance (ii) by obtaining a size-distribution analysis of both the input (i) and output (ii) streams - a collector performance testing procedure, discussed further and illustrated below. This is sometimes referred to as a "reverse grade efficiency" procedure. (3) To determine an input size-distribution (i) by obtaining an output-size distribution (iii) from a known grade-efficiency performance (ii) - a method of particle size analysis, the reverse of the calculations in Example 2.

1. *Collector design* The most important tool needed in collector design calculations is the ability to predict the grade-efficiency function for a given collector operating under specified conditions. Consequently, the central aspect of the theory of collection is the development of mathematical models which will show exactly how grade-efficiency is dependent upon operating parameters such as flow rates, grain-loading, dust properties, collector shape and dimensions, etc. With a complete model in hand an engineer may predict the output to be obtained from a given input. He may test the effect of alternative designs, and the effect of varying the operating parameters in order to arrive at an optimum design. Chapters 6, 7, 8 and 9 are devoted to the development and application of such models.

It happens that whenever the lateral-mixing model, discussed below, is applicable, the resulting grade-efficiency function is of the form of (3.8) or (3.17)

$$\eta = 1 - \exp - Md_p^N$$

Here N (usually $0 < N \leq 2$) is given by the model theory and M is a (known) function of the collector design and operating conditions. A common problem is to find a value of M to meet a required overall efficiency. Thus, for (3.1) and (3.4) or for (3.12) and (3.13)

$$P = 1 - \eta_M = \sum_i g_{o_i} \exp - Md_{p_i}^N \qquad (3.23)$$

is to be solved for M, given P, N, and a set of g_{o_i}. The input size distribution data may be arranged so as to select values of d_{p_i} such that all g_{o_i}'s are equal, say $g_{o_i} = g = 0.1$. Letting $x = e^{-M}$, (3.23) becomes

$$P = g\sum x^{d_{p_i}^N} = g(x^{d_{p_1}^N} + x^{d_{p_2}^N} + \ldots) \tag{3.24}$$

Because of the increasing values of d_{p_i}, the higher power terms in this series may become small rather rapidly, and using only the first term:

$$x \approx (P/g)^{1/d_{p_i}^N} \quad \text{or} \quad M \approx -\ln\left(\frac{P}{g}\right)^{1/d_{p_i}^N} \tag{3.25}$$

If necessary, two or more terms of (3.24) may be used in a trial and error solution for M, using (3.25) as a good first approximation. Illustrations of this method will be given in later chapters.

2. *Collector testing* It is necessary to carry out experimental measurements of dust size distribution and grain-loading of both the input and the output streams in order to establish the real performance of collectors, as well as to provide data for checking theoretical models. Because of the experimental difficulties in making particle size-distribution measurements upon flowing gas streams [3], many published tests of collector performance report only overall efficiencies. There is a very limited amount of data available which is sufficiently complete to be useful in testing of theoretical models. It is not within the purpose of this book to discuss the details of experimental techniques, but it is important to state what type of data is needed and the principles involved in its use.

The determination of a mass grade-efficiency relation requires the measurement of the mass of aerosol per unit volume of gas of each size-grade in both the input and the output streams, that is c_{o_i} and c_i. Then

$$\eta_i = 1 - c_i/c_{o_i} \quad \text{or} \quad P_i = c_i/c_{o_i} \tag{3.26}$$

However, instruments for performing a size-distribution analysis may only give the percentage of each size present, i.e. g_{o_i} and g_i. In this case, the total mass of all particles per unit volume of gas ("grain-loading") must also be measured. Then

$$\eta_i = 1 - \frac{c\, g_i}{c_o g_{o_i}} \quad \text{or} \quad P_i = \frac{c}{c_o} \frac{g_i}{g_{o_i}} \tag{3.27}$$

Alternatively, the total dust flow rates in and out may be used for $c/c_o = E/Qc_o = P$ from (3.14) and

$$\eta_i = 1 - P_i = 1 - P \frac{g_i}{g_{o_i}} \tag{3.28}$$

It is to be noted that measurement of g_i and g_{o_i} only is insufficient to determine the grade-efficiency function. Example 3 illustrates the use of experimental test data to calculate a grade-efficiency curve, the so called "reverse grade-efficiency" method.

Example 3. Data for the particulate emissions from a coal-fired power plant upstream and downstream from an electrostatic precipitator have been published by Holland and Conway [7]. From their plot of cumulative mass distribution, values equivalent to g_{o_i} and g_i have been determined and are tabulated below.

i	Range of d_{p_i} -microns	Mass-fraction in range g_{o_i} up-stream	g_i down-stream	Calculated $P\frac{g_i}{g_{o_i}}$	η_i
1	0-0.6	(0.020)	(0.070)	- -	- -
2	0.6-0.7	0.004	0.01	0.05	0.95
3	0.7-0.8	0.004	0.02	0.10	0.90
4	0.8-1.0	0.007	0.03	0.086	0.914
5	1-2	0.035	0.14	0.08	0.92
6	2-3	0.060	0.16	0.053	0.947
7	3-4	0.240	0.29	0.024	0.976
8	4-5	0.130	0.01	0.002	0.998
9	5-6	0.020	0.0	0	1
10	6-8	0.020	0.02	0.02	0.98
11	8-10	0.030	0.025	0.017	0.983
12	10-20	0.110	0.085	0.015	0.985
13	20-30	0.080	0.070	0.018	0.982
14	30	(0.24)	(0.07)	- -	- -

The overall mass efficiency of the precipitator was not reported, but for purposes of illustration here may be reasonably assumed, in line with Fig. 2, to be 98%, i.e. P = 0.02. Determine the grade-efficiency function.

Using (3.23), values of P_i and η_i may be calculated and tabulated, as in the last two columns. The values of g_{o_i} and g_i and η_i are plotted in Fig. 6. Care must be used in discussing these graphs because the raw data were taken from a small-scale plot which is difficult to read precisely. However, a few comments may be appropriate. Since the dust is the fly-ash from a coal-fired boiler, it is undoubtedly a mixture of several kinds of particles. This is further indicated by the presence of two predominant particle size-ranges, one of 3-4 microns, the other between 10-20 microns. It is not surprising, therefore, to find that the grade-efficiency curve has a rather irregular appearance, although this could be due in part to the uncertainties in the raw data. The value of grade-efficiency in any one particle size-

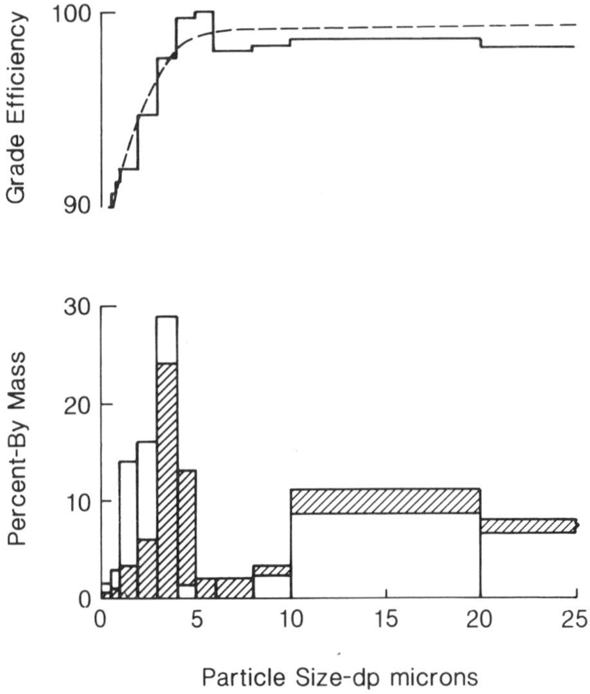

FIG. 6 *Experimental performance of electrostatic precipitator, Example 3.*

range probably represents a composite of the collection action of the precipitator upon the different kinds of particles present. If the overall efficiency of the precipitator is other than the assumed value of 98%, the grade-efficiency graph will be displaced accordingly higher or lower, but its general trend will not be affected. It may be noted that the bar graph for η_i could be approximately represented by a curve such as the dashed one on Fig. 6. This would resemble some of those shown in Fig. 2.

Occasionally the particle-size-distribution of the collected dust may be of interest, and in fact may be easier to obtain from a sample than in the case of the flowing gas streams. From equation (3.14) G_{c_i} may be found from G_{o_i}, G_i and P (on η_M).

$$G_{c_i} = [G_{o_i} - (1-\eta_M)\, G_i]/\eta_M \tag{3.29}$$

where G_{c_i} represents the cumulative per-cent-less-than size d_{p_i} in the collected dust. If the overall collector performance is known, any one of these three size distributions in (3.14) may be determined from the other two, even though the grade-efficiency function is not specifically known.

In principle G could also be used in (3.15) as another way of determining the grade-efficiency

$$\int_0^{G_{o_i}} \eta_i\, dG_{o_i} = (1-P)\, G_{c_i} \tag{3.30}$$

This might be used to infer η from data giving G_o and G_{c_i} as functions of d_{p_i}. This would be done by taking successive intervals of particle size $(d_{p_{i+1}} - d_{p_i})$, and using the corresponding intervals of ΔG_{o_i} and ΔG_{c_i}, calculate η_i for the midpoint of the interval

$$\eta_i \simeq (1-P)\, \Delta G_{c_i}/\Delta G_{o_i} = (1-P)\, g_{c_i}/g_{o_i}$$

E. *MULTIPLE COLLECTORS*

1. *Collectors in parallel* It is frequently necessary and/or desirable to arrange two or more identical collectors to operate in parallel, splitting the gas stream equally among them. This is usually done when the volume of gas flow is too great to be handled by a single collector of reasonable size. Thus two or more (usually an even number) of cyclones may draw feed off of the main gas stream in parallel. A number of small cyclones may be mounted together in parallel, in a device known as a "multiclone". In fabric filters, when the unit collector is a "bag" (size limited to say 8" diameter by 20 ft. length), it is necessary to have many bags taking the flow in parallel.

Electrostatic precipitators are commonly built of a number of collecting plates hung in parallel thus creating a set of containment zones (see Fig. 1) operating in parallel.

So long as the total gas flow is divided uniformly and equally among the collectors in parallel, it may be assumed that each collector is dealing with the same flow of input stream , having the same particle size distribution and grain-loading. If each collector in parallel is identical in size and design, then each will also have the same grade efficiency and same overall efficiency. It is important in this case to insure that no by-passing or short-circuiting of any of the collectors occurs.

2. *Collectors in series* When two or more collectors are connected in series, the emission stream from the first (called "primary" collector) becomes the feed stream to the second (the "secondary" collector) and so forth. Rarely is more than a "tertiary collector used. The volume flow of gas is thus the same through all the collectors, but the feed stream to each becomes successively lower in grain-loading and contains generally finer particles. The successive collectors are not necessarily of the same kind, or if they are, not necessarily of the same sizes and design.

Since the performance of almost all collectors is affected by the grain loading in the feed, it may be desirable to use a primary collector like a cyclone mainly to reduce the loading to a secondary collector such as a filter or electrostatic precipitator in order to get maximum grade-efficiency from the latter. This is a fairly common arrangement. Cyclones are sometimes used in series in order to improve the overall collection of finer particles e.g. a "high-efficiency" design following a "high through-put" design; see Fig. 2 for grade efficiency curves.

For two collectors in series, let $\eta_1(d_p)$ and $\eta_2(d_p)$ represent the mass grade-efficiency functions of each, and $g_o(d_p)$ the particle size distribution in the feed by weight-fraction. Then at any particle size d_p:

$$P_1(d_p) = 1-\eta_1(d_p) \quad (3.4)$$

$$E_1(d_p) = Q_o c_o g_o(d_p) \cdot P_1(d_p) \quad (3.31)$$

$$P_2(d_p) = 1-\eta_2(d_p)$$

$$E_2(d_p) = P_2(d_p) \cdot E_1(d_p)$$

$$= Q_o c_o\, g_o(d_p) \cdot P_1(d_p) \cdot P_2(d_p) \quad (3.32)$$

and, for the combination:

$$P_T(d_p) = E_2(d_p)/Q_o c_o g_o(d_p) \quad (3.4)$$

$$P_T(d_p) = P_1(d_p) \cdot P_2(d_p) \quad (3.33)$$

Since this relationship (3.33) will hold for all values of d_p, it will hold for the overall penetration as well:

$$P_T = P_1 \cdot P_2 = 1 - \eta_T \quad (3.34)$$

whence

$$\eta_T = 1 - (1-\eta_1)(1-\eta_2) = \eta_1 + \eta_2 - \eta_1\eta_2$$

and

$$P_T = 1-\eta_1-\eta_2+\eta_1\eta_2 \quad (3.35)$$

The overall amounts of collection and emission may be computed by integration. Using (3.31) and (3.32)

$$E_1 = E_2 = Q_o c_o \int_o^\infty g_o(d_p) \cdot P_1(d_p) \cdot P_2(d_p)\, dd_p \quad (3.36)$$

$$C_T = C_1 + C_2 = Q_o c_o - E_2$$

and

$$\eta_T = C_T/Q_o c_o = 1-E_2/Q_o c_o = \int_o^\infty g_o(d_p)P_1(d_p)P_2(d_p)dd_p \quad (3.37)$$

Crawford [8] has given a thorough and systematic treatment of the analogous relationships which exist for several other multiple-collector arrangements. These include schemes in which the primary collector is really a "concentrator", both exit streams from which are sent to secondary collectors, and also schemes in which recycling of a stream is introduced, thus involving series-parallel combinations. An example of recycle is discussed in Chapter 6 in connection with cyclone design.

III. BASIC MODELLING CONCEPTS

A. *RESIDENCE TIME OF PARTICLES*

A fundamental factor which influences whether a given aerosol particle will be collected is the length of time during which that particle comes under the influence of the collecting force(s). This time must be sufficient for the particle to travel at the collecting velocity from its point of entrance into the zone of influence of the collecting force to a point of collision, before the particle is swept out of the zone of influence by the air stream velocity.

The length of time a particle is within the zone of influence is called its residence time, t_r. For collision to occur, the residence time t_r must be equal to or greater than the collecting time. The collecting time t may be defined as:

$$t = \frac{\text{distance travelled from point-of-entry to collecting surface}}{\text{collecting velocity}} \tag{3.38}$$

The sketches in Fig. 7 illustrate how the relationship between t_r and t may be affected by: (a) location of point of entry; (b) difference in air stream velocity; and (c) difference in particle size, or mass.

For the purpose of this elementary discussion, the collecting velocity is regarded as constant. In fact, however, when a

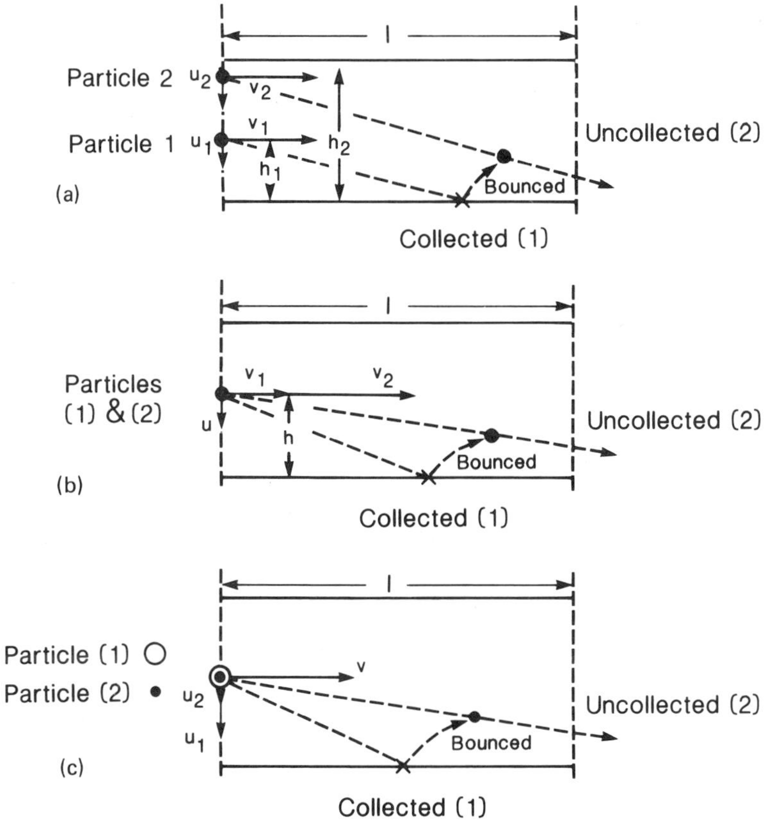

FIG. 7 *Relation between collecting time and residence time. (a) Effect of location of point of entry; (b) effect of air stream velocity; (c) effect of particle size or mass.*

particle enters a collecting zone, it undergoes an initial period of acceleration in the direction of the collecting force. In many instances this period is very short, and for practical purposes the particle may be regarded to have attained a steady collecting velocity almost instantaneously.

An overall average residence time for all particles in a stream may be calculated from the free volume of the collector and the volume flow rate.

$$t_{r_{avg}} = \frac{\text{volume of zone of influence in collector}}{\text{volumetric flow rate of air stream}} = \frac{V}{Q} \tag{3.39}$$

The actual residence time of an individual particle may be quite different from this value, depending upon the nature of the fluid velocity profile.

(a) In the case of plug (or piston) flow all parts of the air stream are moving with the same velocity v, and the residence time for all particles is the same:

$$t_r = \frac{V}{Q} = \frac{\ell}{v} \tag{3.40}$$

where ℓ = length of zone of collection, in direction of flow.

(b) In the case of a real laminar or turbulent flow, there will be a fluid velocity distribution pattern over the cross-section of the collection zone perpendicular to the direction of flow. The precise description of this pattern must be known, in order to calculate the residence times for particles entering along different streamlines.

Sometimes a particle may bounce off after a collision with a collecting surface. It will then be carried farther along by the air stream, at the same or a different air stream velocity depending upon the flow pattern. It may even be carried out of the collecting zone as shown in Fig. 7, which assumes plug flow for the sake of simplicity in the illustration. The particle which is uncollected due to the bounce effect may be regarded as equivalent to one: (a) entering at a larger collecting distance $h_2 > h_1$; or (b) having a lower collecting velocity, $u_2 < u_1$; or (c) traveling with a higher air stream velocity, $v_2 > v_1$, as indicated by the lines marked "Bounced" in Fig. 7. In each case, the effect is to give the particle an apparent collecting time which is greater than its residence time.

A particle may also rest on, or move along the collecting surface for a time and then become re-entrained due to some mechanical action (such as "rapping" the plate of an electrostatic precipitator) or by a current of higher-velocity air. The collecting surface does, in time, have to be cleaned of collected particles in order that it may continue to function, or else it must be renewed in some way. If the aerosol particles are liquid, or the surface is flushed with a liquid, or coated with some oily or adhesive substance, then the effects of bouncing or re-entrainment are greatly reduced or eliminated.

Another effect which must be considered is that which eddy currents in the fluid may have upon suspended particles. Such currents will tend to keep the uncollected particles mixed back into the air stream, and prevent them from moving steadily toward the collecting surface at a theoretical collecting velocity. Instead, their paths would consist of erratic wanderings back and forth due to eddy currents, or in the case of very small particles due to their bombardment by gas molecules (Brownian diffusion, discussed in Chap. 4). There would, however, be a net motion toward the collecting surface, because on the average the eddies would carry the particle in either direction at random with equal probability. The zone of influence of the collecting force is really then only fully effective in the immediate vicinity of the collecting surface, and in the bulk of the nominal collecting zone the particles may be more or less well-mixed.

B. *ELEMENTARY MODELS*

There are three idealized situations which may be identified as basic models to serve as a starting point to describe real collection processes: (1) Plug flow, no radial or axial mixing of uncollected particles; (2) complete radial (lateral) mixing of uncollected particles; (3) complete back-mixing, both radially and axially.

Each of these models may be used as the basis for predicting a corresponding form of grade-efficiency relationship. The nature of this form will also depend to a very significant extent upon the way in which the collecting velocity u_i may depend upon d_{p_i} in the particular application of the model. The possible relationships between u_i and d_{p_i} are presented in the detailed discussion of collecting forces given in Chapters 4 and 5. The three models will be described in general form at this point for application to specific collectors later.

1. *Plug flow, no mixing* This model may be illustrated in terms of a rectangular two-dimensional view of the zone of influence. It is assumed that the particles are distributed uniformly across the section, that all particles are of the same size and kind, d_{p_i}, that the collecting force is operating only in one direction, and that the trajectories of all particles comprise a family of parallel straight lines. Whether a given particle is

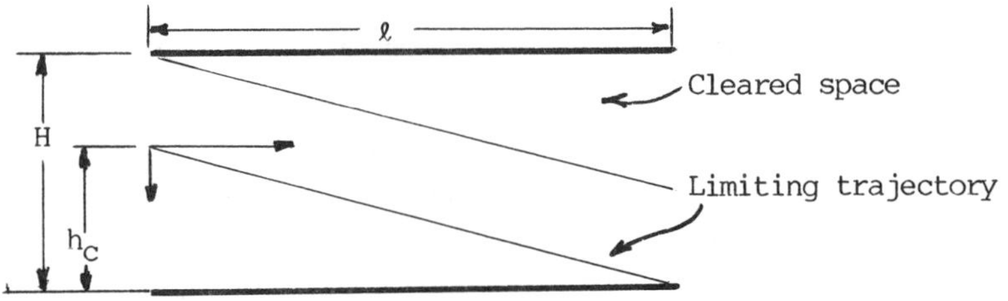

collected then is determined solely by the distance h from its point of entry to the collecting surface. All those particles for which this distance $h < h_c$ will collide with the surface. The shaded space in the sketch becomes completely cleared of particles. The efficiency of collection, by number or by mass, will then be given by

$$\eta_i = h_{c_i}/H = u_i\, \ell/vH \tag{3.42}$$

The grade-efficiency relationship will be determined by the way in which the velocities u_i depend upon d_{p_i}. If $u_i \propto d_{p_i}^2$, which is a common case, then η_i vs. d_{p_i} will be a parabolic graph with $\eta_i = 1.00$ for all particles large enough so that $u_i \geq \frac{vH}{\ell}$ (see Fig. 8).

2. *Lateral mixing* This model assumes that the uncollected particles are well-mixed by lateral turbulence in the gas stream, hence their concentration is always uniform across any lateral or radial section perpendicular to the collecting surface.

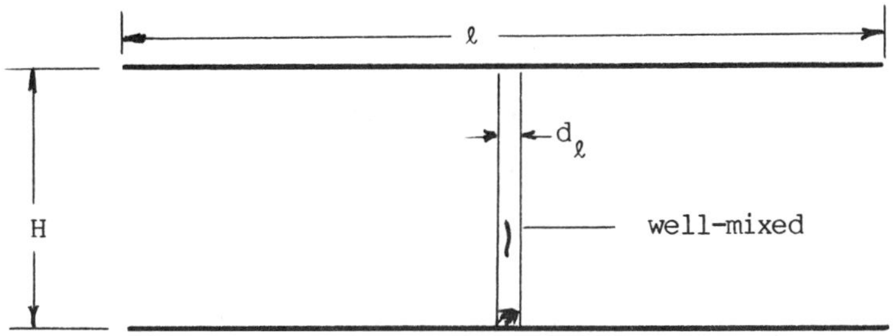

Plug flow is assumed in the axial direction. As the stream flows over a differential distance $d\ell$ during a period of time dt, all particles of the ith grade within a distance $u_i dt$ of the collecting surface will move into that surface. There will occur a fractional reduction in particle concentration which is the same everywhere along the direction of flow. It is given by:

$$-\frac{dn_i}{n_i} = \frac{u_i dt}{H} = \frac{u_i}{Hv} d\ell$$

This may be integrated over the length of the collecting zone to give:

$$-\int_{n_{i_o}}^{n_i} \frac{dn_i}{n_i} = \frac{u_i}{vH} \int_o^{\ell} d\ell = \frac{u_i}{v}\frac{\ell}{H}$$

$$\ln n_i/n_{i_o} = -u_i\ \ell/vH = \ln P = -\text{NTU} \tag{3.43}$$

The collection efficiency is then:

$$\eta_i = 1 - \exp(-u_i\ \ell/vH) = 1 - \exp(-u_i t_r/H) \tag{3.44}$$

where $t_r = \ell/v$ is the residence time. Again, the grade-efficiency relationship will depend upon how u_i is related to d_{p_i}, but in any event the exponential function will require that η_i approach 1.00 asymptotically as d_{p_i} increases. If $u_i \propto d_{pi}^N$, the equation may be rewritten as

$$\eta_i = 1 - \exp(-M d_{p_i}^N) \tag{3.45}$$

where M represents all the factors which do not depend upon d_{p_i}. This is the most commonly used model as has been mentioned in Eqn. (3.17) above, with N ranging between -1 and 2 for the various collectors. It is capable of representing several of the types of curves shown in Fig. 2.

If N = 2 this expression possesses a point of inflection at

$$d_{p_{inf}} = \sqrt{1/2M} \quad \text{and} \quad \eta_{inf} = 0.394 \tag{3.46}$$

The graph will, therefore, be S-shaped as shown in Fig. 8.

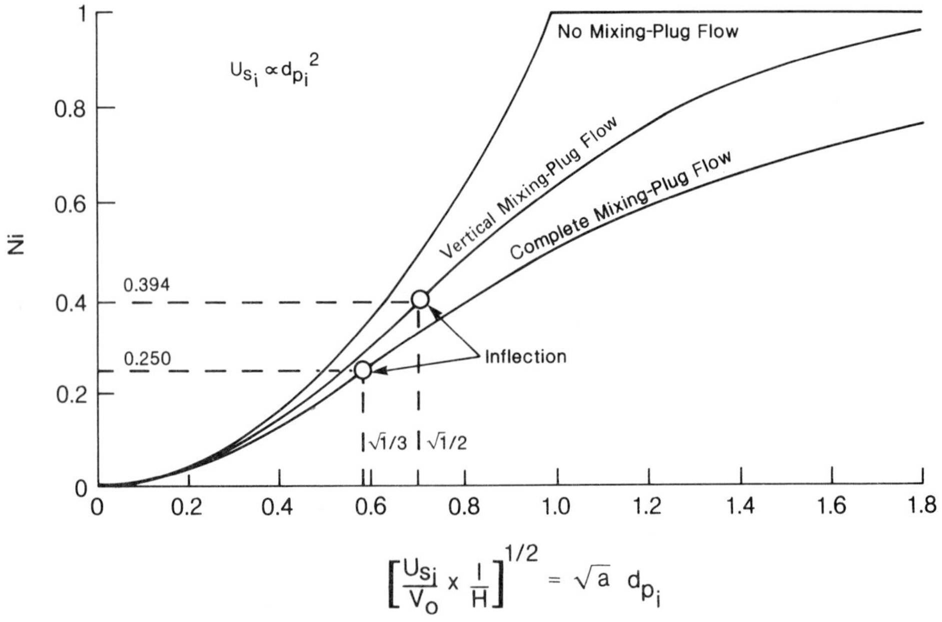

FIG. 8 *Generalized grade efficiency for gravity collector.*

3. *Back mixing* This model assumes that the degree of turbulence in the gas is such that at any instant all uncollected particles are uniformly mixed throughout the entire volume of the collector. During a unit period of time, per unit width of collecting space, dust collected = $(nu_i \ell)$ particles/time; dust leaving (uncollected) = $(nv_o H)$ particles/time. The grade-efficiency is given by

$$\eta_i = \frac{nu_i \ell}{nu_i \ell + nv_o H} = \frac{u_i \ell / v_o H}{1 + u_i \ell / v_o H} \tag{3.47}$$

This functional relationship also has η_i approaching 1.00 asymptotically as d_{p_i} increases. For the case of $u_i \propto d_{p_i}^2$, it may be rewritten as

$$\eta_i = \frac{M\, d_{p_i}^2}{1 + M d_{p_i}^2} \tag{3.48}$$

which has a point of inflection at

$$d_{p_{inf}} = \sqrt{1/3M} \quad \text{and} \quad \eta_{inf} = 0.250 \tag{3.49}$$

and is, therefore, also S-shaped when graphed.

A comparison of all three models is shown in Fig. 8 for the case of a gravity collector where the collecting velocity is proportional to $d_{p_i}^2$.

It should be clear at this point why the grade-efficiency relationship never corresponds to a precise cut-off at a certain critical particle size. Any real collector operation is going to have effects present such as are embodied in each of these models. Either the effect of the position of particle at entry, or the effect of back mixing will be such as to require a gradual change in efficiency with particle size, rather than a step jump from 0% to 100% at a certain size.

4. *Application of models* The basic models of collection efficiency just described may be applied to the performance of specific collecting devices in either of two ways: by the study of the fundamentals of particle motion, or by the empirical analysis of experimental data. The details of the fundamental approach form the subject matter of the later chapters which deal with each type of collector individually.

It is evident that the exact nature of the grade-efficiency relationship corresponding to each model will depend upon the manner in which the collecting velocity is related to particle size. It involves an identification of all the forces acting upon a particle, and the use of Newton's law of motion to predict the path taken by the particle under the combined action of these forces. To apply this information to a given kind of collector requires a detailed knowledge of the nature of the collecting forces operating in that collector. This, in turn, leads to the functional relation between u_i and d_{p_i}. The fundamentals of particle mechanics are presented in Chapters 4 and 5.

These studies show that, in many cases, $u_i \propto d_{p_i}^2$. This corresponds to a range of conditions in which the motion of the particle is governed by what is called Stokes Law. So common is this situation, that it was used above as the basis for the graphical illustration of the models. However, other relationships such as $u_i \propto d_{p_i}$ are possible. A more general treatment of the models can be made by taking

$$u_i \propto d_{p_i}^N$$

where N may range from -1 to +2 in value. There is a range of conditions just outside of those corresponding to Stokes Law in which the value of N may lie between 1.6 and 2.0.

With different values of N, the appearance of the grade-efficiency graphs will change markedly. For example, the model graphs may not exhibit a point of inflection, nor a range of downward concavity. As an aid to identifying possible relationships, general formulas have been derived for the value of η at the point of inflection as follows:

Lateral-mix model: $$\eta_{inflec} = 1 - \exp\,[(1-N)/N] \qquad (3.50)$$

Back-mix model: $$\eta_{inflec} = (N-1)/2N \qquad (3.51)$$

Typical values are tabulated below. Since real values of η must lie between 0 and 1, a negative value of η_{inflec} means that no point of inflection will appear.

N	Grade-Efficiency at Inflection	
	Lateral-Mix	Back-Mix
-1	Negative	1
0.5	Negative	Negative
1	0	0
1.5	0.283	0.167
1.6	0.313	0.188
1.7	0.338	0.206
1.8	0.359	0.222
1.9	0.377	0.237
2.0	0.393	0.250

It is noteworthy that values of η_{inflec} are at the low end of the range.

Furthermore, there is no requirement that the value of N be the same throughout the entire range of operating conditions for a given type of collector. If the nature of action of collecting forces changes with particle size, or with other parameters of operation, N may shift in value from one end of the range to the other. This can give rise to what might be called a "composite model". Such models are observed in practice, for example, the graph for a filter in Fig. 2. Without the insight obtained from the principles of particle mechanics, such behavior would be difficult to understand.

5. *Testing of models* The other approach to the application of the models involves the empirical examination of an experimentally determined grade-efficiency curve. Simple graphical tests may be made to determine whether one of the models is applicable, and if so, the value of N.

a. To test for model (1) - plug flow, no mixing: Plot log η_i vs. log d_{p_i}. If the model fits, a straight line will be obtained with slope equal to N. For if $u_i = a\, d_{p_i}^N$, from (3.42)

$$\log \eta_i = N \log d_{p_i} + \log \left(\frac{a\ell}{vH}\right) \tag{3.52}$$

b. To test for model (2) - lateral-mixing only: Plot log $[-\ln(1-\eta_i)]$ or log NTU vs. log d_{p_i}. If the model fits, a straight line will be obtained with slope equal to N. For, if $u_i \propto d_{p_i}^N$, from (3.44) and (3.45)

$$1 - \eta_i = \exp\left(-\frac{u_i \ell}{vH}\right) = \exp\left(-Md_{p_i}^N\right)$$

$$-\ln(1-\eta_i) = NTU = Md_{p_i}^N$$

$$\log[-\ln(1-\eta_i)] = \log NTU = \log M + N \log d_{p_i} \tag{3.53}$$

c. To test for model (3) - complete back-mixing: Plot $\log[\eta_i/(1-\eta_i)]$ vs. log d_{p_i}. If the model fits, a straight line will be obtained with slope equal to N. For if $u_i \propto d_{p_i}^N$, from (3.47)

$$\eta_i = \frac{Md_{p_i}^N}{1 + Md_{p_i}^N}$$

$$\frac{\eta_i}{1-\eta_i} = \frac{Md_{p_i}^N}{1 + Md_{p_i}^N} \Bigg/ \frac{1}{1 + Md_{p_i}^N} = Md_p^N$$

$$\log \frac{\eta_i}{1-\eta_i} = \log M + N \log d_{p_i} \qquad (3.54)$$

From a glance at the typical performance graphs presented in Figs. 2 and 3, it might appear that none of the models could fit most of these curves because the curves exhibit neither a point of inflection nor a range where they are concave downward. There are several possible explanations to account for this:

(1) The experimental data in the low end of the particle-size range may be either inaccurate or incomplete to the extent that a point of inflection is not revealed even though it may exist. That is, the curve as drawn may simply be roughly extrapolated down to $d_{p_i} \to 0$ from the lowest data point, say in the range of $d_{p_i} = 1.5$ µm. Experimental measurements on very fine particles are usually difficult to make accurately.

(2) One of the models may fit with a value of $N \neq 2$, such that it does not call for an S-shaped curve.

(3) Due to particle bouncing or re-entrainment, the collection efficiency over part of the range of particle sizes may be less than called for by the model.

(4) The value of N, or indeed the nature of the applicable model itself may shift over the range of particle size.

Example 4. Test the graph presented in Fig. 4 for agreement with each of the basic models.

In order to test all three models it is necessary to plot η, $\ln[1/(1-\eta)]$, and $\eta/(1-\eta)$ respectively, against d_p on log-log paper, as given by Eqns. (3.52), (3.53) and (3.54). This may be done conveniently by first preparing a table of these values based upon values of η_i read from Fig. 4 for selected values of d_{p_i}. This table is as follows.

d_{p_i} - µm	η_i	$\ln \frac{1}{1-\eta_i}$	$\frac{\eta_i}{1-\eta_i}$
2	0.50	0.693	1
4	0.65	1.05	1.86
6	0.73	1.31	2.70
10	0.84	1.83	5.25
20	0.93	2.66	13.3
30	0.965	3.35	27.6
40	0.975	3.69	39.0

All three graphs may be made on the same chart by using 2 x 2 cycle log-log paper and defining the ordinate scale appropriately for each function of η. These plots are given as Fig. 9.

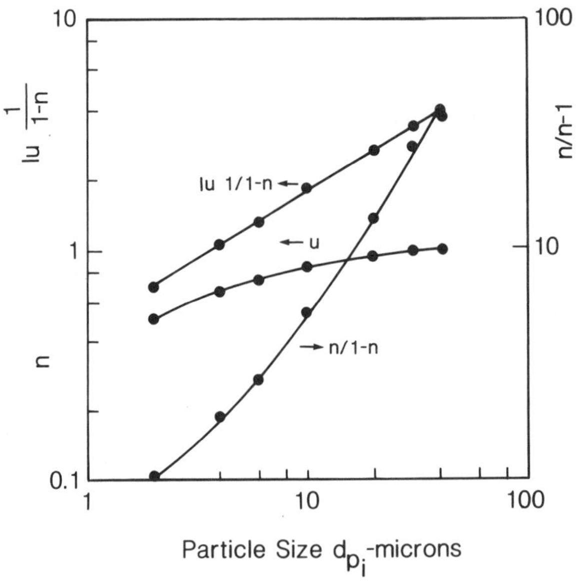

FIG. 9 *Test of models on cyclone data.*

The plot of log $\ln[1/(1-\eta_i)]$ vs. log d_{p_i} gives a good straight line, while the other two plots are curved, thus confirming agreement with the lateral mixing model. The value of N, equal to the slope of this line, may be calculated by selecting two points, say for d_{p_i} = 2μm and 30 μm:

$$N = \text{slope} = \frac{\log 3.35 - \log 0.693}{\log 30 - \log 2} = 0.582$$

No point of inflection is to be expected since from Eqn. (3.50) $\eta_{inflec} = -1.05$.

IV. ENERGY REQUIREMENTS AND EFFICIENCY

The energy required to operate a dust collection system will be mainly of two kinds: (a) work required to move the dust-laden air stream through the equipment; (b) additional energy of a form unique to the type of collection force being employed. The first kind is due to the friction losses incurred by the flow of the stream through the equipment. The second kind may be electrical, as in an electrostatic precipitator; heat, as in a thermal depositor; other mechanical work, as in pumping a scrubbing fluid, or in shaking a fabric filter to clean it of collected dust.

The second kind should never be overlooked. However it may be completely absent in certain mechanical collectors, such as cyclones, and even when present may in some cases be small relative to the first kind, although in other cases may be the major energy requirement.

The first kind is always present. Hence it is always important to be able to estimate the pressure drop in the gas stream as it flows through the system, and to calculate the corresponding power loss which must be provided by the fan or blower which moves the stream. Thus,

$$\text{Power loss} = Q \times \Delta P/33{,}000 \quad \text{hp} \tag{3.55}$$

where Q = flow rate, in actual cubic feet per minute (expressed as acfm), and ΔP = total pressure drop, in lbf per square foot.

Since Q may be very large, it is important to strive for low ΔP, which is then usually expressed in inches of water. Values of 10 inches water or less are rather common, although in certain types of collectors ΔP may range upwards to ten times this value. For these units:

$$\text{Power loss} = 1.58\text{x}10^{-4}\ Q\text{x}\Delta P''\ \text{hp.} = 1.18\text{x}10^{-4}\ Q\ \Delta P''\ \text{kw} \qquad (3.56)$$

where $\Delta P''$ = inches of water. A stream of 10,000 cfm at 10 inches water will thus require 15.8 hp or 11.8 kw.

The total pressure drop due to frictional losses will be made up of losses in the duct work plus loss in the collector itself. The first may be calculated by the standard methods of fluid mechanics, in which it is usually assumed that the gas behaves in a noncompressible manner. The latter requires special treatment which may be different for each type of collector. Thus, along with the modelling for collector efficiency, it is equally important to develop a model for pressure drop, i.e. frictional loss, for each collector. Both must be considered in the design.

The scheme which is used to estimate frictional energy losses is based upon a mechanical energy balance across the collecting device.

$$\Delta\ (\frac{v^2}{2g_c}) + \frac{\Delta P}{\rho_f} + g\ \frac{\Delta Z}{g_c} = W_s - F \qquad (3.57)$$

Each term represents energy per unit mass of fluid stream. The delta (Δ) terms are taken as the difference between a point in the exit duct immediately beyond the collector and a point in the inlet duct just upstream of the collector. If these ducts have approximately the same cross-sectional area, $\Delta(v^2) \approx 0$, and if they are at the same (or nearly the same) elevation, $\Delta Z \approx 0$. Furthermore there is usually no work exchanged between the gas stream and its surroundings between these two points so that $W_s = 0$. The equation usually reduces to $\Delta P/\rho_f = -F$.

It is then convenient to evaluate F as a multiplier of the inlet velocity head $v_o^2/2g_c$, say $F = N_H\, v_o^2/2g_c$ where N_H is called the "number of inlet velocity heads" frictional loss. Finally, the pressure drop is expressed as

$$\Delta P = -\rho_f N_H\, v_o^2/2g_c \tag{3.58}$$

Most applications of energy loss calculations are thus resolved into a method of predicting N_H for a given collector operating under a given set of conditions.

In a wet-scrubber type of collector, there is also the energy required to pump the scrubbing fluid. This is calculated in a manner similar to Eqn. (3.55) except that the units are usually different. Commonly they are given as ΔP in psi (lbf/in^2) and Q in gpm (gallons/min), and the calculation becomes:

$$\text{power loss} = 5.83\text{x}10^{-4}\ Q\ \Delta P_p\ \text{hp} = 4.35\text{x}10^{-4}\ Q\ \Delta P_p\ \text{kw} \tag{3.59}$$

when ΔP_p is the pressure drop involved in pumping the liquid. If some liquid is lost in the process, this may be expressed as equivalent to a power consumption:

$$\text{Equivalent power loss} = 60\ Q_c\ \left(\frac{\$}{\text{gal}}\right) \text{x} \left(\frac{\text{kw-hr elec.}}{\$}\right)\ \text{kw} \tag{3.60}$$

where Q_c = water consumed, gpm; \$/gal = cost of scrubbing liquid; and \$/kw-hr = cost of electricity. The total power consumption for the scrubber would then be given by the sum of (3.56), (3.59) and (3.60).

Some typical power consumption values for common dust collecting equipment are given by Sargent [1], Stairmand [9], Vandergrift [10], and Benson [11]. They range from 0.2 kw/1000 cfm to 4 kw/1000 cfm for various kinds of equipment, capacities and collection efficiency. In every case it is found that the higher the overall collection efficiency, the higher the power requirement. As a very rough rule of thumb, an average value may be taken as 1 kw/1000 cfm.

Example 5. Estimate the total electrical power required to process 1000 acfm of gas in each of the following cases, assuming the energy efficiency of pumps and fans to be 75%: (a) cyclone having 6-in water pressure drop; (b) Venturi scrubber having 20-in water pressure drop in the gas, 20 psi pressure drop for 7 gpm scrubbing water, with 4% loss of water at $1.05/1000 gal, electricity at $0.03/kw-hr; (c) electrostatic precipitator having 0.5-in water pressure drop and using 0.3 kw of electricity.

Solution: with reference to equations given above:

(a) Equation (3.56):

$$\text{power consumed by fan} = \frac{1.18 \times 10^{-4} \times 1000 \times 6}{0.75} = 0.94 \text{ kw}$$

(b) Fan-power (Eqn. 3.56):

$$= \frac{1.18 \times 10^{-4} \times 1000 \times 20}{0.75} = 3.15 \text{ kw}$$

Pump power (Eqn. 3.59)

$$= \frac{4.35 \times 10^{-4} \times 7 \times 20}{0.75} = 0.08 \text{ kw}$$

Water loss (Eqn. 3.60):

$$= 60 \times 7 \times \frac{0.04}{1000} \times \frac{1.05}{0.03} = 0.59 \text{ kw}$$

Total equivalent power consumed = 3.15 + 0.08 + 0.59 = 3.82 kw

(c) Fan power (Eqn. 3.46)

$$= \frac{1.18 \times 10^{-4} \times 1000 \times 0.5}{0.75} = 0.08 \text{ kw}$$

Electrical power for collection = 0.3 kw

Total power consumed = 0.08 + 0.3 = 0.38 kw

The energy efficiency of the collection process is an important consideration to which little attention has been given. Stukel and Rigo [12] have taken as an approximation to the theoretically minimum amount of work required to separate the particles from the gas, simply the kinetic energy of the particles removed. Per unit time:

$$W_{min} = \eta_M (Q_o c_o) v^2/2g_c \tag{3.61}$$

For typical conditions of $\eta_M = 0.95$, $c_o = 5$ gr/ft^3, $v = 50$ ft/s this will give

$$W_{min} = \frac{0.95}{32.2} \times \frac{1000}{60} \times \frac{5}{7000} \times \frac{(50)^2}{2} \times \frac{0.745}{550} = 5.95\times10^{-4} \text{ kw/1000 cfm}$$

This corresponds to an energy efficiency of about 0.059%. The entire range of cases considered by Stukel and Rigo fall within energy efficiency values of 0.008 to 0.1%. Similar calculations by Soo [13] also show the efficiency of energy utilization in common collectors to be well below 0.5%. There is a need to learn ways of separating particulates from gases with a much improved energy efficiency. Stukel and Rigo remark that these efficiences are "intolerably low."

V. GENERAL SURVEY OF COLLECTORS

A. *COMPARISON OF PERFORMANCE*

It is helpful, particularly in making a preliminary survey of choice of collecting equipment, to have some general method of overall comparison of performance. Several such methods are available and may be used as rough guides.

A convenient method of comparing the grade-efficiency performance of various collectors is to replot data such as given by Fig. 2 as a semi-log plot of log $(1-\eta_i)$ vs. d_{p_i}. As an illustration of this, a few selected cases taken from Fig. 2 are shown in Fig. 10.

This method of plotting gives all curves the common point in the upper left corner where $d_{p_i} = 0$ and $\eta = 0$ or $P = 1.00$. It is really a plot of log P vs d_p. The less efficient performance is represented by a curve with a smaller slope i.e. extending farther to the right. For the cases shown, performance could be ranked in descending order, as

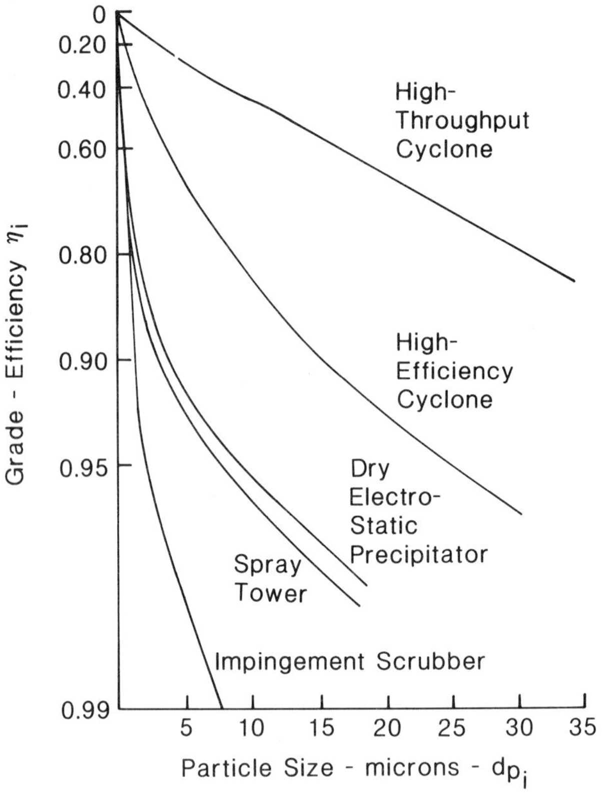

FIG. 10 *Comparison of collector performance.*

1 - High throughput cyclone
2 - High efficiency cyclone
3 - Dry electrostatic precipitator
4 - Spray tower
5 - Impingement scrubber

which is very readily observed by a glance at the chart. Note that the curve for a fabric filter (not shown) would not fit the general pattern. It would be close to the vertical axis, but would have an up and down trend which could hardly be seen on the scale of this Fig. 10.

One should be careful however only to draw very rough general conclusions from such comparisons. It must be remembered that the precise performance curve for any kind of collector depends upon operating parameters. The curves shown are examples only. They could each be shifted to some extent in either direction by a variation in gas flow rate, nature of dust, temperature, or other conditions. One cannot say, for instance, that all scrubbing collectors are more efficient (or less efficient) than all electrostatic precipitators.

There is no reason to expect that such plots should be straight lines. It may easily be shown that a straight line would correspond to a special case of the lateral-mix model with N = 1, i.e. $u_i \propto d_{p_i}$. For from Eqn. (3.45)

$$\ln (1 - \eta_i) = \ln P_i = - Md_{p_i} \tag{3.62}$$

or $2.303 \log (1 - \eta_i) = - Md_{p_i}$, so that $- M/2.303$ would be the slope of this line, always negative. Such a case as this rarely occurs. If $N \neq 1$, the lines will be curved.

Stairmand [14] has provided some very useful charts which take into account costs as well as efficiency. As a basis for his comparison of the performance of collectors he defined three typical dusts designated as "coarse," "fine," and "superfine".

TABLE 1 *Particle size distributions of standard dusts*

Particle Size microns (μm)	% by weight less than size Superfine Dust	Fine Dust	Coarse Dust
150	-	100	-
104	-	97	-
75	100	90	46
60	99	80	40
40	97	65	32
30	96	55	27
20	95	45	21
10	90	30	12
7.5	85	26	9
5.0	75	20	6
2.5	56	12	3

Source: Ref [14]

The size distributions of these are given in Table 1. He then selected typical grade-efficiency performance for each of 18 different types of collectors at three different particle sizes: 50 μm, 5 μm, 1 μm. The overall mass efficiency of collection by each collector was then calculated for each of the three types of dust. This made it possible to list the collectors in order of increasing overall efficiency on each kind of dust. The list is in Table 2.

Next an estimate of the cost of each kind of collecting equipment was made on the basis of a set of assumed conditions. The Total Annualized Cost (TAC), including capital charges, per $1000m^3$ of gas treated was plotted against the overall collection efficiency on a log-probability chart as shown in Figure 11 for the fine dust, and on Figure 12 for all three dusts. It is extremely interesting to note that a straight line correlating Total Annualized Cost with overall efficiency is fairly well defined on this type of plot. These lines could be treated in the same manner as those for particle size distribution based upon Eqn. (2.22), using a median TAC for 50% collection efficiency and a standard geometric dispersion σ_g as in Eqn. (2.31).

TABLE 2 *Efficiency of various gas-cleaning equipment on three typical dusts*

Dust Collector	Collection Efficiencies (%) on		
	Coarse Dust*	Fine Industrial Dust*	Super-fine* Industrial Dust (Fume)
Inertial collector	81.2	58.6	14.3
Medium-efficiency cyclones	84.6	65.3	22.4
Low resistance cellular cyclones	91.3	74.1	29.9
High efficiency cyclone	93.9	84.2	52.3
Impingement scrubber (Doyle type)	95.7	87.9	60.8
Self-induced spray deduster	97.6	92.3	70.3
Void spray tower	97.9	94.5	83.7
Fluidised-bed scrubber	99.0	96.2	82.7
Irrigated-target scrubber (Peabody type)	99.3	97.9	91.8
Electrostatic precipitator	99.5	98.5	94.8
Irrigated electrostatic precipitator	99.6	99.0	97.4
Flooded-disc scrubber-low energy	99.85	99.5	98.3
Flooded-disc scrubber-medium energy	99.90	99.7	99.0
Venturi-scrubber-medium energy	99.94	99.8	99.3
High-efficiency electrostatic precipitator	99.96	99.85	99.4
Venturi-scrubber-high energy	99.97	99.90	99.6
Shaker-type fabric filter	99.97	99.92	99.6
Reverse-jet fabric filter	99.98	99.95	99.8

Source: Ref [14]   *For size distribution see Table 1

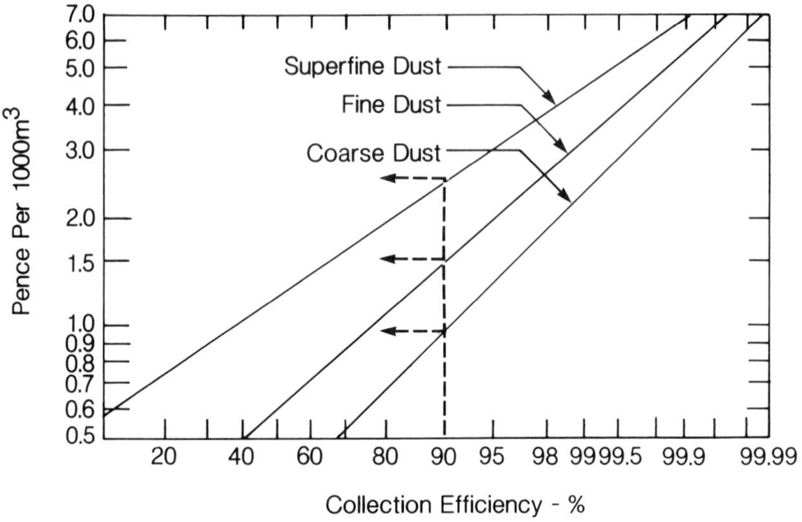

FIG. 11 *Cost of cleaning equipment. Treating fine dust* [14].

1. *Inertial collector*
2. *Medium efficiency cyclone*
3. *Low resistance cyclone*
4. *High efficiency cyclone*
5. *Impingement scrubber*
6. *Self induced spray*
7. *Void spray tower*
8. *Fluidized bed scrubber*
9. *Irrigated target*
10. *Electrostatic precipitator*
11. *Irrigated electrostatic precipitator*
12. *Flooded disc-low energy*
13. *Flooded disc-medium energy*
14. *Venturi-medium energy*
15. *High efficiency ESP*
16. *Venturi-high energy*
17. *Shaker fabric filter*
18. *Reverse-jet fabric filter*

Reproduced by permission from Filtration/Separation.

Stairmand's charts as originally published express costs in pence of equipment available in Great Britain as of the period of the late 1960's. Because of fluctuating relationships between British and U.S. currency, and because of increase due to inflation since that time, these cost numbers must be regarded only as relative values. Methods of making cost estimations at current prices for equipment available in the U.S. are discussed in Sec. B below.

These charts may be useful in preliminary screening studies of collector selection, as will be illustrated below. However

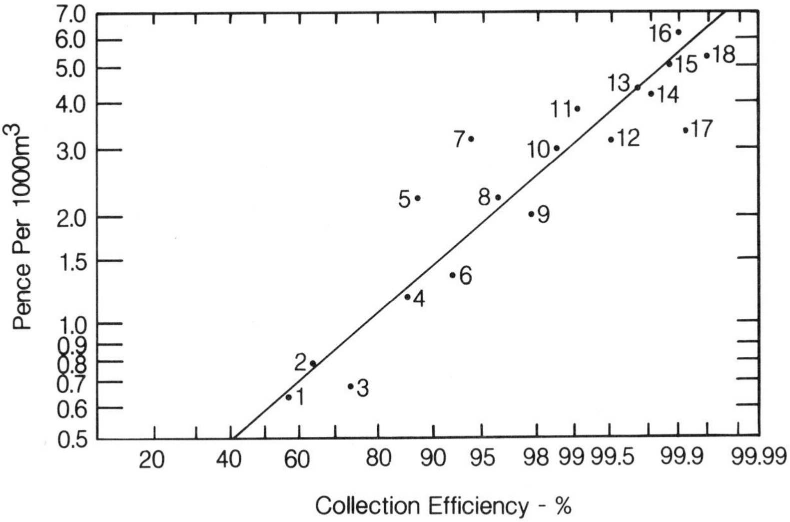

FIG. 12 *Cost of cleaning equipment [14]. Reproduced by permission from Filtration/Separation.*

much care must be observed in using them, particularly with regard to the conditions assumed for the basis of comparison. In specific cases, deviations from one or more of the parameters assumed could make considerable difference in actual performance or cost. The assumptions made in computing the TAC for comparison are:

Gas flow rate: 100,000 m^3/hr at 20°C (58,800 scfm)
Dust loading: 10 gm/m^3 (4.37 grains/ft^3)
Dust Size: as given in Table 1
Combined fan and motor efficiency: 60%
Annual operating time: 8000 hr.
Amortization rate: 10%

Capital cost includes auxiliaries such as fans, pumps, and motors, and allowance for site clearance and erection as follows: 100% of cost of equipment, except for electrostatic precipitators; 50% of cost of equipment for electrostatic precipitators.

One must also be very careful in reading the scales of Figs. 11 and 12, especially the probability scale used for the efficiency axis. Only a few tenths, or even hundredths, of a percentage point lie across the spread of the performance of collectors numbered 12-18, i.e. those above 99.5%. A very small change in some operating condition could easily result in shifting the position of one collector in this group. One could not, for instance, make a sweeping generalization such as "A reverse jet fabric filter (No. 18) is always the most efficient collector" just by glancing at Fig. 11. Finally, it must be pointed out that there are other considerations which will enter into the choice of a collector for a given duty besides its efficiency and its cost, although these are certainly very important. Some of these are discussed below.

B. *COST ESTIMATION*

Rather comprehensive methods are available for making current estimations of the installed cost of particulate collecting equipment together with the operating and maintenance costs to be expected. An excellent treatment of Control Costs has been given by Vatavuk in the "Handbook of Air Pollution Control Technology" [15] from which much of the following has been drawn. This work is also based upon a series of articles by Neveril et al in JAPCA [16], and by Vatavuk and Neveril in Chemical Engineering [17].

The approach outlined here is intended only for making preliminary study estimates of costs. In order to keep the methodology rather simple it is of necessity rather approximate, giving cost estimates within say $\pm$ 30% of detailed cost studies. The data used in developing these methods are taken to be valid as of 1981 in the United States. To update costs to current values, Vatavuk [15] recommends using the Chemical Engineering Plant Cost Index and its components, which are published in Chemical Engineering magazine and revised monthly.

TABLE 3 *Average cost factors for selected air pollution control equipment*

Items	Cyclone	Electrostatic Precipitator	Fabric Filter	Venturi Scrubber
Direct Costs				
Purchased eqpt.				
Collector	A	(These are the same for all kinds of collectors)		
Auxiliaries	B			
Instr. & Control	0.10 (A+B)			
Taxes	0.30 (A+B)			
Freight	0.50 (A+B)			
Base Price	X = total of above			
Installation	0.40X*	0.67X	0.72X	0.56X
Indirect Costs				
Installation	0.35X*	0.57X	0.45X	0.35X
T.I.C.	1.75X*	2.24X	2.17X	1.91X

Source: Ref [15] except * estimated by the author.

For preliminary studies, first the Total Installed Cost (TIC) of the equipment (including collector and all auxiliaries) must be estimated. Table 3 lists the items which go into TIC, broken down into Direct Costs and Indirect Costs, of Installation. Note that the calculation must begin with an estimate of the purchased cost (A) of the collector, and (B) of all auxiliaries. (B) must include allowance for hoods, ducts, fan system, pumps, stack, and perhaps cooling chambers, screw conveyors, and other items peculiar to the system. The value of (A) depends, of course, upon the required size and design of the collector. Specification of these is the major emphasis of each of the chapters in Part II dealing with a particular type of collector. Cost estimation of (A) is also given there.

Direct Installation Costs include: foundation and supports, handling and erection, electrical, piping, insulation, painting, and perhaps site preparation, facilities and buildings. Indirect Installation Costs include: engineering and supervision, construction and field, construction fee, performance test, model study, and an allowance for contingencies. Detailed factors for all these items are given in reference [15], but in Table 3 they are all lumped together into the Multipliers, as shown.

The Total Annualized Cost (TAC) is defined as

$$TAC = D + I - R \tag{3.63}$$

where D = direct operating costs

I = indirect operating costs

R = recovery credits

All items are expressed in \$/year. Direct operating costs include utilities, labor, maintenance, replacement parts, waste disposal, and overhead. The estimation of utilities (electricity, water, etc.) requires a knowledge of the energy needs of the collector, which are discussed in Section IV above. Indirect operating costs cover taxes, insurance, administration and capital recovery as determined by depreciation over the useful life of the equipment. Recovery credits would be taken for the value (if any) of the particulate recovered, in case it could be used. Details of estimating all of these items will be found in the references cited.

C. *CONSIDERATIONS IN SELECTION OF COLLECTORS*

In addition to efficiency, energy requirements, and costs, there are other matters which need to be taken into account when a collector system is to be selected for a given application. Among these are the physical and chemical nature of the particulate, its ultimate disposal, questions of hazards and safety in operation, cleanliness and sanitation, ease of maintenance, etc.

The subject of hazards in dust collection systems is so important as to require special emphasis, particularly with respect to the possibility of explosions and fires. A concise summary of the problem has been given by Daveloose [18], and a detailed treatise by Palmer [19]. The generation of static electricity, or other sources providing an electrical discharge, may lead to ignition or explosion especially when certain conditions are present. These include: the formation of a dust cloud composed of a combustible material, generally finer than 200μm, in concentrations ranging from 15-40 gm/m^3 (6.5-17.5 gr/ft^3) up to 2000-3000 gm/m^3 (870-1300 gr/ft^3), a range which is almost certain to occur somewhere in every collection system. (Below the lower limit of this range flame is not propagated; above the upper, it is smothered for lack of oxygen.) Low humidity in the gas stream promotes the formation of a stable cloud, and also prevents the "drainage" of electrostatic charges. Various mechanical actions, such as are present in filter systems, may also promote cloud formation or cause sparks. In addition to elimination of sources of sparks and flames, prevention of fires may also be enhanced by using anti-static or conducting materials of construction and by careful grounding of all equipment.

Each type of collector has it advantages and disadvantages with respect to all of these items under consideration. Tabulations of the pros and cons are given along with the theory of operation for each of the collectors in Part II. Details of these matters are not within the scope of this text, but some of them are illustrated in the following example of a preliminary screening study. There is also an excellent example of such a study, including cost estimates, given in Ref. [17].

Example 6. A powdered milk product produced by spray drying has the following size-distribution as determined by a Bahco analysis.

Size - microns	% by weight
> 50	62.0
20 - 50	22.0
15 - 20	6.0
10 - 15	4.0
5 - 10	3.6
2 - 5	1.7
< 2	0.7

The residual air stream to be treated will flow at the rate of 34,300 acfm at 220°F, carrying 385 lb/hr of powder. The emission rate allowed by the State E.P.A. regulations is 9.49 lb/hr. Carry out a preliminary study survey to identify possible collectors which might be used in meeting this requirement.

Collection efficiency required = 1 - 9.49/385 = 0.975

Grain-loading of stream = 385 x 7000/60 x 34300 = 1.31 gr/acf

Examination of Table 1 shows that this powder corresponds reasonably well with Stairmand's Coarse Dust. Table 2 then indicates that collectors lying above No. 5 (impingement scrubber) would be required to meet the overall efficiency. The choice lies among some type of scrubber, an electrostatic precipitator, and a fabric filter. The grain-loading is so low as to cause no special problem.

This product is a flammable organic material which is a foodstuff. Although the initial grain-loading is below the lower flammable limit mentioned above, concentration in the neighborhood of a collecting surface will rapidly move into the flammable range as collection procedes. If recovery of some 375 lb/hr is economically worthwhile it must be done in a clean and sanitary manner, and without risk of dust explosion. This rules out the use of an electrostatic precipitator because of the danger of dust ignition by sparking. In a wet (scrubber) recovery system, the solution obtained would have to be concentrated and redried by recycling it to the spray drier system. Otherwise it would be discarded into a drain, and the consequences of this form of emission (water pollution ?) from the plant would have to be considered.

With a fabric filter, recovery of clean powder might be difficult as it might tend to become contaminated by the fabric. The air stream is probably rather humid, so that care would have to be taken that it did not cool below its dewpoint and deposit a damp, sticky cake on the fabric. Under humid conditions mold might tend to grow in the residual cake on the fabric as well. A fabric filter might be satisfactory if the cake is to be discarded.

While the efficiency values clearly show that a single cyclone could not work, there is the possibility of using two cyclones in series. Table 2 shows that a medium-efficiency cyclone ($\eta_M = 0.846$) in series with a high-efficiency cyclone ($\eta_M = 0.939$) would have an overall $P_T = (1 - 0.846)(1 - 0.939) = 0.0094$ (see Eqn. 3.34) which would be sufficient. For ease in cleaning all equipment would need to be made of stainless steel.

While obviously more information is needed about this problem, certain steps are clear from the preliminary study: it is critical to know whether it is necessary or desirable to recover the collected powder; accordingly consider scrubber, fabric filter, and perhaps two cyclones in series; grain-loading is so low that a primary collector would not otherwise be needed.

D. SUMMARY

The concepts discussed in this chapter may be summarized by a general system chart, Figure 13. This shows that the principal inputs to the collector system performance are: (a) the type of collector and its specific dimensions, (b) the operating conditions (gas flow, temperature, grain-loading etc.), and (c) the particle-size distribution of the particulate to be collected. The interaction of (a) and (b) produces the grade-efficiency which interacts with (c) to determine the outputs of the system: particulate emitted and particulate collected (quantities and size distribution). Finally (a) and (b) also determine the energy requirements, including the pressure drop.

The chapters in Part II have as their objective the application of this general system model to the specific collectors discussed. Each has a detailed system chart pertaining to the particular forces, design, and conditions involved.

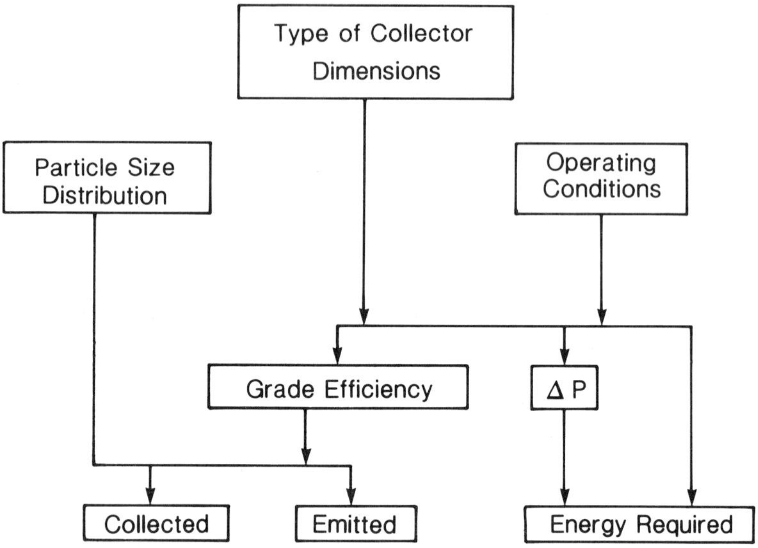

FIG. 13 *General systems analysis of particulate collector performance.*

REFERENCES

1. C.D. Sargent, Chem. Eng., 130 (Jan. 27, 1969).
2. J. H. Abbot and D.C. Drehmel, Chem. Eng. Prog., 72: No. 12, 47 (Dec. 1976).
3. J.D. McCain, K.M. Cushing, and W.B. Smith, Paper 74-117 67th Annual Meeting of A.P.C.A., Denver, Colorado, June 1974.
4. Y.S. Cheng, C.L. Carpenter, E.B. Barr, and C.H. Hobbs, Aerosol Sci. and Technol., 4: 175 (1985).
5. R.E. Sundberg, J. Air Poll. Cont. Assoc., 24: 758 (1974).
6. W.M. Vatavuk, EPA - 450/2 73-002, Aug. (1973).
7. W.D. Holland and R.E. Conway, Chem. Eng. Prog., 69: 93 (1973).
8. M. Crawford, Air Pollution Control Theory, McGraw-Hill, New York (1976), pp. 145-154.
9. C.J. Stairmand, J. Inst. Fuels (London), 29: 58 (1956).
10. A.E. Vandergrift, L.J. Shannon, and P.G. Gorman, Chem. Eng., 80: Deskbook Issue, 107 (June 18, 1973).

11. J.R. Benson and M. Corn, J. Air Poll. Cont. Assoc., 24: 340 (1974).

12. J.J. Stukel and H.G. Rigo, Atmos. Environ., 9: 529 (1975).

13. S.L. Soo, Environ. Sci. and Tech., 7: 63 (1973).

14. C.J. Stairmand, Filtration and Separation, 7: 53 (1970).

15. S. Calvert, H.M. Englund, ed, Handbook of Air Pollution Technology, Chap. 14 Control Costs, W.M. Vatavuk, John Wiley & Sons Inc., New York (1984).

16. R.B. Neveril, J.V. Price, and K.L. Engdahl, J. Air Poll. Cont. Assoc., 28: 829, 963, 1069, 1171 (1978).

17. W.M. Vatavuk, and R.B. Neveril, Chem. Eng., 91: 97 (Apr. 2, (1984) (References to 16 preceding Parts of this series are given also).

18. F. Daveloose, Filtration and Separation 22: 182 (1985).

19. K. N. Palmer, Dust Explosions and Fires, Chapman and Hall, London (1973).

PROBLEMS

1. The particle size distribution of a certain dust, as obtained by an analysis conducted partly by a Coulter Counter and partly by an Anderson Impactor, may be represented by two straight lines on a log-probability plot. These lines intersect at 5.5 μm and 19.5% finer-than, with the Coulter portion having a $\sigma_g = 0.205$, and the Anderson portion a $\sigma_g = 11.1$, with the Anderson covering the finer range. This dust is to be collected by a device which has a grade efficiency performance given by the following equation:

$$\eta_M = 1 - \exp -2\ [1.58 \times 10^{-2}\ d_p^2]^{0.315}$$

where d_p is in microns. For this operation, find:

a. the "cut diameter";

b. the overall efficiency;

c. the particle size distribution of the dust emitted;

d. the rate of emission, per 100 kg of dust fed.

2. Suppose the collection process described by the data of Problem 1 is to be improved to produce an overall collection efficiency of 90%. To what new value would the constant 1.58×10^{-2} have to be changed, assuming the exponent 0.315 is not changed?

3. Check the test results of Example 3, as given in Fig. 6, to see whether any of the elementary models may reasonably be used to describe the grade efficiency. Do the same for the impingement scrubber of Fig. 2.

4. The grade efficiency of a certain gravity (settling chamber) collector was found to be 20% on particles of a certain size and 81% on particles twice as large. Assuming the particles obey Stokes Law, what type of model might represent this collector performance? Would your answer be the same if the grade efficiencies were 25% and 50% respectively?

5. The power consumption of a certain collector was measured as 15 kw. It was processing a stream of gas having an average molecular weight of 32, at 300°F and 15.2 psia, through a duct 18" by 36" in cross-section at an inlet velocity of 50 ft/sec.
Find: (a) the pressure drop across the collector;
(b) the number of inlet velocity heads of frictional energy loss.

6. It is desired to make a preliminary survey for a method (or system) to collect dust from a plant under the following conditions:

Sample of dust:	50% by weight below 30 μm
	88% by weight below 70 μm
	density = 1 gm/cm
Gas flow rate:	30,000 ± 5,000 acfm
Grain loading:	0.5 -4.0 gr/ft^3; average 1.5 gr/ft^3
Emission limitation:	1.5 lb/hr.

Investigate and compare possible collector systems which might do the job, considering both efficiency and relative cost.

7. It has become necessary to control the emission of cement dust from the kiln of a Portland Cement plant in which the operating conditions are as follows: temperature = 250°F; pressure = 1 atm; feed rate to kiln = 5 tons/hr; emission rate of dust (uncontrolled) = 230 lb/ton of feed; air flow = 159,600 acf/ton of feed. The dust may be regarded as equivalent of Stairmand Fine. The emission regulations are given in Chapter 1.

a. Select some possible kinds of collection equipment which might be considered in order to meet this requirement. Indicate their relative costs and power consumption.
b. What will be the grain-loading in the feed to the collection system?
c. Could a cyclone collector be used in any way? If so, or if not, assuming the inlet duct to be 2.28 ft by 1.09 ft, and the value of N = 9 inlet velocity heads, estimate the pressure drop across the cyclone, and the power consumption for the operation.

4 Elementary Particle Mechanics: Movement of Aerosol Particles in Still Gas

I. PARTICLE TRAJECTORIES

The study of particulate collection devices from a fundamental viewpoint requires a knowledge of the principles of mechanics which govern the motion of aerosol particles. The path followed by a particle while it is in the zone of influence of the collecting force(s) determines whether it will collide with the collecting surface or target. It is necessary to be able to predict this path, or trajectory, in order to develop models of collector performance. The methods of doing this will be presented in two parts: first, for motion in still air, and then for particles which are in a stream of moving air.

In certain types of collectors e.g., scrubbers and filters, there may also be a need to determine interaction between the fluid and individual target elements. For example, in a scrubber the target elements are drops of scrubbing liquid in motion relative not only to the dust particles, but also to the air stream. Certain aspects of target motion are treated by the same fundamental principles as apply to the determination of dust particle trajectories. In what follows, wherever the word particle is used it may, in general, be understood to refer either to dust particles or to target elements.

A. *NEWTON'S LAW*

The motion of an individual particle, under the influence of various forces, is governed by Newton's Law of motion: The sum of all forces acting upon a particle = mass of particle x acceleration of particle. This is a vector differential equation: the various forces are vectors, and their sum is a vector addition. The resulting acceleration is a vector. In symbols it is usually written:

$$m_p \, d\bar{u}/dt = \sum \bar{F} \tag{4.1}$$

where the acceleration is represented as the derivative of the particle velocity vector.

A particle trajectory is determined by identifying all of the forces acting in a given situation and representing each of them by the appropriate function. Then the differential equation is solved (in principle) by integration in two stages: first to give the velocity as a function of time, and then, replacing $\bar{u}$, by ds/dt, to give distance (or position) as a function of time. This last equation is that of the trajectory of the particle. The solution is carried out by rewriting the vector equation in terms of its scalar component equations in one, two, or three dimensions as may be required.

Each of the kinds of force which may be encountered in aerosol motion will be studied separately, and then in combination. Solutions will be presented for the various trajectories which are usually involved.

B. *FORCES ON PARTICLES*

1. *Drag force* a. Drag Coefficients - Steady Motion. Whenever there is motion of a particle relative to the surrounding air,

this motion will be resisted or opposed by a "drag force." This is the most fundamental force to be considered for it is always present, even though no other force may be acting appreciably. In particulate collector design work, there is a need to determine drag forces acting upon (a) dust particles themselves, and (b) target elements of target type collectors.

Drag arises out of two phenomena. Due to the shape of the particle (or target), fluid must be displaced around it as it moves. This sets up a greater pressure on the front of the particle than on the rear. The resulting force is called "form-drag." In addition, there is friction between the particle and the adjacent fluid, or between nearby layers of fluid. This results in "frictional-drag" force. The two are usually considered together as producing one overall effect, or force. This is obviously determined by factors such as shape and size of particle, nature of particle surface, speed of motion, nature and properties of the fluid.

For a full treatment of the fluid mechanics of drag, the reader is referred to any standard text on fluid mechanics. In addition, there is an excellent little paperback by Shapiro called "Shape and Flow" [1]. This presents a well-illustrated account of experiments which show the fundamental phenomena involved, and the way in which drag forces depend upon the parameters noted.

The results of theoretical and experimental studies of drag lead to a fairly simple method of calculating it in any particular case. The total drag force may be represented in terms of a drag coefficient C_D which is defined by this equation:

$$F_D = C_D \times A_{proj} \times \rho_f u^2/2 \tag{4.2}$$

This is a scalar equation to give the magnitude of the vector drag force; the direction is exactly opposite to that of the velocity vector. In this equation the term A_{proj} represents the cross-sectional area of the particle (or target) image as projected in the direction of motion. The drag force per unit of projected area is seen to be proportional to the kinetic energy represented by a unit volume of fluid displaced at the particle velocity.

Values of the drag coefficient have been determined by applying Eqn. (4.2) indirectly to experimental data in which the equivalent of F_D, A_{proj} and other terms have been measured. An orderly scheme has emerged for the correlation of these values with the other parameters. For particles of a given kind of shape, C_D may be correlated with the particle Reynold's number, defined as

$$Re_p = x_p \rho_f u/\mu = x_p u/\upsilon_f$$

where x_p is some appropriate linear dimension of the particle (or target), and υ_f is the kinematic viscosity of the fluid. Figure 1 shows this correlation for several shapes. It is apparent that there are several regimes of flow which may be identified by a corresponding range of values of Re. These data apply only to a particle in steady motion with a constant velocity relative to the fluid.

Although drag coefficients have been studied up to very large values of Re_p, as indicated in Fig. 1, applications to conditions where $Re_p > 800$ seldom arise in dust collection work. Further discussion, therefore, will be limited to the lower range of Re_p values. This range may be subdivided into several portions according to the nature of the drag phenomena operative in each portion.

From this point on the particle will be regarded either as being spherical in shape with x_p replaced by d_p, the spherical diameter, or as behaving in a manner equivalent to that of a spherical particle. A particle of irregular shape may thus be characterized by a "size" which is the diameter of an "equivalent" sphere in the sense of having the same drag force.

Equivalent here is used in the sense of the motion of the sphere being the same as that of the particle. For material of unit density (1 gm/cm^3) the size is referred to as the

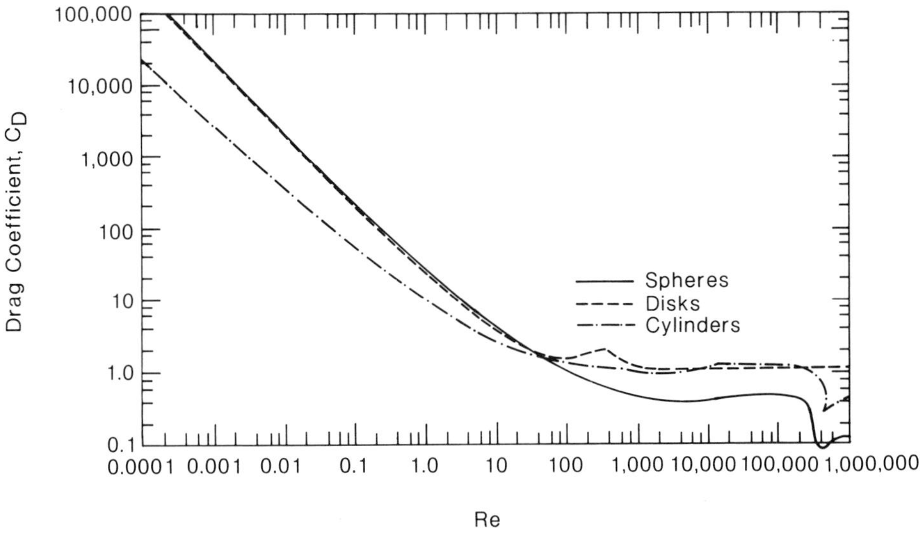

FIG. 1 *Drag coefficients for spheres, disks, and cylinders* [2]. *Reproduced by permission from Perry's Chemical Engineers Handbook, 5th Edition.*

"aerodynamic diameter" of the particle. In this context, the drag coefficient for spherical shape only may be examined in detail as shown in Fig. 2. In the range of Re_p below approximately 0.1 there is evidently a simple relationship for C_D. This has been shown theoretically and confirmed experimentally to be

$$C_D = 24/Re_p \tag{4.3}$$

It corresponds to what is called Stokes Law for drag. Thus

$$F_D = \frac{24}{Re_p} \times A_{proj} \times \frac{\rho_f u^2}{2}$$

For spherical particle $Re_p = d_p u/\upsilon$ and $A_{proj} = \pi d_p^2/4$. Then

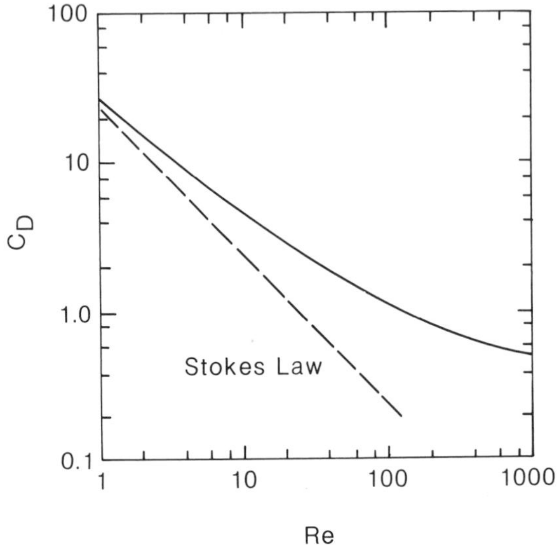

FIG. 2 *Drag coefficients for spheres.*

$$F_D = \frac{24\mu}{d_p u \rho_f} \times \frac{\pi d_p^2}{4} \times \frac{\rho_f u^2}{2} = 3\pi d_p \mu u \tag{4.4}$$

which is Stokes Law [3]. It applies to a regime of pure laminar flow.

As the value of Re_p increases above 0.1, there is a gradual but definitely increasing deviation from the simple relationship of (4.3). The drag coefficient becomes greater than would be anticipated by Stokes Law. This effect is attributed to the gradual onset of turbulence in the fluid surrounding the particle, and especially in the wake of the particle.

A number of equations, both semi-theoretical and empirical, have been proposed for the calculation of C_D from Re_p to give reasonable agreement with the experimental values. The following are recommended to be used up to $Re_p = 800$, in order to give agreement within ± 2% over this range, and at the same time to be relatively simple for calculation.

$Re_p < 0.1$	$C_D = 24/Re_p$	Stokes	[3]
$0.1 < Re_p < 0.5$	$C_D = 24/Re_p + 4.5$	Oseen	[4]
$0.5 < Re_p < 3.0$	$C_D = 24/Re_p + 3.60/Re_p^{0.313}$	Schiller & Naumann	[5]
$400 < Re_p < 800$			
$3.0 < Re_p < 400$	$C_D = 24/Re_p + 4/(Re_p)^{1/3}$	Klyachko	[6]

The Klyachko equation may be used over the entire range of Re_p from 0.5 to 800 with an error generally within ± 3-4%. It is superior to that of Dickinson and Marshall [7] which is stated to be valid within ± 7% for $Re_p < 3000$:

$$C_D = 0.22 + 24(1 + 0.15\ Re_p^{0.6})/Re_p.$$

b. Applicability of Stokes Law: Cunningham Correction. In a great majority of applications to the motion of dust particles, such as are encountered in air pollution control problems, conditions fall within the realm of Stokes Law. For example, consider motion in air at 20°C where $\upsilon_f = 0.151\ cm^2/sec$. In order for Re_p to be ≤ 0.1, it is necessary only that $d_p \leq 151/u$, d_p in microns, u in cm/sec. Figure 3 shows the maximum particle size which will be in the Stokes Law range at various relative velocities and air temperatures.

So widespread is the applicability of this range that it is customary to assume Stokes Law applies as a basic principle in much of dust collection theory. This greatly simplifies the calculation of drag forces. However, one should always check this assumption if errors of more than 2% arising from incorrect drag calculations are to be avoided. On occasion it may be necessary to use one of the more complicated methods of calculation of C_D according to the range of Re_p, instead of simply using Stokes Law. For example, this situation arises in calculating the motion

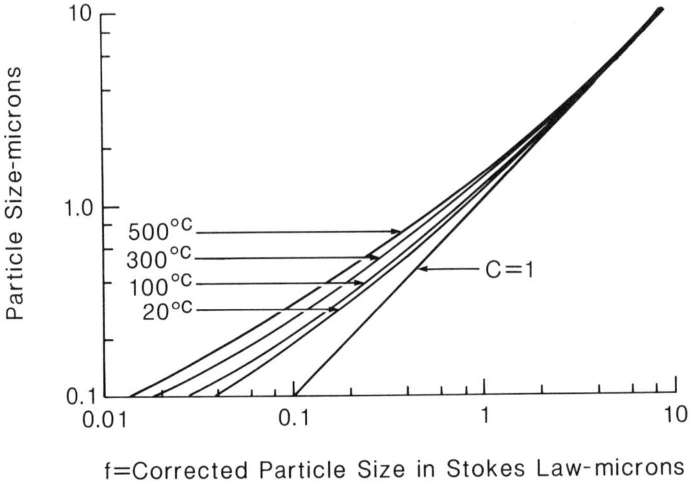

FIG. 3 *Range of applicability for Stokes Law. Conditions of velocity, particle size, and temperature within triangular zone to left of slanting solid line and above dashed line.*

of liquid droplets in scrubber operation, where values of d_p of the order of 200 microns and u of more than 6000 cm/sec are not uncommon.

Another correction must be taken into account whenever the particle size is small enough to be of the order ot the mean free path (λ) of the gas molecules. Under this condition, the gas no longer behaves as a continuous medium with respect to the particle and its drag effect is reduced. To correct for this so-called "slip" condition, a correction factor known as the Cunningham factor C, is introduced into Stokes Law.

$$F_D = \frac{3\pi\mu d_p u}{C} = 3\pi\left(\frac{d_p}{C}\right)\mu u \qquad C > 1 \tag{4.5}$$

Both theoretically and experimentally it is found that the value of C depends upon the ratio of the mean free path of the gas molecules to the size of the particle, expressed as the Knudsen Number, $Kn = 2\,\lambda/d_p$. Knudsen and Weber [8] showed that

$$C = 1 + A\ Kn \tag{4.6}$$

with

$$A = \alpha + \beta \exp(-\gamma/Kn)$$

At low values, $Kn \leq 0.1$, the gas behaves as a continuum with respect to the particle, the value of A becomes constant at α and the value of C approaches 1. For $Kn > 10$ the gas is in the free molecular regime, A again approaches a constant value $(\alpha + \beta)$ asymptotically and C becomes very large. There is a transition range for $0.1 < Kn < 10$.

Until recently the experimental evaluation of α, β, and γ has depended largely upon the work of Milliken and his students (1910 - 1923), in the famous oil-drop experiments in which liquid droplets were used as the particles. Davies proposed [9] an equation now widely used to correlate these data, as

$$C = 1 + Kn\ [1.257 + 0.400 \exp(-1.10/Kn)] \tag{4.7}$$

in which $A \rightarrow 1.257$, and $A \rightarrow \alpha + \beta = 1.657$, at the lower and higher extremes respectively.

To use this equation Kn is calculated by taking λ = mean free path of gas molecules, based upon the Chapman-Enskog equation.

$$\lambda = \upsilon/0.499\bar{u} \tag{4.7a}$$

in which $\bar{u}$ = mean molecular velocity, cm/sec, given according to the kinetic theory by

$$\bar{u} = \sqrt{8RT/\pi M} \tag{4.7b}$$

from which, for a perfect gas,

$$Kn = 0.0226 \frac{\mu}{d_p p} \sqrt{\frac{T}{M}} \tag{4.7c}$$

where μ = gas viscosity, poises; T = absolute temperature °K; P = gas pressure, atmos.; d_p = particle diameter, cm.; M = molecular weight; R = gas constant. 8.31×10^7 ergs/°K mole.

This correction depends upon temperature, pressure, and particle size, being greater at higher temperatures, lower pressures, and for smaller particles. Figure 4 indicates the magnitude of the Cunningham correction to the drag force F_D by giving the value of d_p/C as a function of d_p at several temperatures, for dry air under 1 atmosphere pressure.

As a rough guide to the magnitude of the correction at 20°C, 1 atm: $C = 1 + 0.165/d_p$ (d_p in microns). For particles 8 microns and smaller the correction will be more than 2%. Figure 3 indicates, at each temperature, the minimum particle size for which the correction will be less than 2% at atmosphere pressure. On this graph the triangular zone above and to the left of the lines marked with a given temperature indicates the range of

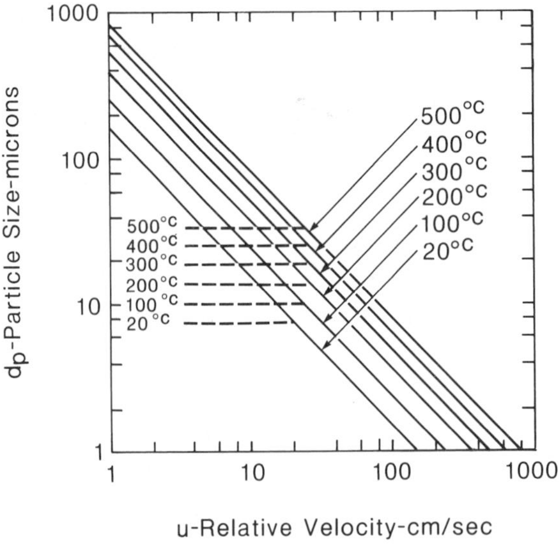

FIG. 4 *Slip factor correction on Stokes Law. Dry air at 1 atmosphere;* $F_D = 3\pi f \mu u$; $C = d_p/f$ = *Cunningham factor.*

velocity and particle size within which Stokes Law may be used without correction, at that temperature.

The validity of Eqn. (4.7) rests upon rather limited experimental data obtained only for air at temperatures up to 80°C and at pressures of 1 atmosphere and below. To apply it to other gases at elevated temperatures and pressure, must involve an extrapolation of an unknown degree of uncertainty. However, it is obvious that the mean free path should tend to be smaller at higher pressures and lower temperatures.

Willeke [10] has considered this problem from a theoretical point of view and derived formulas to be used at elevated temperatures. They are:

$$Kn = \frac{6.052 \times 10^{-4}}{d_p P} \frac{T}{(1 + 110/T)} \tag{4.8}$$

$$C = 1 + Kn\,[1.246 + 0.42 \exp - 0.87/Kn] \tag{4.9}$$

in which T = °K, d_p = microns, P = atmospheres. Here $A \to 1.246$, and $A \to \alpha + \beta = 1.666$ at the lower and higher extremes respectively. These equations are probably better than (4.7) and (4.7c) but no data are available for checking.

Recently Allen and Raabe [11] have performed a new series of experiments, very carefully using an improved Milliken-type apparatus and solid spherical particles. They correlate a large number of their experimental results with

$$C = 1 + Kn\,[1.142 + 0.558 \exp(-0.999/Kn)] \tag{4.10}$$

in which $A \to 1.142$, and $A \to \alpha + \beta = 1.700$ at the lower and higher extremes respectively. They associate the differences in the values of C between their results and Milliken's, with the difference in the fraction of gas molecules undergoing specular reflections between liquid drops and between solid particles.

Although the values of α, β, and γ seem to differ quite a bit among these three correlations, the net result on values of C is not very great. Agreement is generally within 2%, being much better at low values of Kn and a little poorer at very high Kn. In this text the Davies correlation is used whenever values of C are quoted. If more precision is required for solid particles, Eqn. (4.10) may be used.

The concept of a "Stokes Law Particle" is sometimes used to refer to an equivalent spherical particle moving under conditions such that Stokes Law applies and the Cunningham factor is equal to 1, within a combined accuracy of say less than ± 1%. A whole set of special equations for various calculations may be developed for Stokes Law Particles. These will be indicated as they arise.

A more general definition of "aerodynamic diameter", incorporating the Cunningham factor, has been given by Calvert [12] as

$$d_{p_a} = d_p \sqrt{C\rho_p}$$

This will be utilized in scrubber calculations, Chapter 9.

The procedure for determining an appropriate value for C_D and F_D is illustrated in the following example.

Example 1. Determine the drag force on a spherical particle moving through still, dry air in each of the following cases:

(a) d_p = 100 μm, u = 100 cm/sec, t = 20°C, P = 1 atm.
(b) d_p = 1 μm, u = 10 cm/sec, t = 100°C, P = 1 atm.
(c) d_p = 40 μm, u = 30 cm/sec, t = 20°C, P = 1/2 atm.
(d) d_p = 0.1 m, u = 10 cm/sec, t = 300°C, P = 1/2 atm.

For each case the procedure will be the same. First, calculate the Re_p. If $Re_p > 0.1$, Stokes Law will not apply and C_D must be calculated from an appropriate expression as listed above, or read from Fig. 2. The drag force will then be calculated from

Eqn. (4.2). If $Re_p \leq 0.1$, Stokes law applies and the drag force may be calculated directly from Eqn. (4.4) provided no Cunningham correction is needed. To determine this, calculate Kn from Eqn. (4.7c) and if Kn > 0.016, determine C from (4.7). Then use (4.5) for the drag force. Whenever the pressure is equal to 1 atm., Figs. 3 and 4 may be used to shorten the work.

(a) At 20°C, 1 atm., $\mu = 1.81 \times 10^{-4}$ poises, $\rho_f = 1.205 \times 10^{-3}$ gm/cm^3, $\upsilon = 0.151$ cm^2/sec. Figure 4 indicates no Cunningham correction is needed.

$$Re_p = \frac{100 \times 10^{-4} \times 100}{0.151} = 6.62 >> 0.1$$

Using Klyachko formula $C_D = 5.75$ (Fig. 2 reads $C_D = 5.6$):

$$F_D = 5.75 \times \frac{\pi(100 \times 10^{-4})^2}{4} \times \frac{1.205 \times 10^{-3}(100)^2}{2}$$

$$= 2.72 \times 10^{-3} \text{ dynes}$$

(b) At 100°C, 1 atm, $\mu = 2.17 \times 10^{-4}$ poises, $\rho_f = 0.940 \times 10^{-3}$ gm/cm^3, $\upsilon = 0.231$ cm^2/sec. Figures 3 indicates that Stokes Law applies, but a Cunningham correction is needed. Figure 4 reads f = 0.82 μm (C = 1/0.82 = 1.22) for d_p = 1 μm at 100°C.

$$F_D = 3\pi \ f \ \mu \ u = 3\pi \times 0.82 \times 10^{-4} \times 2.17 \times 10^{-4} \times 10$$

$$= 1.68 \times 10^{-6} \text{ dynes}$$

(c) At 20 C, 1/2 atm. $\mu = 1.81 \times 10^{-4}$ poises, $\rho_f = 0.603 \times 10^{-3}$ gm/cm^3, $\upsilon = 0.302$ cm^2/sec. Since P ≠ 1 atm. Figs. 3 and 4 cannot be used.

$$Re_p = \frac{40 \times 10^{-4} \times 30}{0.302} = 0.397 > 0.1, \text{ Stokes Law does not apply}$$

Using Oseen formula:

$$C_D = \frac{24}{0.397} + 4.5 = 65. \text{ (off-scale Fig. 2)}$$

For a particle of 40 μm, a Cunningham correction is probably unnecessary:

$$Kn = \frac{0.0226 \times 1.81 \times 10^{-4}}{40 \times 10^{-4} \times 0.5} \sqrt{\frac{293}{28.9}} = 6.5 \times 10^{-3} \qquad C = 1.008$$

$$F_D = 65 \times \frac{\pi(40 \times 10^{-4})}{4} \times \frac{0.603 \times 10^{-3}(30)^2}{2} = 2.22 \times 10^{-4} \text{ dynes}$$

(d) At 300 C, 1/2 atm, $\mu = 2.93 \times 10^{-4}$ poises, $\rho_f = 0.307 \times 10^{-3}$ gm/cm^3, $\upsilon = 0.954$ cm^2/sec. Since $P \neq 1$ atm. Figs. 3 and 4 cannot be used.

$$Re_p = \frac{0.1 \times 10^{-4} \times 10}{0.954} = 1.05 \times 10^{-4} < 0.1$$

Stokes Law applies but undoubtedly a Cunningham correction is needed.

$$Kn = \frac{0.0226 \times 2.93 \times 10^{-4}}{0.1 \times 10^{-4} \times 0.5} \sqrt{\frac{573}{28.9}} = 5.89 \qquad C = 10.36$$

$$F_D = 3\pi \times \frac{0.1 \times 10^{-4}}{10.36} \times 2.93 \times 10^{-4} \times 10 = 2.66 \times 10^{-8} \text{ dynes}$$

c. Deceleration Due to Drag. When no force other than drag is operating, it is obvious that a particle cannot maintain a steady motion relative to the air and that a deceleration must occur. For a spherical particle moving linearly in still air, Newton's law in one-dimension becomes

$$\frac{\pi d_p^3}{6} \rho_p \frac{du}{dt} = - F_D = - C_D \frac{\pi d_p^2}{4} \times \frac{\rho_f u^2}{2} \tag{4.11}$$

This indicates that the deceleration due to drag is

$$\frac{du}{dt} = - \frac{3}{4} C_D \frac{u^2}{d_p} \times \frac{\rho_f}{\rho_p} \tag{4.12}$$

Before proceeding with the integration of (4.12), consideration must be given to whether the values of C_D obtained for steady motion will also apply to accelerating (or decelerating) motion. According to Fuchs [13], for Reynolds' numbers not exceeding a few hundred, it is possible, without making a large error, to assume that the drag resistance of the air does not depend upon the acceleration. The experimental evidence for this will be discussed below. At this point, the effect of acceleration upon C_D (if any) will be neglected in order that Equation (4.12) may be solved.

(1) Time of travel. The time of travel required for the velocity to decrease from u_o at t = 0, to u may then be found:

$$t = - \frac{4}{3} \frac{\rho_p}{\rho_f} d_p \int_{u_o}^{u} \frac{du}{C_D u^2} \tag{4.13}$$

To evaluate the integral it must be transformed in order to utilize the relationship between C_D and Re_p. Replacing u by $Re\, \mu / d_p \rho_f$, (4.13) becomes:

$$t = - \frac{4}{3} \frac{\rho_p d_p^2}{\mu} \int_{Re_o}^{Re} \frac{d(Re)}{C_D Re^2} \tag{4.14}$$

(2) Distance of travel. The linear distance x traversed by a particle while the velocity is decreasing from u_o to u may also

be found by placing u = dx/dt, replacing dt/u in (4.12), and integrating. Thus

$$x = -\frac{4}{3}\frac{\rho_p}{\rho_f} d_p \int_{u_o}^{u} \frac{du}{C_D u} \tag{4.15}$$

Again, transforming the integral into terms of Re,

$$x = -\frac{4}{3}\frac{\rho_p}{\rho_f} d_p \int_{Re_o}^{Re} \frac{d(Re)}{C_D Re} \tag{4.16}$$

The relationship between distance and time of travel is then

$$\frac{x}{t} = \frac{u}{\rho_f d_p} \frac{\int_{Re_o}^{Re} \frac{d(Re)}{C_D Re}}{\int_{Re_o}^{Re} \frac{d(Re)}{C_D Re^2}} \tag{4.17}$$

Evaluation of the integrals in (4.14), (4.16) and (4.17) depends upon the relationship between C_D and Re. If the effect of acceleration upon drag may be neglected, then the integrals may be calculated using the C_D-Re relationship for steady motion. For this purpose, it is useful to have plots of $1/C_D Re^2$, and $1/C_D Re$, vs. Re to evaluate the integrals. These are shown in Figs. 5 and 6.

(3) Drag coefficient-accelerating motion. Torobin and Gauvin [14] have presented an extensive review of the investigations reported on the effect of acceleration upon drag coefficient. They define a modified drag coefficient for accelerating particles as

$$C_{DA} = \frac{F_{DA}}{\frac{1}{2}\rho_f u^2 \times A_{proj}}$$

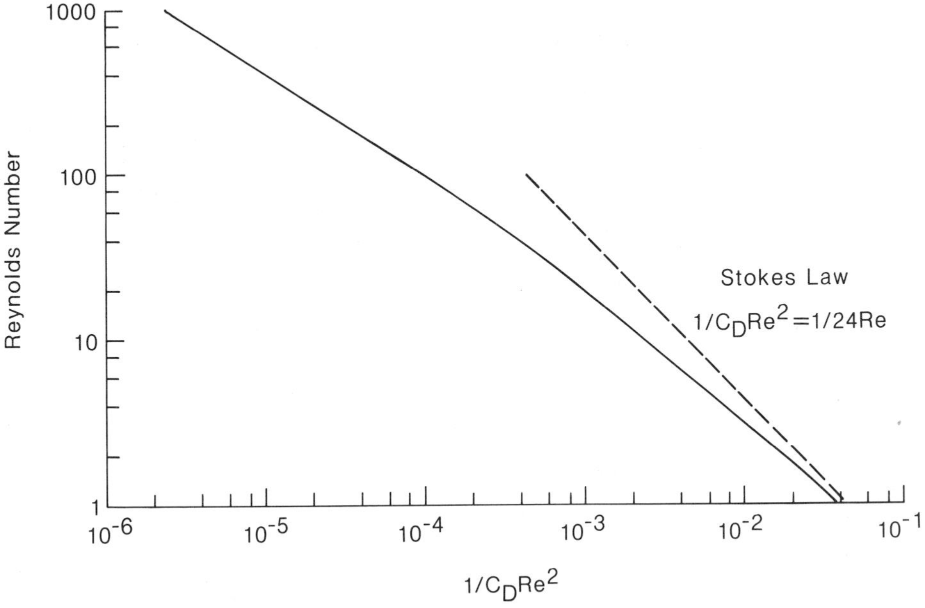

FIG. 5 *$1/C_DRe^2$ for Equations (4.14) and (4.17).*

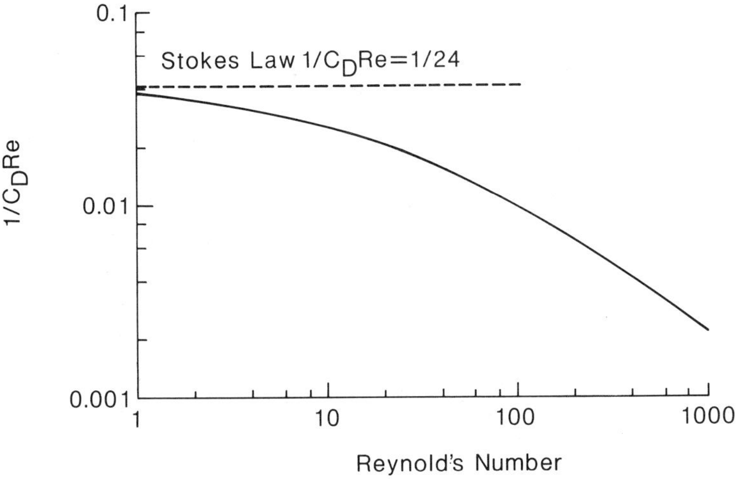

FIG. 6 *$1/C_DRe$ for Equations (4.16) and (4.17).*

where F_{DA} is the drag force acting upon the accelerating particle. At low values of Re_p, experiments indicate that $C_{DA} = C_D$, but at higher values of Re_p some experiments indicate that $C_{DA} > C_D$, others that $C_{DA} < C_D$. There is some indication that $C_{DA} > C_D$ for accelerating particles, and that $C_{DA} < C_D$ for decelerating particles.

For example, the work of Ingebo [15] gives, for deceleration:

$$C_{DA} = 27\ Re_p^{-0.84} \qquad 6 < Re_p < 500$$

in which $C_{DA} < C_D$ throughout the range. However, the experimental evidence is not conclusive, and other effects such as the roughness of the particle and the rate of the acceleration also seem to be involved. Until definitive work has been done on this subject, it is probably satisfactory to follow Fuchs' recommendation, except perhaps in the higher range of Re values encountered in scurbber systems. This question will be considered further in Chapter 9.

(4) Relaxation time and stopping distance. Two very useful concepts emerge from consideration of the deceleration of a Stokes Law particle. Eqn. (4.12) becomes

$$\frac{du}{dt} = -\frac{18\mu}{\rho_p d_p^2} u = -\frac{u}{\tau} \tag{4.18}$$

The quantity $\tau = \rho_p d_p^2/18\mu$ is a basic property of the particle-air system which is called the particle <u>relaxation time</u>. From Eqns. (4.14) and (4.16),

$$t = \tau \ln u_o/u \tag{4.19a}$$

$$u = u_o e^{-t/\tau} \tag{4.19b}$$

$$x = \tau(u_o - u) = u_o\tau(1 - e^{-t/\tau}) \tag{4.19c}$$

and from (4.17)

$$x/t = \frac{(u_o - u)}{\ln(u_o/u)} = u_{\text{log mean}} \tag{4.19d}$$

The physical significance of τ may be expressed as the time required for the velocity to be reduced by drag to (1/e) (or about 36.8%) of its initial value.

The time required for the particle to be brought to rest is infinite, but the distance it travels until u = 0 is finite. It is called the stopping distance of the particle and is given by

$$x_s = \rho_p d_p^2 u_o/18\mu = \tau u_o \tag{4.20}$$

The distance traveled may be expressed as a fraction of the stopping distance, from (4.19d) in dimensionless form:

$$x/x_s = (1 - e^{-t/\tau}) \tag{4.21}$$

These concepts may be extended beyond the Stokes Law range. The relaxation time may be defined more broadly from Eqn. (4.14) by setting $Re = Re_o/e$ in the integration. The stopping distance may be calculated more generally by setting Re = 0 in the integral of Eqn. (4.16). Fuchs [13] has shown how a general chart may be prepared by graphical integration of (4.16) using the data given in Fig. 6, resulting in

$$\frac{3}{4}\frac{\rho_f}{\rho_p}\frac{x_s}{d_p} = \int_o^{Re_o} \frac{d(Re)}{C_D Re} = f\ (Re_o)$$

The quantity $3\rho_f x_s/4\rho_p$ is plotted as a function of $Re_o = u_o d_p/\upsilon$., as shown in Fig. 7. In the Stokes Law range $f(Re_o) = Re_o/24$.

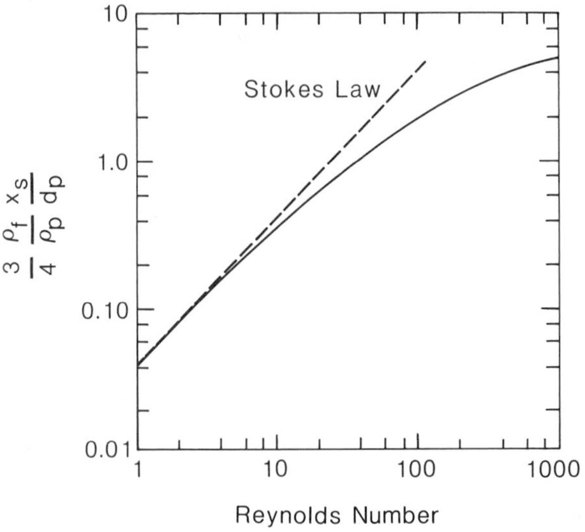

FIG. 7 *Stopping distance of particles.*

Example 2. A spherical particle of lead compounds, 100 microns in diameter, density = 6.0 gm/cm^3, is projected horizontally from the tail-pipe of a stationary automobile into still air at 20°C with an initial velocity of 100 cm/sec. How far will it have travelled horizontally, and for how long, when its velocity is 10 cm/sec? What is its stopping distance?

It will be useful to answer the last question first. The stopping distance may be found most readily by use of Fig. 7.

$$Re_o = \frac{d_p u_o \rho_f}{\mu} = \frac{100 \times 10^{-4} \times 100 \times 1.205 \times 10^{-3}}{1.816 \times 10^{-4}} = 6.64$$

This is well out of the Stokes Law range. Hence, read from Fig. 7: $f(Re_o) = 3\rho_f x_s/4\rho_p d_p = 0.24$, corresponding to $Re_o = 6.64$.

$$x_s = 0.24 \times \frac{4}{3} \times \frac{6 \times 100 \times 10^{-4}}{1.205 \times 10^{-3}} = 16.0 \text{ cm (stopping distance)}$$

When the velocity of the particle has reached 10 cm/sec, the value of Re will be 0.664. Obtain a value of $f(Re_o)$ corresponding to an $Re_o = 0.664$ from Fig. 7 and calculate an x_s for it. Subtract this from the value of x_s found above. Thus $3\rho_f x_s/4\rho_p d_p = 0.027$ when $Re_o = 0.664$ (Fig. 7 extended into Stokes range) and $x_s = 1.80$ cm. Distance traveled = 16. - 1.8 = 14.2 cm.

Finally, the time of flight will be calculated using (4.14) by performing a numerical integration based upon the data shown in Fig. 5. A plot of $1/C_D Re^2$ vs. Re is constructed and the area under it between Re = 6.64 and Re = 0.664 obtained.

$$\int_{6.64}^{0.664} \frac{d(Re)}{C_D Re^2} = -\ 0.0769$$

$$t = -\frac{4}{3} \times \frac{6.0(100 \times 10^{-4})}{1.816 \times 10^{-4}} (-0.0769) = 0.34 \text{ sec.}$$

Note that in this example only the horizontal motion is considered. At the same time the particle will also be falling under the influence of gravity, which is considered next. However, gravity will have no effect upon the horizontal components of motion.

2. *Gravity force* a. Galileo Number. Because a particle has weight there will always be some tendency for it to move vertically downward. It will also be subjected to a buoyancy force equal to the weight of air it displaces. Although for small particles weight and/or buoyancy may be quite negligible, it is important to consider as a fundamental case of particle motion that in which the forces involved are drag, gravity, and buoyancy.

Newton's Law, Eqn. (4.1), is written

$$m_p \frac{d\vec{u}}{dt} = \vec{F}_g + \vec{F}_b + \vec{F}_D \tag{4.22}$$

where $\vec{F}_g = g\, m_p$, the weight of the particle: "gravity force";
$\vec{F}_b = g\, m_f$, the weight of displaced fluid: "buoyancy force."
Since $\vec{F}_g$ and $\vec{F}_b$ act only in the vertical direction, the component equations become

$$m_p \frac{du_x}{dt} = - F_{D_x} \quad \text{horizontal, along x-axis} \tag{4.22a}$$

$$m_p \frac{du_y}{dt} = g(m_p - m_f) - F_{D_y} \quad \text{vertical, along y-axis taking downward to be the positive direction} \tag{4.22b}$$

The equation for the horizontal component of motion is the same as that dealt with in the previous section on drag. The second equation, for the vertical component of motion, must now be solved. For spherical particles only:

$$m_p = \pi d_p^3 \rho_p/6 \quad m_f = \pi d_p^3 \rho_f/6$$

and (4.22b) becomes

$$\frac{du_y}{dt} = g\left(\frac{\rho_p - \rho_f}{\rho_p}\right) - \frac{F_{D_y}}{m_p} = g\left(\frac{\rho_p - \rho_f}{\rho_p}\right) - \frac{3}{4} C_D \frac{\rho_f}{\rho_p} \frac{u_y^2}{d_p} \tag{4.23}$$

To place (4.23) in a dimensionless form, and at the same time to introduce the C_D-Re relationship, replace u_y by $\mu \mathrm{Re}/\rho_f d_p$ and rearrange, giving

$$\frac{4}{3} \frac{\rho_p d_p^2}{\mu} \frac{d\mathrm{Re}}{dt} = \frac{4}{3} g \frac{(\rho_p - \rho_f)\rho_f d_p^3}{u^2} - C_D \mathrm{Re}^2 \tag{4.24}$$

The first term on the right has been called the Galileo Number. It is a property of the fluid-particle system, independent of velocity. It may be defined as:

$$Ga = \frac{4}{3} g \frac{(\rho_p - \rho_f)\rho_f d_p^3}{u^2} \approx \frac{4}{3} g \frac{\rho_p \rho_f}{u^2} d_p^3 \quad \text{for } \rho_f << \rho_p \tag{4.25}$$

The differential equation for vertical motion may finally be written:

$$\frac{dRe}{dt} = \frac{Ga - C_D Re^2}{24\tau} \tag{4.26}$$

Taking $u_y = u_{y_o}$, and $Re = Re_o$, at $t = 0$, the solution will involve this integration:

$$\frac{t}{24\tau} = \int_{Re_o}^{Re} \frac{dRe}{Ga - C_D Re^2} \tag{4.27}$$

A companion equation for the distance travelled vertically downward, may be obtained by placing $dt = dy/u_y$ in (4.23) and again using $u_y = \mu Re/\rho_f d_p$. Then (4.23) becomes

$$\frac{dRe}{dy} = \frac{(Ga - C_D Re^2)}{Re} \times \frac{3\rho_f}{4\rho_p d_p} \tag{4.28}$$

The solution of (4.28) may be indicated, using the limits $y = 0$ when $u_y = u_{y_o}$, and $Re = Re_o$.

$$y = \frac{4}{3} \frac{\rho_p}{\rho_f} d_p \int_{Re_o}^{Re} \frac{Re\, d(Re)}{Ga - C_D Re^2} \tag{4.29}$$

Time of travel and vertical distance of travel may thus be related through (4.27) and (4.29).

Both of the integrals involved in these equations must be evaluated by substituting the proper C_D-Re relationship, according to the range of Re encompassed by the limits of integration. Except in the case of Stokes Law this is sufficiently complex that the integrals are best computed by numerical methods. However, as will be shown below, the need to have these integrals evaluated rarely arises in most particle-gas systems of interest in dust collection problems. For the sake of completeness, however, the results are presented for Stokes Law particles, where $C_D Re^2 = 24\ Re$

$$\frac{t}{\tau} = -24 \int_{Re}^{Re_o} \frac{dRe}{Ga - 24\ Re} = \ln \frac{Ga - 24\ Re_o}{Ga - 24\ Re} \tag{4.30}$$

$$y = \frac{4}{3} \frac{\rho_p}{\rho_f} d_p \int_{Re_o}^{Re} \frac{Re\ d(Re)}{Ga - 24\ Re}$$

$$y = \frac{\rho_p d_p}{18 \rho_f} (Re_o - Re) - \frac{Ga}{24} \ln \frac{Ga - 24\ Re}{Ga - 24\ Re_o} \tag{4.31}$$

b. Terminal Settling Velocities. An extremely important aspect of the basic differential equation for vertical motion is the condition when $du_y/dt = 0$. This corresponds also to $dRe/dt = 0$, and represents a state of steady motion. Physically this may be understood as a particle reaching a steady vertically downward velocity, which is called u_s. If the initial downward velocity of the particle $u_{y_o} > u_s$, the particle will decelerate, and if $u_{y_o} < u_s$ it will accelerate, until u_s is attained. The value of u_s is, therefore, called the terminal settling velocity of the particle.

Inspection of Eqns. (4.26) and (4.27) reveals that the condition for terminal settling velocity to be attained is that

$$C_D Re^2 = Ga \tag{4.32}$$

and also reveals that the time required will be infinite. However, as will be shown, for systems of practical interest the limiting condition may be approached very closely within a very short time. For all practical purposes, many such particles may be regarded as settling at their terminal velocities all the time.

To relate the terminal settling velocity to the particle size, the following scheme is used. It is seen that

$$\sqrt[3]{Ga} = \sqrt[3]{C_D Re^2} = \left[\frac{4}{3} g \frac{(\rho_p - \rho_f)\rho_f}{\mu^2}\right]^{1/3} d_p \tag{4.33}$$

is a dimensionless number which is directly proportional to the particle size and does not involve the terminal settling velocity. Similarly,

$$\frac{C_D Re^2}{Re^3} = \frac{Ga}{Re^3} = \frac{C_D}{Re}$$

also satisfies the condition for zero vertical acceleration and produces a dimensionless number

$$\sqrt[3]{\frac{Re_s}{C_D}} = \frac{u_s d_p \rho_f}{\mu} \times \left[\frac{3\mu^2}{4g(\rho_p - \rho_f)\rho_f}\right]^{1/3} d_p = \left[\frac{3\rho_f^2}{4g(\rho_p - \rho_f)\mu}\right]^{1/3} u_s \tag{4.34}$$

which is directly proportional to the terminal settling velocity and does not involve the particle size.

From the fundamental data of Fig. 1 or 2 a plot of

$$(C_D Re^2)^{1/3} \quad \text{vs.} \quad (Re/C_D)^{1/3}$$

may be prepared and is given as Fig. 8. Here the ordinate is directly proportional to the value of d_p. For a given fluid-solid

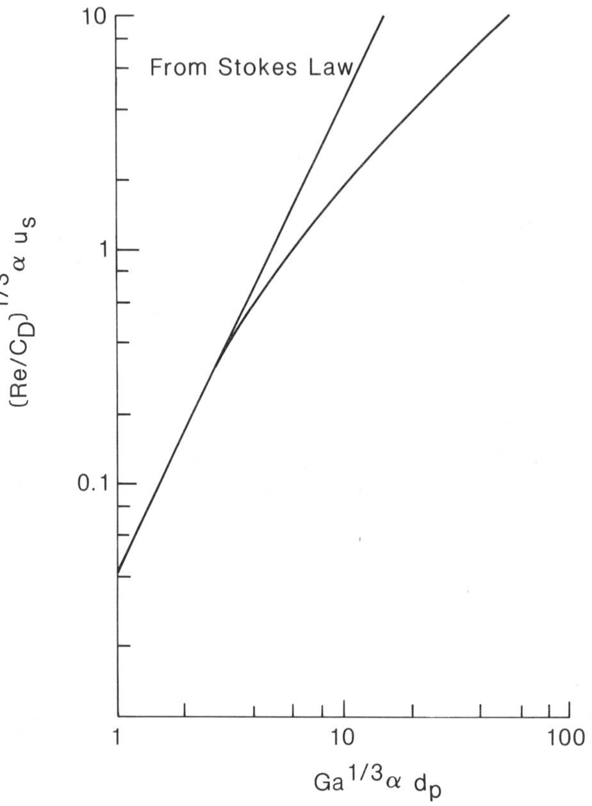

FIG. 8 *Terminal settling velocity as related to particle size.*

system at a specified temperature, both proportionality factors may be calculated and the graph used: (a) to predict the terminal settling velocity of a particle of given size; (b) to determine the size of a particle having a given, or measured, terminal settling velocity. The latter may be used as the basis of an experimental method of determining the "sedimentation diameter" or equivalent spherical diameter, of a particle of unknown or equivalent shape by measuring its terminal settling velocity.

If greater precision is desired, then the following equations may be used [16]:

$$\log (Ga^{1/3}) = 0.7841 + 0.7107 \log (\frac{Re}{C_D})^{1/3} + 0.1898 \left[\log (\frac{Re}{C_D})^{1/3}\right]^2$$

$$+ 0.06289 \left[\log (\frac{Re}{C_D})^{1/3}\right]^3 \tag{4.35}$$

$$\log (\frac{Re}{C_D})^{1/3} = - 1.387 + 2.153 \log(Ga^{1/3}) - 0.548 (\log Ga^{1/3})^2$$

$$+ 0.05665(\log Ga^{1/3})^3 \tag{4.36}$$

Both of these equations may readily be programmed for solution on a pocket calculator. This makes it possible to solve directly for d_p, given u_s [Eqn. (4.35)], or to find u_s, given d_p [Eqn. (4.36)].

For a Stokes law particle the results are simply

$$u_s = \frac{\upsilon Ga}{24 d_p} = \frac{gC}{18} \frac{(\rho_p - \rho_f)}{\mu} d_p^2 = gC\tau \qquad \text{when } \rho_f << \rho_p \tag{4.37}$$

or

$$d_p = \sqrt{\frac{18\mu\, u_s}{gC(\rho_p - \rho_f)}}$$

where the Cunningham factor is placed as shown, when needed. This equation for u_s (4.37) is sometimes referred to as Stokes Law also.

It must be emphasized that the range of particle sizes for which (4.37) is accurate, is not the same as that for which Fig. 3 applies. The use of Stokes Law for drag calculation, as given in Fig. 3, is independent of particle density, but the use of (4.37) to calculate a terminal settling velocity is not independent of ρ_p. The criterion to be applied here is that $Re_s = d_p u_s/\upsilon \leq 0.1$ or $Ga \leq 2.4$ for accuracy within 2%. Taking $\rho_f << \rho_p$, this corresponds to

$$\frac{4}{3}\,\frac{g\rho_p\rho_f d_p^3}{24\mu^2} = \frac{g\tau d_p^3}{\nu} \leq 0.1$$

In dry air at 20°C and 1 atm., for example, this reduces to

$$\frac{4 \times 980 \times 1.205 \times 10^{-3}}{3 \times 24\ (1.81 \times 10^{-4})^2}\,\rho_p d_p^3 = 2.002 \times 10^6 \times \rho_p d_p^3 \leq 0.1$$

or $\rho d_p^3 < 49.9 \times 10^{-9}$ gm. Table 1 lists values determined from this relationship, and also values for air at higher temperatures.

Example 3. Find the terminal settling velocity of a spherical particle under each of these conditions:

(a) d_p = 100 μm, ρ_p = 2.60 gm/cm^3, dry air at 100°C, 1 atm.
(b) d_p = 60 μm, ρ_p = 1.0 gm/cm^3, dry air at 300°C, 2 atm.
(c) d_p = 8 μm, ρ_p = 0.6 gm/cm^3, dry air at 100°C, 1 atm.

In each case the value of $(Ga)^{1/3}$ should be calculated and used on Fig. 8 or with Eqn. (4.36) to obtain the value of $(Re/C_D)^{1/3}$ from which u_s may be calculated. Table 1 may be used as a quick check to see whether Stokes Law is followed. If so Eqn. (4.37) may be used instead of Fig. 8.

(a) From Table 1, Stokes Law does not apply.

$$Ga^{1/3} = \frac{4}{3} \times 980 \times \left[\frac{(2.6)0.940 \times 10^{-3}}{2.17 \times 10^{-4})^2}\right]^{1/3} \times 100\text{x}10^{-4} = 4.08$$

$(Re/C_D)^{1/3}$ = 0.55 from Fig. 8 or Eqn. 4.36, and Eqn. (4.34) gives

$$u_s = 0.55 \times \left[\frac{4 \times 980\ (2.6) \times 2.17 \times 10^{-4}}{3 \times (0.940 \times 10^{-3})^2}\right]^{1/3} = 51.8 \text{ cm/sec.}$$

TABLE 1 *Maximum size of sphere for which terminal settling velocity may be calculated by Stokes Law (4.37)*

	Air Temp. 20°C	Air Temp. 100°C	Air Temp. 300°C
	$Re_s \leq 0.1$ or $Ga \leq 2.4$ for accuracy within 2%		
	Dry air, at 1 atmos. pressure		
ρ_f =	1.205×10^{-3} gm/cm^3	0.940×10^{-3} gm/cm^3	0.613×10^{-3} gm/cm^3
μ =	1.816×10^{-4} poises	2.17×10^{-4} poises	2.93×10^{-4} poises
$\rho_p d_p^3$ =	49.9×10^{-9} gm	92.0×10^{-9} gm	$257. \times 10^{-9}$ gm
ρ_p(gm/cm^3)	d_p(microns)	d_p(microns)	d_p(microns)
0.2	63	77	109
0.3	55	67	95
0.4	50	61	86
0.6	44	54	75
0.8	40	49	68
1.0	37	45	64
1.5	32	39	56
2.0	29	36	50
3.0	26	31	44
4.0	23	28	40
6.0	20	25	35
*Minimum size	8	11	18

*Note: For diameters less than the minimum size listed (taken from Fig. 3), a Cunningham correction factor will be needed.

(b) From Table 1, Stokes Law applies, even though P = 2 atm, because the value of ρ_f has a negligible effect on Ga. Hence, from (4.37), C = 1, and

$$u_s = 980 \times 1 \times \frac{1.0 \times (60 \times 10^{-4})^2}{18 \times 2.93 \times 10^{-4}} = 6.7 \text{ cm/sec.}$$

(c) Table 1 indicates Stokes Law may be used but a Cunningham correction is needed. Figure 4 gives f= 7.62, hence C = 8.0/7.6 = 1.05 and

$$u_s = 980 \times \frac{1.05 \times 0.6 \times (8 \times 10^{-4})^2}{18 \times 2.17 \times 10^{-4}} = 0.10 \text{ cm/sec.}$$

Example 4. (a) Assuming a raindrop behaves like a rigid sphere what is the size of a drop which is observed to fall at the rate of 5 ft/sec.? (b) A high-volume sampler takes in 60 ft.^{3} of air per minute vertically upward through an area of 1 sq. ft. What is the largest sized particle of density 2.6 gm/cm^3 which it can collect?

(a) Here u_s = 5 x 30.5 = 152 cm/sec. Take the air to be at 20°C, 1 atm.

$$\left(\frac{Re}{C_D}\right)^{1/3} = \left[\frac{3 \times (1.205 \times 10^{-3})^2}{4 \times 980 \times 1.0 \times 1.816 \times 10^{-4}}\right]^{1/3} 152 = 2.78$$

From Fig. 8 or Eqn. (4.35), $(Ga)^{1/3}$ = 14.2, and from (4.33)

$$d_p = \left[\frac{3 \times (1.816 \times 10^{-4})^2}{4 \times 980 \times 1.0 \times 1.205 \times 10^{-3}}\right]^{1/3} 14.2 = 0.039 \text{ cm}$$

(b) $$u_s = \frac{60}{1 \times 60} \times 30.5 = 30.5 \text{ cm/sec.}$$

$$\left(\frac{Re}{C_D}\right)^{1/3} = \left[\frac{3 \times (1.205 \times 10^{-3})^2}{4 \times 980 \times 2.6 \times 1.816 \times 10^{-4}}\right]^{1/3} 30.5 = 0.405$$

From Fig. 8 or Eqn. 4.35, $(Ga)^{1/3}$ = 3.4, and from (4.33):

$$d_p = \left[\frac{3 \times (1.816 \times 10^{-4})^2}{4 \times 980 \times 2.6 \times 1.205 \times 10^{-3}}\right]^{1/3} 3.4 = 6.8 \times 10^{-3} \text{ cm}$$

or 68 μm

c. Gravitational Motion of Stokes Law Particles. The equations for time and distance of settling may be reformulated to advantage in terms of the terminal settling velocities for Stokes Law particles. Differential equation (4.23) will become

$$\frac{du_y}{dt} = g - \frac{u_y}{\tau} \tag{4.38}$$

Applying the condition that Ga = 24 Re_s and noting that $Re/Re_s = u_y/u_s$, or $Re_o/Re_s = u_{y_o}/u_s$, Eqn. (4.30) for time becomes:

$$\frac{t}{\tau} = \ln \frac{1 - u_{y_o}/u_s}{1 - u_y/u_s}$$

which may be arranged to solve for u_y/u_s:

$$\frac{u_y}{u_s} = 1 - (1 - \frac{u_{y_o}}{u_s})e^{-t/\tau} \tag{4.39}$$

For the case of a particle starting to fall from rest $u_{y_o} = 0$, and

$$\frac{u_y}{u_s} = 1 - e^{-t/\tau} \tag{4.40}$$

Eqn. (4.31) for distance becomes

$$y = u_s\,\tau \left[\frac{u_{y_o}}{u_s} - \frac{u_y}{u_s} - \ln \frac{1 - u_y/u_s}{1 - u_{y_o}/u_s}\right] \tag{4.41}$$

This may be expressed in terms of time rather than velocity by using (4.39) to replace u_y/u_s. The result is, in dimensionless form:

$$\frac{y}{u_s\tau} = \frac{u_{y_o}}{u_s} + (\frac{t}{\tau} - 1) + (1 - \frac{u_{y_o}}{u_s})e^{-t/\tau} \tag{4.42}$$

When the particle starts from rest (4.42) becomes

$$\frac{y}{u_s \tau} = \frac{t}{\tau} - (1 - e^{-t/\tau}) \tag{4.43}$$

These equations show clearly that when a particle has been falling for a period of time $t \approx 5\tau$ it will have attained over 99.3% of its terminal velocity starting from rest. For most practical purposes, it may be assumed that the acceleration period of fall is essentially completed at $t = 5\tau$, and thereafter the particle continues to fall at steady terminal settling velocity. The additional distance traveled after that point will be virtually equal to $(t - 5\tau)u_s$.

The relaxation time τ is quite small for small particles in air. For example, at 20°C, $\mu = 1.81 \times 10^{-4}$ poises, and for $d_p = 1$ micron $= 10^{-4}$ cm, and $\rho_p = 1$ gm/cm^3

$$\tau = \frac{1 \times (10^{-4})^2}{18 \times 1.81 \times 10^{-4}} = 3.07 \times 10^{-6} \text{ sec.}$$

and

$$5\tau = 1.54 \times 10^{-5} \text{ sec.}$$

This period of time is so small that for most purposes the accelerating period may be neglected entirely.

d. Dimensionless Forms for Stokes Law Particles. Dimensionless parameters are used extensively in modelling dust collection devices. Several useful ones may be obtained by putting Eqn. (4.23) into other interesting dimensionless forms for Stokes Law particles. This is done by selecting various sets of reference values as follows:

(i) Select a set of "reference" values, say v_o = an appropriate fixed or constant velocity; D = an appropriate fixed linear dimension; $t_o = D/v_o$, a reference time. Then define dimensionless variables as $\tilde{u}_y = u_y/v_o$ dimensionless velocity; $\theta = t/t_o = tv_o/D$ dimensionless time. Substitute these variables into the differential equation which then becomes:

$$\frac{d\tilde{u}_y}{d\theta} = \frac{g(\rho_p - \rho_f)}{\rho_p} \frac{D}{v_o^2} - \frac{18\mu D}{\rho_p v_o d_p^2} \tilde{u}_y = G - \tilde{u}_y/\text{Stk} \tag{4.44}$$

Two dimensionless groups appear automatically, and may be defined as

$$G = \frac{g(\rho_p - \rho_f)}{\rho_p} \frac{D}{v_o^2} \simeq \frac{gD}{v_o^2} \tag{4.45}$$

called "gravity" parameter, which is a dimensionless acceleration; and

$$\text{Stk} = \frac{\rho_p v_o d_p^2}{18\mu D} = \frac{\tau v_o}{D} = \frac{x_s}{D} \tag{4.46}$$

the "inertial" parameter, called Stokes Number which is the ratio of the particle stopping distance to D. Their product is a dimensionless velocity:

$$G \cdot \text{Stk} = \frac{g(\rho_p - \rho_f) d_p^2}{18\mu v_o} = \frac{u_s}{v_o} \tag{4.47}$$

which is the ratio of the particle terminal settling velocity to v_o. These three dimensionless parameters will be useful in discussing deposition of particles from moving air in the next chapter, where specific meanings will be assigned to v_o and to D.
(ii) Another approach to a dimensionless form may be made by selecting the reference time to be τ, in conjunction with the reference velocity v_o. In this case, the resulting equation is

$$\frac{d\tilde{u}_y}{d\theta} = \frac{g(\rho_p - \rho_f)}{\rho_p}\frac{\tau}{v_o} - \tilde{u}_y = G \cdot Stk - \tilde{u}_y \tag{4.48}$$

No new groups are obtained.

(iii) Finally, what is perhaps the neatest dimensionless form of Eqn. (4.23) may be obtained by selecting τ as reference time $\theta = t/\tau$ and u_s as reference velocity, $\tilde{u}_y = u_y/u_s$. When these substitutions are made (4.23) is transformed into

$$\frac{d\tilde{u}_y}{d\theta} = 1 - \tilde{u}_y \tag{4.49}$$

which is readily integrated to give

$$\tilde{u}_y = 1 - (1 - \tilde{u}_{y_o})\, e^{-\theta}$$

This is precisely the same as Eqn. (4.39) but it is obtained much more easily in this way. This technique of normalizing is often useful in simplifying the solution of differential equations.

3. *Other forces* a. Effect of Other Forces. If forces other than drag, gravity and buoyancy are acting upon the particle, they may easily be incorporated into Newton's Law:

$$m_p \frac{d\vec{u}}{dt} = \vec{F}_D + \vec{F}_g + \vec{F}_B + \sum \vec{F}_{other} \tag{4.50}$$

The nature of the "other forces(s)" must be specifically identifed, and a functional relationship to all other pertinent variables must be supplied in order to solve the equation and to find the particle trajectory. The treatment here will be limited to the case where other forces are "steady", i.e., independent of particle velocity or time.

The calculation of the trajectory where an additional constant force is acting is rather simple and depends mainly upon the direction of action of the additional force. Three general cases will be considered first, and then the specific cases of an electrostatic force and a thermal force will be shown.

(a) One-dimensional, vertical motion, F_o vertical; only one component equation exists:

$$m_p \frac{du_y}{dt} = (F_G - F_B) - F_{D_y} \pm F_{o_y} \qquad \begin{array}{l} + \text{ if } F_{o_y} \text{ acts downward} \\ - \text{ if } F_{o_y} \text{ acts upward.} \end{array} \tag{4.51}$$

Since F_{o_y} is constant it may merely be added to $(F_G - F_B)$, and Eqns. (4.22b), (4.23), (4.24) etc. will be essentially the same as before with an added constant term. For steady motion:

$$(F_G - F_B) \pm F_o = F_{D_y} \tag{4.52}$$

(b) Two-dimensional, vertical and horizontal, F_o is in the same direction as horizontal motion; two component equations are needed:

$$\text{Horizontal:} \quad m_p \frac{du_x}{dt} = F_{o_x} - F_{D_x} \tag{4.53}$$

(This equation would also cover the case of one-dimensional horizontal motion in which F_{o_x} and F_{D_x} are the only forces acting.)

$$\text{Vertical:} \quad m_p \frac{du_y}{dt} = (F_G - F_B) - F_{D_y} \tag{4.54}$$

(c) Three-dimensional motion, two F_o forces acting horizontally at right angles; three component equations are needed:

Horizontal: x-direction
$$m_p \frac{du_x}{dt} = F_{o_x} - F_{D_x} \tag{4.55}$$

Horizontal: z-direction
$$m_p \frac{du_z}{dt} = F_{o_z} - F_{D_z} \tag{4.56}$$

Vertical:
$$m_p \frac{du_y}{dt} = (F_G - F_B) - F_{D_y} \tag{4.57}$$

Each of these equations is similar in form to one which has been solved above, with the simple addition of the appropriate constant term representing F_o. The condition for steady motion in any one direction is readily obtained by setting the corresponding acceleration term equal to zero.

b. Electrostatic Force. Where a particle is carrying a constant electrical charge and is placed in an electrostatic field of constant potential gradient, an electrostatic force is developed in the direction of the gradient and is given by

$$F_E = q \cdot E \tag{4.58}$$

where q = charge on particle, coulombs; E = field strength, potential gradient, volts/cm; F_E = electrostatic force, in volt-coulombs/cm, or joules/cm. Wherever F_o appears in the above sets of equations, it would be replaced by F_E where E must be the gradient in the appropriate direction x, y, or z as indicated. Then a new constant term $F_E/m_p = q \cdot E/m_p$, representing acceleration in direction of E, will appear in the corresponding differential equations. The solutions previously obtained may readily be adapted to include F_E. The following example illustrates how this may be done.

Example 6. A Stokes Law particle is projected into still air with an initial horizontal velocity u_{o_x}. It immediately enters an electrostatic field of constant horizontal potential gradient E_z, perpendicular to the direction of original motion, and it immediately acquires a charge of q coulombs. Develop equations for determining the location of the particle at any instant after entry. Show the nature of the particle trajectory. Take u_{o_x} along x-axis, gravity along y-axis, and $F_{E_z} = q \cdot E_z$ along z-axis, also $u_{o_y} = u_{o_z} = 0$.

Motion in x-direction: given by Eqns. (4.18) and (4.19c)

$$x = u_{o_x} \tau (1 - e^{-t/\tau})$$

Motion in y-direction: given by Eqns. (4.37) and (4.43)

$$y = g\tau^2 \left[\frac{t}{\tau} - (1 - e^{-t/\tau})\right]$$

Motion in z-direction: given by solving Eqn. (4.56)

$$\frac{du_z}{dt} = \frac{q_p E_z}{m_p} - \frac{u_z}{\tau}$$

which is in precisely the same form as (4.38), with qE_z/m_p replacing g. This will give:

$$z = \frac{6q_p E}{\pi \rho_p d_p^3} \tau^2 \left[\frac{t}{\tau} - (1 - e^{-t/\tau})\right] = \frac{q_p E \rho_p d_p}{54\pi\mu^2} \left[\frac{t}{\tau} - (1 - e^{-t/\tau})\right] \tag{4.59}$$

The trajectory will be an exponential-type of curve in three dimensions until $t \approx 5\tau$. After that time steady conditions will prevail: there will be no further motion in the x-direction, $x = x_s$, and the trajectory will be essentially a straight line in the y-z plane located at $x = x_s$, with

$$u_z = \frac{z}{t} = \frac{6q_p E\tau}{\pi \rho_p d_p^3} = \frac{qE}{3\pi d_p \mu} = u_{E_s}, \text{ and } u_y = \frac{y}{t} = g\tau = u_s \tag{4.60}$$

The quantity $q/3\pi d_p\mu$ is sometimes called the "mobility" (b) of the charged particle in this field. As we have seen $t = 5\tau$ will be a very short period of time. The initial curved portion of the trajectory will be very short, and the overall path will be essentially a straight line corresponding to the limiting velocity components u_S and u_{E_S} as given.

Attention should be called to the velocity u_{E_S} obtained in the above example. This is a steady-state electrostatic "drift" velocity, which may be obtained by setting the right-hand side of (4.53) equal to zero. The electrostatic drift velocity will appear again in Chapter 7.

Measurement of the electrical mobility of particles having a known charge is the principle of a method of particle size analysis applicable to fine particles in the range of 0.003-1 μm. Details of the operation are given by Allen [17].

Example 7. Take conditions as in example 6 to be as follows. Particle: size 10μm, density 2.6 gm/cm^3, charge 100 electrons. Field: 2000 volts/cm, air at 20°C; initial velocity: 100 cm/s. Determine the position of the particle with respect to its point of entry into the field, 10 sec. later. Note: charge on a single electron is 1.601×10^{-19} coulombs, and force in volt-coulombs/cm (same as joules/cm) is equal to 10^{-7} times the force in dynes.

The relaxation time is $\tau = 7.98 \times 10^{-4}$ sec., hence $t/\tau = 10/8 \times 10^{-4} \simeq 10^4$ is very large and the acceleration period may be neglected. No Cunningham correction is needed for a 10μm particle. The motion is analyzed by components:

Horizontal motion, x-direction drag force only:

$$Re_o = \frac{100 \times 10 \times 10^{-4} \times 1.205 \times 10^{-3}}{1.814 \times 10^{-4}} = 0.664$$

Fig. 8 shows this is in Stokes law range. Equation (4.20) gives:

$$x = x_S = u_o\tau = 100 \times 7.98 \times 10^{-4} = 0.08 \text{ cm}$$

Vertical motion, y-direction, gravity force: from (4.25)

$$Ga = \frac{4 \times 980 \times 2.6 \times 1.205 \times 10^{-3} \times (10 \times 10^{-4})^3}{(1.814 \times 10^{-4})^2} = 0.124$$

$$(Ga)^{1/3} = 0.499$$

Figure 8 shows this is in Stokes law range. Equation (4.37) gives:

$$u_s = gC\tau \qquad C=1$$

$$= 980 \times 7.98 \times 10^{-4} = 0.78 \text{ cm/sec.}$$

$$y = u_s t = 0.78 \times 10 = 7.8 \text{ cm}$$

Horizontal motion, z-direction, electrostatic force: from (4.60)

$$u_z = u_{E_s} = \frac{qE}{3\pi dp\mu} \text{ (drift velocity)}$$

$$= \frac{100 \times (1.601 \times 10^{-19}) \times 2000 \times 10^7}{3\pi \times (10 \times 10^{-4}) \times 1.814 \times 10^{-4}} = 0.187 \text{ cm/sec}$$

$$z = 0.187 \times 10 = 1.87 \text{ cm}$$

c. Thermal Force-Thermophoresis. It has long been observed that a particle suspended in a gas having a temperature gradient across it will be subjected to a thermal force F_T which causes it to move away from the higher temperature region and toward the lower temperature region of the gas. The process is called thermophoresis. F_T would be one of the forces represented by F_o in Eqns. (4.51) through (4.57), according to the direction of the temperature gradient.

In the absence of other forces F_T will be opposed only by F_D. Where $F_T = -F_D$ a steady thermophoretic velocity u_T will be reached. Particles will move at this velocity toward a colder surface bounding a hot gas, and be deposited upon it. Thermal deposition is a well-known phenomenon associated with the soiling of walls in the vicinity of hot pipes, and with the fouling of "cold" surfaces in gas heat exchangers.

An excellent review of theoretical and experimental studies of thermophoresis and thermal deposition has been given by Gieseke [18]. The thermophoretic velocity may be correlated in terms of the thermal dimensionless group

$$Th = -\frac{u_T \rho_f T}{\mu_f \frac{dT}{dx}} = -\frac{u_T T}{\nu_f \frac{dT}{dx}} \tag{4.61}$$

in which dT/dx is the temperature gradient (always < 0) in the direction of motion. For Stokes law particles the value of the thermal force will be given by

$$F_T = \frac{3\pi\mu_f^2 d_p}{C\rho_f T}\left[-Th\,\frac{dT}{dx}\right] = \frac{m_p u_T}{\tau} \tag{4.62}$$

The number Th is found to be essentially constant for small particles, characterized by $Kn > 1$, indicating that u_T is independent of particle size (and particle properties), directly proportional to the temperature gradient, and only very weakly dependent upon temperature for a given gas. The value of Th according to various theories seems to lie between 0.42 and 1.5, with experimental values falling in the neighborhood of 0.5.

In this case the force is attributed simply to gas molecules impinging on the particle from opposite sides with different mean velocities. The resulting momentum exchange depends in part upon what fraction of the collisions are perfectly elastic or specular, and what fraction are diffuse. Experiments indicate that about 80% are diffuse, which leads to a value of Th = 0.57.

For coarser particles, Kn $\leq$ 1, the situation is more complex and is treated by the thermal creep theory. There is a creeping flow of gas from the colder to the warmer regions along the surface of the particle which in turn experiences a force in the cold direction. Th becomes a function of Kn, as well as of the ratio of the thermal conductivity of the gas to the thermal conductivity of the particle. Th decreases with Kn in general, and may fall as low as 0.02 at Kn = 0.01 for low k_g/k_p values. Gieseke [18] presents a graph which summarizes all of these relationships for Th.

Some idea of the relative importance of thermophoresis may be obtained by estimating u_T in the case of a particle 0.1 micron in diameter in dry air at 300°C and P = 1 atm. From Eqn. (4.7c) Kn = 2.94. Taking Th = 0.5 for illustration, (4.61) gives:

$$u_T = \frac{\mathrm{Th} \times \nu_f}{T} \frac{dT}{dx} = \frac{0.5 \times 2.93 \times 10^{-4}}{0.613 \times 10^{-3} \times 573} \frac{dT}{dx} = 4.2 \times 10^{-4} \frac{dT}{dx} \frac{\mathrm{cm}}{\mathrm{sec}}$$

It is evident that unless the temperature gradient is very high, say of the order of 1000°C/cm or more, the thermal velocity will not be large enough to be significant. At least in most situations of ordinary temperature gradients, F_T is not important in comparison with other forces acting.

Therefore to use thermal force as a mechanism for particulate collection on a large scale will require a source of heat. This is a major energy requirement. Estimates indicate that the power required may be an order of magnitude greater than that required for other modes of gas cleaning, unless the gas is already hot. In this case, if the gas is cooled by proper heat transfer surfaces thermal deposition will occur. Such deposition from turbulent flows inside tubes and in annular spaces has been studied by Byers and Calvert [19] and by Singh and Byers [20]. Collection efficiencies up to 30% were observed. Appropriate mathematical models were developed. Thermal force may therefore be regarded as a contributing mechanism to particle collection in some situations. Effective thermal precipitators are available to use for collecting small samples of fine (0.01-5μm) aerosols from flowing gas streams [17].

Because of the relative unimportance of thermal deposition as a mechanism for industrial particle collection, it will not be discussed in further detail. However one should be aware of large temperature gradients wherever they occur, in order not to overlook the possibility of thermal deposition occuring.

II. MOTION DUE TO DIFFUSION

A. *BROWNIAN MOTION*

Aerosol particles suspended in an isothermal gas are constantly subjected to bombardment by the molecules of the gas due to their random thermal motion. At each collision an exchange of momentum occurs. If the aerosol particles are very small these collisions may cause the particles to move about in a random fashion. This is called Brownian motion.

As a consequence of Brownian motion very small aerosol particles tend to diffuse through a gas from regions of higher particle concentration toward regions of lower concentration. The process is analogous to true diffusion in gases and is governed by a differntial equation of the same form. In rectangular coordinates, this is

$$\frac{\partial n}{\partial t} = \mathcal{D} \left[\frac{\partial^2 n}{\partial x^2} + \frac{\partial^2 n}{\partial y^2} + \frac{\partial^2 n}{\partial z^2} \right] \tag{4.63}$$

This equation also has the same form as the well known equation for heat conduction and can be treated by the same mathematical techniques.

Particles in Brownian motion will strike a confining surface of the aerosol and may be deposited there. If so, this creates a region of lower concentration adjacent to the surface, and a concentration gradient toward it thus promoting diffusion toward the surface. Diffusive deposition from either stationary or moving gas may, under some conditions, be a significant mechanism of particle collection. It will be discussed further below and in later chapters.

The quantity $\mathcal{D}$ in Eqn. (4.63), called the diffusivity of the particle, is a basic physical parameter which depends upon the nature and temperature of the gas, and upon the particle size. The value of $\mathcal{D}$ may be estimated in two theoretical ways.

(a) For particles of a size order the same as or greater than the mean free path of the molecules ($Kn \lesssim 0.5$), according to Einstein [21]

$$\mathcal{D} = \frac{CkT}{3\pi\mu d_p} \text{ cm}^2/\text{sec.} \tag{4.64}$$

where k = Boltzman's constant = 1.37×10^{-16} ergs/°K· molecule.

(b) For particles much smaller than the mean free path of the gas molecules ($Kn >> 0.5$), but still very large compared to the molecules themselves, according to Langmuir [22]

$$\mathcal{D} = \frac{4kT}{3\pi d_p^2 p} \quad \frac{8RT}{\pi M} \tag{4.65}$$

Symbols are as defined for Eqn. (4.64), but here consistent units must be used.

Illustrative values of $\mathcal{D}$ are shown below for dry air at 20°C and 1 atm. (1.013×10^6 dynes/cm^2). For (4.64), the values of C were computed by Eqn. (4.7).

Particle Diffusivities

d_p–μm	Kn	$\mathcal{D}$ – cm^2/sec. Einstein (4.64)	$\mathcal{D}$ – cm^2/sec. Langmuir (4.65)
10	0.0131	2.3×10^{-8}	---------
1	0.131	2.7×10^{-7}	---------
0.216	0.5	1.7×10^{-6}	1.7×10^{-6}
0.1	1.31	----------	7.8×10^{-6}
0.01	13.1	----------	7.8×10^{-4}
0.001	131	----------	7.8×10^{-2}

For very small particles, d_p less than a few tenths of a micron, the Brownian effect becomes significant. It plays a role in particle collection, particle deposition, and particle coagulation. Brownian deposition in aerodynamic capture will be discussed in Chapter 5, and in subsequent chapters where it arises in connection with specific collectors. Particle deposition and coagulation in still air are discussed below.

1. *Diffusive deposition* Equation (4.63) has been applied to a number of cases of diffusive deposition of aerosol particles from a stationary gas onto confining surfaces in the absence of significant effects from other forces. Fuchs [13, pg. 193] presents results for the following cases: (a) a plane vertical wall in contact with an infinitely large volume of aerosol; (b) a plane horizontal surface, likewise; (c) two parallel vertical planes of infinite extent; (d) two parallel horizontal planes of infinite extent; (e) aerosol situated inside a spherical vessel; (f) aerosol contained inside an infinitely long cylinder; (g) the surface of a sphere surrounded by an infinitely large volume of aerosol; (h) the surface of an infinitely long cylinder, likewise.

In each case diffusion in one dimension only (normal to the surface) is considered, using only the first term on the right side of (4.63), or its counterpart in spherical or cylindrical coordinates. It is assumed that the aerosol is initially at a uniform concentration n_o of particles all of the same size, and that all undeposited particles remain at the original size, i.e., there is no agglomeration occuring.

The general theory of solution of (4.63) given by Pich [23] yields first the concentration of remaining particles as a function of location (distance from surface) and time. This is then used in turn to give either the number of particles deposited in time t, or the average concentration $\bar{n}$ remaining undeposited after time t.

The rigorous solutions of (4.63) are expressed as an infinite series of terms in the dimensionless group $Fo = \mathcal{D} t/R^2$. For example in case (e) above, the average remaining concentration is given by Pich as:

$$\bar{n} = \frac{6}{\pi^2} n_o \sum_{k=1}^{\infty} \frac{1}{k^2} \exp - \pi^2 k^2 \text{ Fo} \tag{4.66}$$

However, for small values of Fo which frequently arise in practice such solutions are not suitable for practical calculations, nor for the inverse determination of Fo from a given value of $\bar{n}/n_o$.

For such cases Pich [23] developed a theory of deposition at small values of Fo which, as modified by Slinn [24], leads to the following simplified approximate solutions for cases (e) and (f).

Case (e): Deposition internal to a sphere of radius R

$$\frac{\bar{n}}{n_o} = 1 - 6(\text{Fo}/\pi)^{1/2} + 3 \text{ Fo} \tag{4.67}$$

Case (f): Deposition internal to a cylinder of radius R

$$\frac{\bar{n}}{n_o} = 1 - 4(\text{Fo}/\pi)^{1/2} + \text{Fo} + \frac{11}{24} \frac{(\text{Fo})^{3/2}}{\sqrt{\pi}} \tag{4.68}$$

For example, particles of 0.1 μm diameter, $\mathcal{D} = 7.8 \times 10^{-6}$ cm^2/sec, depositing inside a sphere of 1 mm radius, will be reduced to half their initial concentration (Fo = .0305 from Eqn. 4.67) after

$$t = \frac{0.305 \times (0.1)^2}{7.8 \times 10^{-6}} = 39.1 \text{ sec}$$

Calculations of this kind are generally not involved in particle collection work, where Brownian motion (if any) is taking place in a moving fluid. However, they may be of interest in estimating the deposition on the walls of a container in which a dilute aerosol is stored, or in estimating deposition in small vessels such as those making up the lungs, during periods of time in which there is little or no flow and in which coagulation of the particles is negligible.

2. *Thermal coagulation of aerosols* Particles in Brownian motion may collide with one another and in so doing adhere to form a new larger particle. This process, called thermal coagulation, is always present in aerosols and will also reduce the concentration of particles as time goes on. If successive collisions occur to the same particles, eventually much larger ones form and may become subject to the action of other forces. Thus coagulation may aid indirectly in the particle collection process. Coagulation may also be promoted by the presence of other forces, e.g., electrostatic, external to the particles and superimposed upon the Brownian movement. An extensive review is given by Zebel [25].

An elementary theory of thermal coagulation, adapted from Smoluchowski [26,27], is useful to review here. Assuming all particles to be spherical, initially of the same size, and of concentration n per unit volume at any instant, he estimated the total number of collisions per second to be $4\pi d_p \mathcal{D} n^2$. Assuming that each collision results in adherence and therefore reduces the number of particles by one

$$-\frac{dn}{dt} = 4\pi d_p \mathcal{D} n^2 = K_o n^2 \tag{4.69}$$

Upon integration, assuming K_o to be independent of time,

$$\frac{n}{n_o} = \frac{1}{1 + K_o n_o t} \tag{4.70}$$

or

$$\frac{1}{n} - \frac{1}{n_o} = K_o t$$

where $n = n_o$ at $t = o$. When $n/n_o = 0.5$, $t = 1/K_o n_o$ is called the half-life of the aerosol. Theoretical values of K_o, the coagulation rate constant, in air at 20°C may be calculated from (4.69) using the Einstein value for $\mathcal{D}$ from Eqn. (4.64). Results are:

Particle size d_p-microns	Coagulation rate constant K_o-cm^3/sec.
10	2.9×10^{-10}
1	3.4×10^{-10}
0.1	8.4×10^{-10}
0.01	6.5×10^{-9}

It is evident that K_o could not be strictly independent of time, because the average particle size in the aerosol is becoming larger as collisions occur. Nevertheless the reciprocal relationship shown by (4.70) is found to hold experimentally, albeit with values of K_o some 30% higher than the theoretical. This is attributed to the counteracting influence of the greater dispersity of particle sizes as coagulation proceeds.

Extension of the theory to the case of particles of two different sizes present initially shows that if one size is appreciably larger than the other, the coagulation rate constant is much greater than that given above. In such aerosols the smaller particles disappear rapidly and are said to be "eaten up" by the larger ones. This is the case with ambient urban aerosols.

Example 8. For particles 0.1 micron in diameter, initially present in air at 20°C, in concentration $10^5/cm^3$, how frequently do collisions occur initially, what is the half-life, and what will the concentration be after 20 minutes?

For $d_p = 0.1$, $K_o = 8.4 \times 10^{-10}$, the initial rate of collision is $K_o n_o^2 = (8.4 \times 10^{-10})\,(10^5)^2 = 8.4$ per sec.

The half-life is $t_{1/2} = 1/8.4 \times 10^{-10} \times 10^5 \times 3600 = 3.31$ hr.

After t = 20 min.:

$$n = \frac{10^5}{1 + \frac{20}{3.31 \times 60}} = 9.08 \times 10^4 \text{ per cm}^3.$$

B. *DIFFUSIOPHORESIS*

If a gas is a mixture of two kinds of molecules A and B such that there is a difference in concentration between two points, the process of gaseous diffusion will take place according to Fick's Law. The mass rate per unit area flow of A will be in the direction of decreasing concentration of A, and proportional to the concentration gradient of A and the diffusion coefficient of A-B. Similarly for B, except that the concentration gradient (and flux) will be in the opposite direction.

If the molecules of A are heavier than B, then small particles (Kn>>1) suspended in the gas will be subjected to bombardment by a greater proportion of A on one side than the other. The particles will take on a net motion in the same direction as the diffusion of A. This process is called diffusiophoresis.

For the case where the average molecular velocity of the mixture is zero (A and B counter-diffusing through each other), according to Waldmann and Schmitt [28], the particle takes on a diffusion velocity u_D in the x-direction, given by

$$u_D = \left[\frac{\delta_A\sqrt{m_A} - \delta_B\sqrt{m_B}}{\delta_A n_A\sqrt{m_A} + \delta_B n_B\sqrt{m_B}}\right] D_{AB}\,\frac{\partial n_A}{\partial x} \tag{4.71}$$

in which m = mass of molecule, proportional to molecular weight; n = concentration of molecule, or mole-fraction; $\delta = 1 + \frac{\pi}{8}a$; a = accomodation coefficient; D_{AB} = mutual diffusion coefficient for the two gases. The velocity is in the direction of the gradient of A ($\partial n_A/\partial x < 0$), the heavier molecule. Since $\delta_A \simeq \delta_B$, (4.71) simplifies to

$$u_D = -\left[\frac{\sqrt{m_A} - \sqrt{m_B}}{m + \sqrt{m_A m_B}}\right] D_{AB} \frac{\partial n_A}{\partial x} \tag{4.72}$$

in which $m = n_A m_A + n_B m_B$ is proportional to the average molecular weight of the mixture. Note that in air where $m_{O_2} \simeq m_{N_2}$, the quantity in brackets is very small (0.0062) and the concentration is uniform so that u_D is ordinarly zero. In a mixture for which $m_A = m_B$, u_D would always be zero, regardless of concentration gradients.

For the special, but important, case where one gas is stationary overall, another phenomenon comes into play. Consider the surface of an evaporating liquid which is sending vapor molecules to diffuse into the surrounding resting gas. To compensate for the tendency of the gas molecules to diffuse toward the surface, there must be a (hydrodynamical) flow of gas-vapor mixture away from the surface at a velocity u_{St} given by

$$n_A u_{St} = D_{AB}\, \partial n_A / \partial x \tag{4.73}$$

where x is taken as a positive direction away from the surface. This velocity is called a Stefan flow. It will be superimposed upon the diffusion velocity, so that the net velocity is $(u_D + u_{St})$.

Adding (4.72) and (4.73), in which A refers to gas molecules and B to vapor, and using the fact that $\partial n_A/\partial x = -\,\partial n_B/\partial x$ it may be shown that

$$u_D + u_{St} = -\left[\frac{\sqrt{m_B}}{n_A\sqrt{m_A} + n_B\sqrt{m_B}}\right] \frac{D_{AB}}{n_A} \frac{\partial n_B}{\partial x} \tag{4.74}$$

The net result is always a particle velocity away from the evaporating surface, as $\partial n_B/\partial x$ is < 0. On the other hand, if the vapor is condensing from the gas onto a liquid surface the particle motion will be toward the surface, as $\partial n_B/\partial x > 0$.

A diffusion force F_D, analogous to the thermal force F_T given by Eqn. (4.62), may be calculated for Stokes law particles in the same way.

$$F_D = 3\pi\mu d_p(u_D + u_{St})/C \tag{4.75}$$

Well-known observations confirm these predictions. A clear space forms in a dusty gas above the surface of an evaporating drop of liquid. Experiments confirm the Stefan-Waldmann theory quantitatively [29].

The phenomenon is of importance in wet scrubbing systems of particulate collection where drops of colder liquid are used as target collectors. If the surrounding gas is saturated with liquid, condensations will occur and thus help to drive dust particles onto the target drops. This is referred to as force-flux-condensation (FFC). Conversely, if the gas is unsaturated and warmer drops are evaporating, the particle collection process is hindered. The role of FFC will be discussed further in Chapter 9.

REFERENCES

1. A. H. Shapiro, Shape and Flow, Anchor Books, Doubleday and Co., Inc., New York, (1961).
2. J. H. Perry ed., Chemical Engineers Handbook, 5th edition McGraw Hill, New York, (1976).
3. G. G. Stokes, Trans. Cambridge Phil. Soc., 9: Part II, 51 (1851).
4. C. Oseen, Neure Methoden und Ergebnisse in der Hydrodynamik, sec. 16, Leipzig, (1927).
5. L. Schiller and L. A. Naumann, Zeit. Vor. Deut, Ing., 77: 318 (1933).
6. L. Klyachko, Otopl. i Ventil.: No. 4 (1934).
7. D. R. Dickinson and W. R. Marshall, AIChE Journ., 14: 541 (1968).
8. M. Knudsen and S. Weber, Ann. Phys. 36: 981 (1911).
9. C. N. Davies, Proc. Phys. Soc., 57: 259 (1945).

10. K. Willeke, J. Aerosol Sci., 7: 381 (1976).

11. M. D. Allen and O. G. Raabe, Aerosol Sci. and Tech. 4: 269 (1985).

12. S. Calvert in Air Pollution-Vol. IV, 3rd Ed. (A. Stern, ed.), Academic Press, New York (1977).

13. N. A. Fuchs, The Mechanics of Aerosols, Pergamon Press, New York (1964).

14. L. B. Torobin and W. H. Gauvin, Canad. J. Chem. Eng., 37: 224 (1959).

15. R. Ingebo, NASA: Tech. Note 3762 (1956).

16. W. Koch, Personal Communication (1977).

17. T. Allen, Particle Size Measurement, 3rd ed. 81-83, Chapman and Hall, London, (1981).

18. J. A. Gieseke, in Air Pollution Control-Part II (W. Strauss, Ed.) Wiley Interscience, New York (1972).

19. R. L. Byers and S. Calvert, Ind. Eng. Chem. Fund., 8: 646 (1969).

20. B. Singh and R. L. Byers, Ind. Eng. Chem. Fund., 11: 127 (1972).

21. A. Einstein, Z. Electrochemie, 14; 235 (1908).

22. I. Langmuir, OSRD Report No. 865 (1942).

23. J. Pich, Atmos. Environ. 10: 131 (1976).

24. W. G. N. Slinn, Atmos. Environ. 10: 789 (1976).

25. G. Zebel, in Aerosol Science (C. N. Davies, Ed.) Chap. II, Academic Press, New York (1966).

26. M. von Smoluchowski, Phys. Z., 17: 557, 585 (1916).

27. M. von Smoluchowski, Z. Phys. Chem. 92: 129 (1918).

28. L. Waldman and K. H. Schmitt in Aerosol Science (C. N. Davies, Ed.) Chap. VI, Academic Press, New York (1966).

29. P. Goldsmith and F. G. May in Aerosol Science (C. N. Davies, Ed.) Chap. VII, Academic Press, New York (1966).

PROBLEMS

1. A particle 15μm in diameter, having a density of 3.1 gm/cm^3, was observed to have a stopping distance of 4.0 mm in still air at 20°C and 1 atm pressure. With what initial velocity was it propelled?

2. Spherical particles of density 1.3 gm/cm^3 are projected into dry air at 100°C and 1 atm with an initial horizontal velocity component of 10 cm/s. Determine the size of particle which will have a stopping distance of 15 cm.

3. A droplet of water 200µm in diameter is formed by condensation in a flue gas, at just below 100°C, as the gas is rising in a stack with a velocity of 1m/s. Which way will this drop move? How long will it take to travel vertically 1 m of stack height?

4. A certain aerosol sample (density of particles = 1.3 gm/cm^3) was found to follow a log-normal probability size distribution with a number median size of 35µm and a value of $\sigma_g = 3.50$. Determine the velocity of a rising air current (20°C, 1 atm) which would prevent:

(a) at least 1% of the particles from settling out;

(b) at least 16% of the mass of the particles from settling out.

5. An aerosol having an MMD of 5 µm and a $\sigma_g = 3.2$, particle density = 1.8 gm/cm^3, was placed in a current of air at 100°C and 1 atm flowing upward with a velocity of 0.300 cm/s. What fraction by weight of the particles will be carried upward?

6. What would be the electrostatic drift velocity of a particle 1 µm in diameter carrying 75 electronic charges in an electric field potential gradient of 2000 volts/cm in dry air at 20°C and 1 atm?

7. A drop of water 1000 µm in diameter falls from rest in air at 20°C and 1 atm carrying a charge of 100 electrons. There is an electrostatic field acting vertically downward with a potential gradient of 2000 volts/cm.

(a) What would be the initial acceleration of the drop, that is at the moment it starts to fall?

(b) What fraction of this acceleration is due to the electrostatic field?

8. (a) Under the same conditions as given in problem No 6 above, how long would it take a particle 0.40 μm in diameter to travel 1 inch?

(b) What thermal gradient would be needed in order to produce a velocity equal to that in (a) if Th = 0.42 for Kn $\geq$ 1.00?

9. An aerosol composed of particles 0.1 μm in diameter, having a density of 2.0 gm/cm^3, is flowing in air parallel to a flat horizontal surface maintained at 20°C. Determine the minimum temperature gradient at the surface necessary to prevent particle deposition by settling. (Neglect Brownian diffusion).

10. Two parallel vertical plates are spaced 1 mm apart and maintained at a temperature difference of 100°C and a voltage difference of 100 volts, the hotter plate being at the lower potential. Estimate how many electronic charges a particle 1 μm in diameter must have in order that it will not move toward either plate.

11. It has been stated in the literature that particles 0.02 μm in diameter, initially present in the atmosphere (20°C, 1 atm) at a concentration of 10^5 particles per cm^3, are reduced to 1/2 their number in one hour by the process of coalescence or agglomeration. Does this statement agree with the Smoluchowski theory?

12. Suppose particles 0.01 μm in diameter to be initially uniformly suspended in air at 20°C, 1 atm inside a sphere of diameter 1 cm at a concentration of 10^6 particles/cm^3. During the first one second of elapsed time how would the reduction in

concentration by thermal coalescence compare with that due to diffusional deposition on the wall, assuming that either of these processes could take place independently of the other?

13. The following data were determined on a smoke of cadmium oxide particles suspended in a large volume of still air, in which the concentration of particles was measured as a function of time.

t - min	8	24	43	62	84
$n/cm^3 \times 10^6$	0.92	0.47	0.33	0.24	0.21

(a) Why is the concentration of particles decreasing?

(b) What was the initial concentration, $particles/cm^3$?

(c) What is the rate constant for this process?

5
Trajectories of Particles in Moving Gas—Aerodynamic Capture

I. PARTICLE TRAJECTORIES

This chapter deals in detail with the methods of predicting whether a particle being carried in a moving stream of gas will collide with a collecting surface or with a target collector. It is based upon the general discussion of Mechanisms of Collection given in Chapter 3. The basic approach is to calculate the trajectory or path followed by a given particle from its point of entrance into a zone of collection under whatever forces may be present to act upon it. The information presented in Chapter 4 about forces and motion of particles in a still gas will be applied here to a flowing gas. The results obtained will in turn form the basis for the modelling of the various kinds of collectors which are treated in the subsequent chapters.

A. *THE ROLE OF THE GAS VELOCITY*

The prediction of particle trajectories in a gas which is in motion is more complex than for a gas which is at rest. The velocity of the gas stream with respect to the particles, as well as with reference to the stationary surroundings, must be taken into account. A differential equation must be written for the streamlines of the gas as well as for the aerosol particle, and these two equations considered simultaneously. Velocity vectors for the gas, and for the particle, must be determined.

Whenever the velocity vector for the fluid at a point differs in direction and/or magnitude from that of the particle at the same location, there is said to be "slip" between particles and gas. The vector difference of these velocities is called the "slip," or "relative," velocity. The magnitude of this relative velocity determines the magnitude of the drag force. The equations of Chapter 4 for calculating drag are applicable so long as relative velocity is substituted for particle velocity.

In Newton's Law the expression for the drag force will become

$$\vec{F}_D = -\frac{C_D A_{proj} \rho_f}{2} (\vec{u} - \vec{v}) \cdot |\vec{u} - \vec{v}| \tag{5.1}$$

$$|\vec{F}_D| = \frac{C_D A_{proj} \rho_f}{2} |\vec{u} - \vec{v}|^2 \tag{5.2}$$

where $\vec{u}$ = velocity vector of particle; $\vec{v}$ = velocity vector of fluid. Figure 1 illustrates these vectors and their components, in two dimensions, and shows that $(\vec{u} - \vec{v})$ = velocity vector of particle relative to fluid and

$$|\vec{u} - \vec{v}| = \sqrt{(u_x - v_x)^2 + (u_y - v_y)^2} \tag{5.3}$$

In calculations this drag force will have to be solved into its components:

$$F_{D_x} = \vec{F}_D \cos \alpha = \frac{C_D A_{proj} \rho_f}{2} (v_x - u_x) \; |\vec{u} - \vec{v}| \tag{5.4}$$

$$F_{D_y} = \vec{F}_D \sin \alpha = \frac{C_D A_{proj} \rho_f}{2} (v_y - u_y) \; |\vec{u} - \vec{v}| \tag{5.5}$$

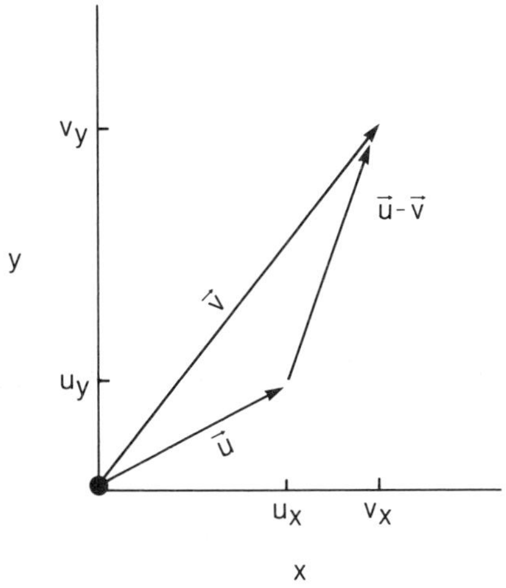

FIG. 1. *Fluid velocity $\vec{v}$, particle velocity $\vec{u}$, and drag $(\vec{u} - \vec{v})$ vectors.*

Several basic cases may be examined, with reference to the concepts of collecting zones, collecting velocity, and residence time described in Chapter 3. The simplest is that of plug flow (constant $\vec{v}$) for the fluid, and a constant collecting velocity $\vec{u}$ perpendicular to $\vec{v}$. As illustrated in Fig. 7, Chapter 3, the particle velocity is a constant and the trajectory is a straight line.

Consider the same case but with laminar flow of the gas stream. The fluid will have a symmetrical parabolic velocity distribution profile across the zone. As the particle moves across the fluid it will constantly encounter a different value of v, hence its velocity vector will vary in length and direction. The path of the particle will no longer be straight. The actual trajectory may be visualized readily, Fig. 2, by comparing it with the straight-line path the particle would have followed in a plug-flow air stream at $v = v_o$, where v_o is air velocity at the point of entry of the particle. If it can be assumed that the particle instantaneously adjusts its velocity u_x to be equal to v_x at every

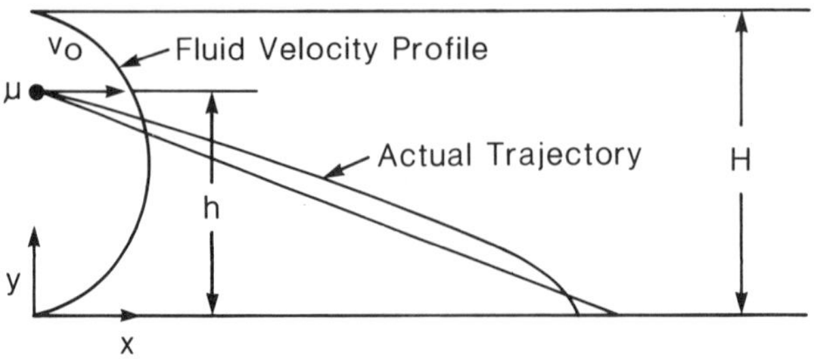

FIG. 2. *Trajectory of particle between parallel plates, laminar flow.*

point, i.e., no slip in the x-direction, the curved path may be calculated from a knowledge of the actual fluid velocity profile and the nature of the collecting force. This is done below for certain cases.

The precise shape of the fluid velocity profile will vary with the value of Reynold's number based upon the duct size. When this number becomes sufficiently large, there will be an onset and development of turbulent flow. Not only will the gross velocity profile become flatter and less parabolic, but also there will now be eddies in the fluid to affect the particle motion.

Another case of importance is that of the flow of the gas stream around a collector target element. (See Fig. 1 (b), Chapter 3). Here a more complex pattern of fluid stream lines will be set up around the curvature and into the wake of the target. The exact nature of the pattern will vary with fluid Reynold's number based upon the size of the target, i.e., $Re_f = vD/\nu_f$. Where these streamlines begin to curve upstream of the target, the particle may not be able to follow owing to its inertia. The particle path will be a curve cutting across the fluid streamlines toward the target surface, as shown in Fig. 3. There will be "slip," and the drag force must be reckoned accordingly. A knowledge of the fluid mechanics of flow around whatever shape the target is, will be needed in order to place the fluid streamlines.

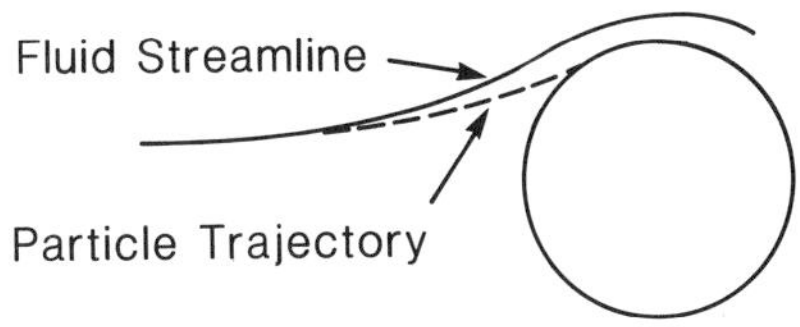

FIG. 3. *Trajectory of particle approaching target surface.*

These examples illustrate the importance of the fluid velocity in determining the particle trajectory. Particle trajectories will next be examined in detail with the objective of determining deposition upon collector surfaces (parallel planes, and cylinders) and upon collector targets (cylinders and spheres).

B. *TRAJECTORIES WITHIN CONTAINMENT ZONES*

1. *Parallel Plane Boundaries* If the collection zone is bounded by parallel plane surfaces as in Fig. 1a, Chapter 3, and the collecting velocity $\vec{u}$ is constant, the particle trajectory will depend upon the velocity profile of the gas flow between the planes. There are three cases of interest: (a) plug flow, $\vec{v}$ constant; (b) laminar flow, $\vec{v}$ parabolic profile; (c) turbulent flow, $\vec{v}$ logarithmic profile, eddy currents. Some examples are: (1) gravity settling between horizontal planes, where $\vec{u}$ acts vertically downward and is equal to u_S, the terminal settling velocity of the particles; Eqns. (4.33) or (4.37).
(2) electrostatic attraction toward a plane bounding a potential difference across a gas, with $\vec{u}$ acting in the direction of the potential gradient, and equal to u_E; Eqn. (4.60).
(3) thermophoresis between two planes of different temperatures, $\vec{u}$ acting from hotter toward colder plane and equal to u_T, the thermophoretic velocity of the particle; Eqn. (4.62).
(4) diffusiophoresis from the horizontal surface of an evaporating liquid (Stefan flow), $\vec{u}$ acting vertically upward equal to u_D; Eqn. (4.72) or (4.73).

These and similar examples may all be analyzed in one general treatment, making these assumptions:

(i) The aerosol composed of various sized particles (i-grade) is homogeneous and uniformly distributed between planes at the point of entrance into the zone.

(ii) The particles behave individually, acted upon by the collecting force(s) and drag; buoyancy is neglected.

(iii) There is no slip in the direction of flow.

(iv) Particles accelerate rapidly to the collecting velocity, which is constant in the direction perpendicular to flow and acts only toward one plane.

(v) The planes may be horizontal or vertical; if horizontal, gravity will have to be considered as part of the total collecting force (action toward the lower plate only).

As shown in Fig. 2, the coordinate system is taken as x in the direction of flow from point of entrance, y perpendicular to flow from plane toward which collecting velocity acts. The spacing between the planes is H, their length in direction of flow is ℓ. The point at which a particle enters the zone is at $x = 0$, $y = h$.

a. Plug Flow of Gas.

Fluid: $v_x = v_o, \quad v_y = 0$

Particle: $u_{x_i} = v_o$ (independent of d_{p_i})

$u_{y_i} = u_{c_i}$, constant but may depend upon d_{p_i}

This corresponds to the no-mixing model presented in Chapter 3, III.B.1, illustrated there with $u_{c_i} = u_{s_i}$ the terminal settling velocity of particles between horizontal planes.

The particle trajectories are straight lines having slopes $= u_{c_i}/v_o$, and the grade efficiency is given by

$$\eta_i = u_{c_i}\ell/v_oH \tag{5.6}$$

All particles for which $u_{c_i} \geq v_oH/\ell$ will be collected with 100% efficiency. This establishes a minimum particle size which will be collected completely.

The relationship between η_i and d_{p_i} will depend upon the nature of the collecting force and its relationship to d_{p_i}. This can be obtained from Chapter 4. For example:

(a) Gravity settling: $u_{c_i} = u_{s_i}$ which depends upon Ga_i and may be obtained from Fig. 8, Chapter 4. For Stokes law particles $u_{s_i} \propto d_{p_i}^2$ and the grade-efficiency plot is as shown in Fig. 8, Chapter 3.

(b) Electrostatic attraction: $u_{c_i} = u_{E_i}$, given by Eqn. (4.60). For Stokes law particles $u_{E_i} \propto d_{p_i}$, and the grade-efficiency plot is linear from $\eta_i = 0$ at $d_{p_i} = 0$, up to $\eta_i = 1$ at d_{p_i} corresponding to $u_{E_i} = v_oH/\ell$.

(c) Thermophoresis: $u_{c_i} = u_{T_i}$, given by Eqn. (4.62) which is independent of d_{p_i} for Kn >> 1. The grade-efficiency is the same for all particle sizes and is given by (5.6) with $u_{c_i} = u_T$.

b. Laminar Flow of Gas

Fluid: $$v_x = v_{max}\left[1 - \frac{4(y - H/2)^2}{H^2}\right] = 4v_{max}\left[\frac{y}{H} - \left(\frac{y}{H}\right)^2\right]$$

$$v_y = 0$$

$$v_{max} = (3/2)v_o \qquad (v_o = \text{average velocity})$$

Particle: $u_{x_i} = v_x$ (independent of d_{p_i})

$u_{y_i} = u_{c_i}$, constant, but may depend upon d_{p_i}

The parabolic fluid velocity profile, illustrated in Fig. 2, holds for $Re_f = v_o H/\nu_f < 890$.

To obtain the particle trajectory, set

$$u_{x_i} = v_x = \frac{dx}{dt} \quad \text{and} \quad u_{y_i} = u_{c_i} = -\frac{dy}{dt}$$

Then

$$\left(\frac{dx}{dy}\right)_i = -\frac{6v_o}{u_{c_i}}\left(\frac{y}{H} - \frac{y^2}{H^2}\right) \tag{5.7}$$

It is convenient to put (5.7) into dimensionless form using H and v_o as reference values. Then $x^* = x/H$; $y^* = y/H$; $u^*_{c_i} = u_{c_i}/v_o$ and (5.7) becomes

$$\left(\frac{dx^*}{dy^*}\right)_i = -\frac{6}{u^*_{c_i}}\left(y^* - y^{*2}\right) \tag{5.8}$$

Integrating (5.8) with the boundary condition that at $x^* = 0$, $y^* = h/H = h^*$, gives the equation of the trajectory defined for $0 \leqq y^* \leqq h^*$, i.e., from point of particle entrance at $y = h$, to the point where the particle collides with collecting plane at $y = 0$. The result may be written

$$x^* = \frac{1}{u^*_{c_i}}\left[\left(2y^{*3} - 3y^{*2}\right) - \left(2h^{*3} - 3h^{*2}\right)\right] \tag{5.9}$$

All types of trajectories represented by (5.9) have a point of inflection at $y^* = 1/2$, and are normal to the surface at $y = 0$.

The efficiency may be determined by finding the value of h^* for the trajectory which terminates at $y = 0$, $x = \ell$ or $x^* = \ell/H$. Equation (5.9) gives

$$3h_i^{*2} - 2\,h_i^{*3} = \frac{\ell}{H}\,\frac{u_{c_i}}{v_o} \tag{5.10}$$

All particles entering below this value of h_i^* are collected, hence $\eta_i = h_i^*$ = root of (5.10) lying between 0 and 1. The values of this root corresponding to values of $\ell u_{c_i}/Hv_o$ are tabulated below.

$\ell u_{c_i}/Hv_o$	h_i^*	$\ell u_{c_i}/Hv_o$	h_i^*
0	0	0.6	0.5671
0.1	0.1958	0.7	0.6367
0.2	0.2871	0.8	0.7129
0.3	0.3633	0.9	0.8042
0.4	0.4329	1	1
0.5	0.5		

c. Turbulent Flow of Gas. As the value of $Re_f = v_oH/\nu_f$ for the duct increases above about 890, there will be an onset of turbulence in the fluid. The eddy currents developed will attain root-mean-square velocities in the vertical direction of approximately 0.03 to 0.1 of v_o, the average fluid velocity. For example, at $v_o \approx 1$ m/sec, these will be 3-10 cm/sec, appreciably greater than the terminal settling velocity of say 20-micron particles of unit density, $u_s \approx 1.3$ cm/sec. Hence particles of this size and finer will tend to become uniformly mixed over the vertical cross-section of the duct at velocities $v_o \geq 1$ m/sec.

When such conditions prevail, the collection efficiency may be modelled by the lateral-mixing model discussed in Chapter 3, III.B.2. The velocity profile becomes very flat and $v \approx v_o$. In time interval $dt = d\ell/v_o$ the reduction in number of particles of ith grade by collection from an element of gas $d\ell$ in length and unit width, will be $-dn_i = (nu_{c_i}/H)\, dt = (nu_{c_i}/Hv_o)\, d\ell$. The parameter $k_i = u_{s_i}/Hv_o$, and the collection efficiency is given by

$$\eta_i = 1 - \exp - (u_{c_i}\ell/Hv_o) \tag{5.11}$$

In all of the above equations (5.6), (5.10), and (5.11) for the grade-efficiency of collection by a constant collection velocity, there are two parameters. One is the velocity ratio

$u^*_{c_i} = u_{c_i}/v_o$ which is determined by the properties of particle and fluid, and by operating conditions. The other is the geometric ratio ℓ/H which represents the physical shape of the collector space. It will be seen that these two kinds of parameters will always occur in the model for any collecting device.

Comparison of grade-efficiencies from (5.6), (5.11) and (5.11) shows that for $0 < u_{c_i}\ell/v_oH < 0.5$ they rank in the order LAMINAR > PLUG > TURBULENT, while for $0.5 < u_{c_i}\ell/v_oH$ the order is PLUG > LAMINAR > TURBULENT. At $u_{c_i}\ell/v_oH \geqq 1$, $\eta_i = 1$ for PLUG and LAMINAR, but $\eta_i \to 1$ asymptotically ($\eta_i = 0.9933$ at $u_{c_i}\ell/v_oH = 5$) for TURBULENT.

Example 1. A thermal sampling collector, consisting of parallel horizontal plates, operates under the following conditions: H = 2mm, ℓ = 5 cm; top plate at 100°C, bottom plate at 0°C; v_o = 10.0 cm/s; dry air at 1 atm. Estimate the grade efficiency of collection of particles of unit density ranging from d_p = 0.001 μm to 1 μm .

Assuming the average gas temperature to be 50°C, $\mu = 1.96 \times 10^{-4}$ gm/cm·s, $\rho_f = 1.093 \times 10^{-3}$ gm/cm^3, and $v_oH/\nu_f = 11.1 < 890$, so that flow is laminar.

The value of Kn (Eqn. 4.8) will range from 148 to 0.148, so that Th may be taken as ≈ 0.50 and u_T will be

$$u_T = \frac{0.50 \times 1.96 \times 10^{-4}}{1.093 \times 10^{-3}\ (273 + 50)} \times \frac{(100 - 0)}{(0.2)} = 0.14 \text{ cm/s}$$

In Equation (5.10) u_T will be the same for all particles

$$\frac{\ell}{H}\frac{u_T}{v_o} = \frac{5 \times 0.14}{0.2 \times 10.0} = 0.35$$

and $h^* \approx 0.40$. Particles of all sizes in this range will be collected by a thermal force with an efficiency of 40%. If it is desired to obtain 100% collection efficiency, the air flow rate should be reduced to below

$$v_o = \frac{5 \times 0.14}{0.2 \times 1} = 3.50 \text{ cm/s}$$

There will also be a small effect of gravity, so that the coarser particles will tend to be collected to a slightly greater extent. The terminal settling velocity of 1 μm particles is given by Eqn. (4.39) as

$$u_s = \frac{980 \times 1.18 \times (1)}{18 \times 1.96 \times 10^{-4}} (1 \times 10^{-4})^2 = 0.0033 \text{ cm/s.}$$

Since $u_s << u_T$ the gravity effect may be neglected.

2. *Cylindrical Boundary - Spinning Gas* Centrifugal force may be utilized as an effective collecting force by causing the dust-laden gas stream to flow in a circular manner inside a cylindrical boundary. This is the basis for the operation of cyclone collectors which are discussed in detail in Chapter 6. Fundamental to the modelling is a knowledge of particle trajectories in a spinning gas which is presented here.

Consider a particle entering tangentially at speed u_{T_1} into a horizontal plane of a spinning gas stream at R_1 as shown in Fig. 4. Owing to its inertia, which is represented by a centrifugal force of $F_c = m_p u_T^2/R$ at any instant and to the force of gravity, the particle will follow a helical path downward. Its velocity vector u will have a tangential component (u_T), a radial component (u_R) and an axial component (u_S). The velocity of the spinning gas will be assumed to have only a tangential component v_T, with $v_R = 0$. It may also be assumed that there is no tangential slip, i.e., that $u_T = v_T$. However, v_T is not to be regarded as a constant, but to vary with R. Tests have shown that this variation will likely follow the relationship

$$u_T R^n = \text{constant} = u_{T_1} R_1^n \tag{5.12}$$

For an ideal fluid in such a "vortex" flow n = 1, but in real flows the value of n may range downward to 0.5.

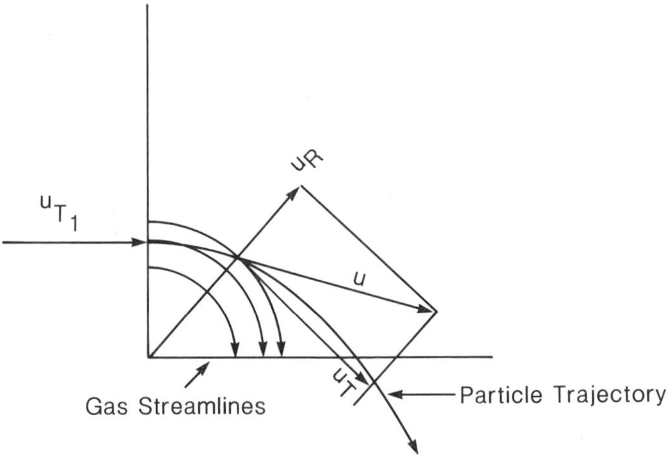

FIG. 4. *Particle trajectory in a spinning gas.*

Newton's Law is applied only to the radial component of motion, with the drag force directed radially inward.

$$m_p \frac{du_R}{dt} = \frac{m_p u_T^2}{R} - F_D \tag{5.13}$$

Since $u_R = \frac{dR}{dt}$, this may be written for a spherical particle:

$$\frac{d^2R}{dt^2} = \frac{u_{T_1}^2 \; R_1^{2n}}{R^{2n+1}} - \frac{3}{4}\frac{C_D}{d_p}\frac{\rho_f}{\rho_p}\left(\frac{dR}{dt}\right)^2 \tag{5.14}$$

This equation is obviously non-linear and is further complicated by the fact that C_D will also vary with u_R.

Applying (5.14) to a Stokes law particle gives some simplification:

$$\frac{d^2R}{dt^2} = \frac{{u_{T_1}}^2 {R_1}^{2n}}{R^{2n+1}} - \frac{1}{\tau}\frac{dR}{dt} \tag{5.15}$$

but (5.15) cannot be solved in general except by numerical approximation methods. Fortunately a good approximation to it is obtained by neglecting the second order term although this is equivalent to saying that the particle moves radially outward at a constant velocity which is obviously inconsistent with the resulting equation

$$\frac{dR}{dt} = \frac{\tau u_{T_1}^2 R_1^{2n}}{R^{2n+1}} = \frac{\tau u_{T_2}^2}{R_2} \left(\frac{R_2}{R}\right)^{2n+1} \tag{5.16}$$

This may be integrated to give the position R at any time t with respect to R_1 when t = 0.

$$t = \frac{R_1^2}{2(n+1)\tau u_{T_1}^2} \left[\left(\frac{R}{R_1}\right)^{2n+2} - 1 \right] \tag{5.17}$$

This equation relating radial position R to elapsed time t enables calculation of the particle trajectory. It will be used in connection with modelling the cyclone collector in Chapter 6. It is useful to restate it in terms of u_{T_2} and R_2, using (5.12)

$$t = \frac{R_2^2}{2(n+1)\tau u_{T_2}^2} \left[\left(\frac{R}{R_2}\right)^{2n+2} - \left(\frac{R_1}{R_2}\right)^{2n+2} \right] \tag{5.18}$$

where R_1 and R_2 designate respectively the innermost and the outermost radial positions of interest, and $R_1 < R < R_2$.

3. *Cylindrical boundary - spinning gas - charged particles in a radial electrostatic field* As an example of a case of a combination of forces acting upon a particle, consider the same situation as 2. above with the addition of an electrostatic field of gradient E in the radial direction, and particles carrying

charge q. Equation (5.13) would be modified by the addition of the electrostatic force term qE

$$m_p \frac{du_R}{dt} = \frac{m_p u_T^2}{R} - F_D + qE \tag{5.19}$$

Here E will vary inversely with the radius and may be written $E = E_o R_1/R$, whence the electrostatic force term will become

$$qE = b'/R = bE_o R_1/R \tag{5.20}$$

where b = electrical mobility as defined in Eqn. (4.60).

Proceeding as above to introduce the velocity relationship (5.12), applying to a spherical Stokes Law particle, and setting $d^2R/dt^2 = 0$, the equation analogous to (5.16) will become

$$\frac{dR}{dt} = \frac{\tau u_{T_1}^2 R_1^{2n}}{R^{2n+1}} + \frac{b'}{R} \tag{5.21}$$

wherein $b' = \dfrac{q}{3\pi\mu d_p} E_o R_1$.

Equation (5.21) is of the form

$$\frac{R^{2n+1}}{C_1 + b'R^{2n}} dR = dt \tag{5.22}$$

where $C_1 = (u_{T_1} R_1^n)^2 \tau = (u_{T_2} R_2^n)^2 \tau$

As before, this is to be solved by integration to give the radial position R of the particle at time t, with respect to its position R, when t=0. In case n=0.5, or n=1 this integration is easily done; otherwise a numerical integration is necessary.

For n=1, ideal vortex flow, the result is

$$t = \frac{R_1^2}{2b'}\left[\left(\frac{R}{R_1}\right)^2 - 1\right] - \frac{C_1}{2b'^2}\ln\frac{R^2 + C_1/b'}{R_1^2 + C_1/b'} \tag{5.23}$$

which may be compared to Eqn. (5.17) with n=1.

For n = 1/2, the result is

$$t = \frac{R_1^2}{2b'}\left[\left(\frac{R}{R_1}\right)^2 - 1\right] - \frac{R_1C_1}{b'^2}\left[\frac{R}{R_1} - 1\right] + \frac{C_1^2}{b'^3}\ln\frac{R + C_1/b'}{R_1 + C_1/b'} \tag{5.24}$$

which may also be compared to (5.17) with n = 1/2. These results can also be adapted to the modelling of an "electrified" cyclone in Chapter 6.

C. *INERTIAL TRAJECTORIES*

The path of a particle moving through a spinning gas as discussed above, is a special case of a trajectory determined by particle inertia as the gas flow is subjected to a curved path. A more general approach will be considered next, which may be used to treat several other cases of importance in the modelling of dust collectors, especially for target collectors.

For this purpose, only the drag force acting upon the particle is considered and Newton's law is written using (5.1)

$$m_p\frac{d\vec{u}}{dt} = \vec{F}_D = \frac{-C_DA_{proj}\,\rho_f}{2}(\vec{u} - \vec{v})\cdot|\vec{u} - \vec{v}| \tag{5.25}$$

and placed in dimensionless form by using set (i) of the reference values discussed with Eqn. (4.44). Here v_o will represent an undisturbed steady velocity at some point upstream of onset of the curved path (upstream of target), D will represent some appropriate dimension related to the size of the curvature in path (size of target), and the reference time will be D/v_o. The dimensionless variables then become

$\theta = \frac{v_o}{D} t$ dimensionless time

$\tilde{x} = \frac{x}{D}$ dimensionless particle position along x-axis

$\tilde{y} = \frac{y}{D}$ dimensionless particle position along y-axis

$\tilde{u}_x = \frac{u_x}{v_o} = \frac{1}{v_o} \frac{dx}{dt}$ dimensionless particle velocity, x-component

$\tilde{u}_y = \frac{u_y}{v_o} = \frac{1}{v_o} \frac{dy}{dt}$ dimensionless particle velocity, y-component

$\tilde{v}_x = \frac{v_x}{v_o}$ dimensionless fluid velocity, x-component

$\tilde{v}_y = \frac{v_y}{v_o}$ dimensionless fluid velocity, y-component

Since there is usually symmetry about the z-axis, it is sufficient to deal with the component equations in two dimensions (x,y) only. Equation (5.25) then is resolved into the scalar equations using (5.4) and (5.5):

$$m_p \frac{du_x}{dt} = - C_D A_{proj} \frac{\rho_f(u_x - v_x)(|\vec{u} - \vec{v}|)}{2}$$

$$m_p \frac{du_y}{dt} = - C_D A_{proj} \frac{\rho_f(u_y - v_y)(|\vec{u} - \vec{v}|)}{2} \tag{5.26}$$

which are reexpressed in dimensionless form:

$$\frac{d\tilde{u}_x}{d\theta} = - \frac{C_D A_{proj} D \rho_f}{2m_p} (\tilde{u}_x - \tilde{v}_x) \frac{(|\vec{u} - \vec{v}|)}{v_o}$$

$$\frac{d\tilde{u}_y}{d\theta} = - \frac{C_D A_{proj} D \rho_f}{2m_p} (\tilde{u}_y - \tilde{v}_y) \frac{(|\vec{u} - \vec{v}|)}{v_o} \tag{5.27}$$

It is convenient to introduce the particle Reynold's number based upon relative velocity defined as:

$$Re_p = \frac{(|\vec{u} - \vec{v}|)\, d_p}{\nu_f}$$

Carrying the development forward just for the x-component, (5.27) may be transformed into

$$\frac{d\tilde{u}_x}{d\theta} = - \frac{C_D\ Re_p}{2} \left(\frac{A_{proj}}{m_p}\right) \frac{D\mu}{v_o d_p} (\tilde{u}_x - \tilde{v}_x)$$

For spherical particles $A_{proj}/m_p = 3/2\rho_p d_p$ and the inertial parameter Stk (Stokes number) as defined in Eqn. (4.46) automatically arises:

$$Stk = \frac{C\rho_p d_p^2}{18\mu} \frac{v_o}{D}$$

$$\frac{d\tilde{u}_x}{d\theta} = - \frac{C_D Re_p}{24\ Stk} (\tilde{u}_x - \tilde{v}_x)$$

or

$$\frac{24\ Stk}{C_D Re_p} \frac{d\tilde{u}_x}{d\theta} + \tilde{u}_x - \tilde{v}_x = 0 \tag{5.28}$$

Since the objective here is to find the particle trajectory, i.e., to find the position of the particle (x,y) as a function of time, it is necessary to substitute $\tilde{u}_x = d\tilde{x}/d\theta$ and $d\tilde{u}_x/d\theta = d^2\tilde{x}/d\theta^2$. This gives

$$\frac{24\ Stk}{C_D Re_p} \frac{d^2\tilde{x}}{d\theta^2} + \frac{d\tilde{x}}{d\theta} - \tilde{v}_x = 0$$

and

$$\frac{24\ Stk}{C_D\ Re_p} \frac{d^2\tilde{y}}{d\theta^2} + \frac{d\tilde{y}}{d\theta} - \tilde{v}_y = 0 \tag{5.29}$$

as the equations which must be solved in order to get the trajectory. These equations may also be written to display Re_D, the Reynold's Number for the target, by writing the coefficient of the first term as

$$\frac{24\ Re_p}{2\ C_D}\left(\frac{2Stk}{Re_p^2}\right) = \frac{4}{3}\left(\frac{Re_p}{C_D}\right)\left(\frac{1}{Re_D}\right)\frac{\rho_p}{\rho_f}$$

noting that

$$\frac{2\ Stk}{Re_p^2} = \frac{\rho_p \mu}{9\rho_f v_o^2 D} = \frac{\rho_p}{9\rho_f\ Re_D}$$

This term is independent of particle size. This form of (5.29) has been found useful particularly when dealing with non-Stokesian particles. In the solution the parameters Stk and Re_D remain constant, but Re_p (and C_D with it, see Chapter 4) varies along the path with the relative velocity scalar $|\tilde{u} - \tilde{v}|$. For Stokes law particles, where $C_D = 24/Re$, (5.29) reduces to

$$Stk\ \frac{d^2\tilde{y}}{d\theta^2} + \frac{d\tilde{y}}{d\theta} - \tilde{\upsilon}_y = 0 \qquad (5.30)$$

and the value of Stk must include the Cunningham factor as needed.

Note that here again is the problem of determining the appropriate drag coefficient for an accelerating particle. It is similar to the problem discussed in Chapter 4, I.B.1, c.3. The usual practice here, as recommended there, is to use the values of C_D obtained for steady motion, assuming that little error will thus be introduced because most applications will be for low values of Re_p.

To solve these equations in any given case, for example as illustrated in Fig. 3 of this Chapter, will require a knowledge of the pattern of fluid streamlines because $\tilde{v}_x$ and $\tilde{v}_y$ are in themselves functions of x and y, although they may be taken as independent of time. It is well-known that such a pattern of streamlines varies with the nature of the fluid regime as characterized by the Reynold's Number for the target: $Re_D = v_o D/\nu_f$. Hence a solution of (5.30) can be worked out only for a specified type of target shape and for a fixed value of Re_D. A particular solution for a particle trajectory starting at a point (x_o, y_o) sufficiently upstream so that the curvature represented by D has no effect (i.e., at which $\tilde{u}_x = \tilde{v}_x = 1$, and $\tilde{u}_y = \tilde{v}_y = 0$), will in general be a function of Stk and Re_D. If the pattern is taken to be that of potential flow, i.e., $Re_D \to \infty$, then Re_D will not enter specifically and the solution will depend only upon Stk. It is evident why Stk is called the inertial parameter.

Solution of (5.29) or (5.30) is complex and usually can only be carried out by approximate numerical methods. Fuchs [1, page 136] and Tuttle [2] outline methods which may be employed. Solutions for several cases (see below) have appeared in the literature: for example, flow about an infinitely long cylinder of diameter D, or about a sphere of diameter D. The flow regimes considered are at the two extremes of: (a) an ideal (inviscid) fluid, potential flow $Re_D \to \infty$, velocity components about a cylinder, in polar coordinates (r, θ; $\theta = 0$ at upstream - centerline)

$$v_r = - v_o\ (1 - D^2/4r^2)\cos\theta$$
$$v_\theta = v_o\ (1 + D^2/4r^2)\sin\theta \qquad (5.31)$$

where Eqns. (5.31) are valid only on the upstream side of the cylinder, and (b) A viscous fluid at low $Re_D < 1$, for a cylinder

$$v_r = C_L\ \left(1 - \frac{D^2}{4r^2} - 2\ \ell n\ \frac{D}{2r}\right)\cos\theta$$
$$v_\theta = - C_L\ \left(1 - \frac{D^2}{4r^2} + 2\ \ell n\ \frac{D}{2r}\right)\sin\theta \qquad (5.32)$$
$$C_L = \frac{v_o}{2\ (2 - \ell n\ Re_D)}$$

valid for $\frac{2r - D}{D} \ll 1$, i.e., close to the surface of the cylinder. Equation sets (5.31) and (5.32) are due to Lamb [3].

Corresponding equations are available for velocity components about a sphere. In polar coordinates (r, θ; $\theta = 0$ at centerline downstream) assuming polar symmetry:

(a) potential flow

$$v_r = v_o\ (1 - D^3/8r^3)\cos\theta$$
$$v_\theta = -v_o\ (1 + D^3/16r^3)\sin\theta \qquad (5.33)$$

(b) viscous (Stokes) flow

$$v_r = v_o\left(1 - \frac{3D}{4r} + \frac{D^3}{16r^3}\right)\cos\theta$$
$$v_\theta = -v_o\left(1 - \frac{3D}{8r} - \frac{D^3}{32r^3}\right)\sin\theta \qquad (5.34)$$

For the purpose of modelling dust collection, such solutions are important insofar as they show whether a particle collides with a target or not. The long cylinder is taken as model for the fibers in a filter, and the sphere for the drops of liquid in a scrubber or for the granules in a granular bed filter. The results of the trajectory calculations must be displayed in terms expressing the proportion of particles in the stream which collide and are collected. This leads to the subject of aerodynamic capture to be discussed next.

II. AERODYNAMIC CAPTURE ON ISOLATED SINGLE TARGETS

A. *SINGLE TARGET EFFICIENCY*

The term aerodynamic capture is used to refer to the collision of particles with a target surface either due to inertia or to some other cause. It is customary to express the extent of this "capture" in terms of an "isolated single target efficiency," defined as that fraction of particles of a given size d_{p_i} in the stream which collide with the target surface. This may also be expressed in terms of the equivalent cross-sectional area of the gas stream from which particles are removed by collision with the target surface:

η_{T_i} = isolated single-target efficiency, particles of grade i

$$\eta_{T_i} = \frac{\text{cross-sectional area of gas stream cleared of particles by single target}}{\text{cross-sectional area of target projected upstream}} \tag{5.35}$$

For example, if the target be a long thin cylinder of diameter D, then there is a depth of stream b such that $\eta_{T_i} = b_i/D$, or in the case of a spherical target, a diameter of a stream tube b such that $\eta_{T_i} = (b_i/D)^2$. The determination of η_{T_i} will be considered first for an isolated single target, and later for a single target which is a member of an array of targets.

In addition to "inertial impaction" (as illustrated in Fig. 3) there are a number of other processes which may lead to collision and capture. Particles may be attracted to the target by electrostatic force caused by charges on the particles, on the collector, or on both. Very small particles may be moved at random by Brownian diffusion, and so chance to hit the target if they pass close enough to it. Other forms of diffusion may be directional toward the target, as when a high temperature gradient exists near the target surface (thermophoresis), or when there is a concentration gradient of gas molecules (diffusiophoresis). In some instances gravitational settling onto a target may be significant. The appropriate force, in each case, must be incorporated in Eqn. (5.25).

Combinations of collecting forces may readily exist with each type present contributing its share to the total target efficiency. A single target efficiency for an isolated target may be assigned to each type of collecting force, e.g., η_{II} = isolated single-target efficiency due to inertial impaction, η_{BD} = isolated single-target efficiency due to Brownian diffusion, η_E = isolated single-target efficiency due to electrostatic charge, and so on. The total overall isolated single target efficiency will then be some combination of these, but not necessarily their sum. It is possible that a given particle might have been collected by any one of several mechanisms. If so, its collection should be counted only once in determining the overall target efficiency. This equation is discussed further in G. below.

Detailed methods of detecting which collecting mechanisms are operative in a given situation, and of determining their respective η values, are required. They are presented in the following sections. It will be found that for each mechanism certain dimensionless parameters may be identified which are uniquely associated with the magnitude of η. The calculation of these parameters will lead rather quickly to a determination of which collecting mechanisms are important in a given situation and which may be neglected. The characteristic parameter, or parameters, for the significant mechanisms may then be used to estimate the value of η for each one.

B. *CAPTURE BY INERTIAL IMPACTION*

1. *General approach* Of all the possible aerodynamic capture mechanisms (especially for particles $d_p \geq 1$ μm), that due to inertial impaction is undoubtedly the most common and has received the greatest amount of study. It is of basic importance in the collection of dust by fibrous filters, and by scrubbers. It is also very important in certain particulate sampling devices (such as impingers, Anderson samplers, etc.) and in probes used for stack sampling.

To determine η_{II} for a given case it is necessary to solve the pair of differential equations as (5.29) or (5.30), subject to the existing flow field given by $v_x = v_x(x,y)$ and $v_y = v_y(x,y)$, and to find the limiting trajectory for collision to occur. This establishes the position of the distance b, as shown in Fig. 5 for impaction on a cylinder or a sphere. Then η_{II} is calculated by (5.35). For a particular shape of target or flow field, the value of η_{II} will be found to be a function of Stk and Re_D in general, or merely of Stk in the case of potential flow. (Banks and Kurowski [4] have recently developed a perturbation method for more rapid computation of b in the case of an axisymmetric target.)

As a typical example of this work, two figures dealing with a cylindrical target are reproduced from Ranz's report [5]. Figure 6 shows η_{II} as a function of $\sqrt{Stk}$, at various values of $Re_p^2/2\ Stk = N_{\rho_g}$, as determined from Eqns. (5.29) for very high values of Re_D. On this figure, the capture of Stokes law particles, as determined by Eqns. (5.30), is presented as the special case identified by the curve for $Re_p^2/2\ Stk = 0$. Use of this parameter in correlating data is convenient because it may be determined without knowing particle size. Figure 7 shows the effect of the fluid flow pattern, as represented by different values for Re_D, all for Stokes law only [5,10,11]. The top curve in both Figures is the same.

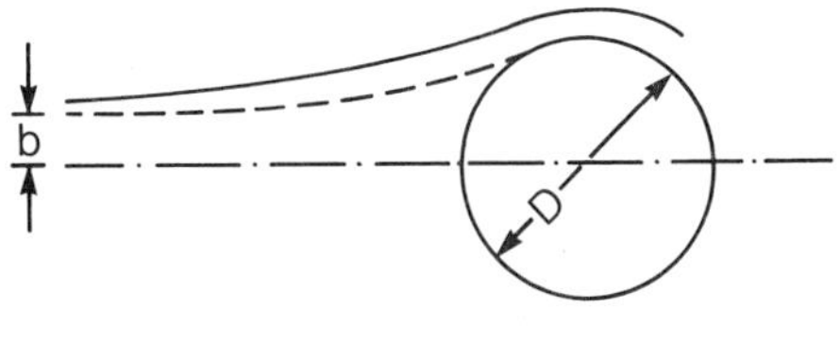

FIG. 5. *Limiting particle trajectory.*

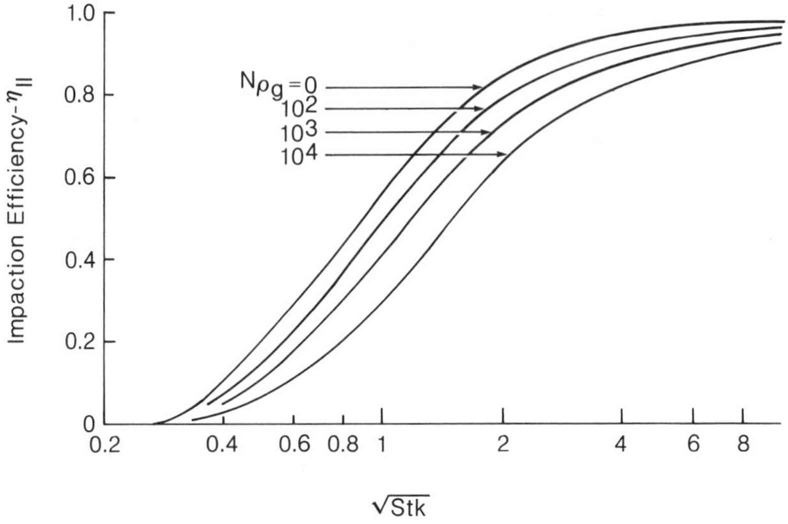

FIG. 6. *Capture efficiency by inertial impaction on spheres, potential flow. Reproduced by permission of Pennsylvania State University.*

On these Figures $\sqrt{Stk}$ is used instead of Stk in order to display the particle size in dimensionless form as the independent variable. Thus

$$\sqrt{Stk} = \frac{d_p}{\left(\frac{18\ \mu D}{C\rho_p v_o}\right)^{1/2}}, \quad \text{a dimensionless particle size.}$$

Bearing in mind, however, that Stk is also equal to x_s/D (see Eqn. 4.46) it is evident that as the stopping distance x_s approaches the same order of magnitude as the diameter of the target, the collection efficiency approaches the order of magnitude of 1.

These results may be generalized to incorporate non-Stokesian particles (i.e. $Re_p > 1$) by using a more general definition of the Stokes number, involving the more general definition of stopping distance implied by Eqn. (4.16). Thus, Israel and Rosner [25] have defined a Stk_{eff} as

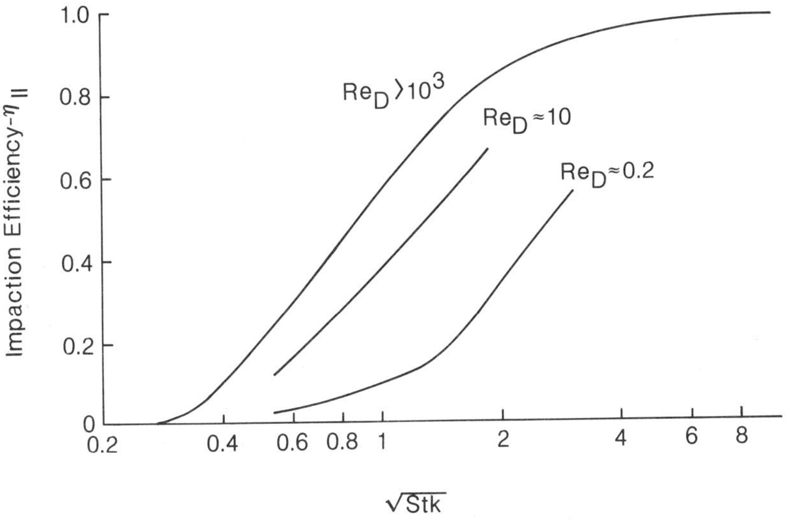

FIG. 7. *Capture efficiency by inertial impaction on spheres, effect of flow regime. Reproduced by permission of Pennsylvania State University.*

$$Stk_{eff} = \frac{8}{3} \frac{\rho_p}{\rho_f} \frac{d_p}{D} \int_0^{Re_p} \frac{d\ Re_p}{C_D\ Re_p} \tag{5.36}$$

which is related to the conventional definition of Stk by

$$Stk_{eff} = \psi\ Stk$$

where

$$\psi = \frac{24}{Re_p} \int_0^{Re_p} \frac{d\ Re_p}{C_D\ Re_p} \tag{5.37}$$

Note that $\psi = 1$ for Stokes law particles. Replotting data such as in Fig. 6 using Stk_{eff} in place of Stk brings all data onto a single curve, i.e. giving $\eta_{II} = f(Stk_{eff})$ for all particles. Israel and Rosner showed [25] that the data for both spheres and cylinders could be successfully treated in this way.

The value of ψ may be determined as a function of Re_p by integrating (5.37) using one of the appropriate expressions for C_D, such as are given in Chapter 4. This has been done [58] using $C_D = 24(1 + 0.158\ Re_p^{2/3})/Re_p$, giving

$$\psi = 19.0\ Re_p^{-2/3} - 47.8\ Re_p^{-1} \tan^{-1}(0.397\ Re_p^{1/3}) \qquad (5.37a)$$

2. *Effect of target shape* Calculations of the type outlined have been made for quite a few target shapes. Usually they have been done either for the case of large Re_D where potential flow (inviscid) fluid may be assumed in order to get the v_x, v_y values, or for the case of laminar flow at very low values of Re_D. Ranz [5] made a survey of these in 1956. More recent summaries are given by Strauss [6], Golovin and Putnam [7], May and Clifford [8], and Tuttle [2] who conducted a comprehensive literature search. Recently some solutions, especially for spheres [15,16] have been computed at intermediate values of Re_D, say between 10 and 400.

Solutions for η_{II} in the form of equations or charts of the type of Fig. 6 or 7, and some experimental results, are available for at least the following shapes in the references listed below. Most of them are not as complete or detailed as those for spheres.

Shape	References
Cylinder	[5,6,7,10,11,20]
Sphere	[5,6,7,12,13,14,15,16,28]
Round jet against flat plate	[5,17,18]
Rectangular jet against flat plate	[5,10,18,19]
Disc collector	[5,7,8]
Ribbon collector	[5,7,8,11]
Elbows	[5,11,21,22]
Wave-shaped channel	[5]
Recessed trough	[5,7]
Rectangular half-body	[5,7,9]
Ellipsoid of revolution	[5,7,12]
Airfoils	[7]
Sampling tube (probe)	[23,24]

The S-shaped appearance of a number of these graphs has suggested that they might be linearized by replotting on log-probability coordinates in a manner similar to that utilized for particle size distribution. Such plotting does give reasonably straight lines for a number of η_{II} vs $\sqrt{Stk}$ charts available, particularly through the middle range of efficiencies. Whitby [26] presented a list of values for median $\sqrt{Stk}$ and for σ_g which may be used to reconstruct such plots. These are given in Table 1 and plotted on Fig. 8.

Only a few equations are available to represent some of these results. For inertial impaction on cylinders at $Re_D = 10$, Pich [27] gives

$$\eta_{II} = \frac{Stk^3}{Stk^3 + 0.77\ Stk^2 + 0.22} \tag{5.38a}$$

For inertial impaction on spheres, the calculations of Herne [13] for potential flow have been fitted by Knettig and Beeckmans [56] to

TABLE 1 *Inertial impaction parameters for various collector target shapes*

Collector	$\sqrt{Stk_M}$	σ_g
Jet on plate, round	0.373	1.24
Jet on plate, rectangular	0.55	1.24
Cylinder	$1.253\ Re_D^{-0.0685}$ $0.2 < Re_D < 150$	1.65
Sphere	0.77	1.91
Ribbon	0.54	2.22
Trough or cup (Impinger)	0.46	1.91
Rectangular half body	0.54	2.32
Sweeping Bend	0.71	1.80
Focusing away	0.14	2.9

Source: Ref. [26]

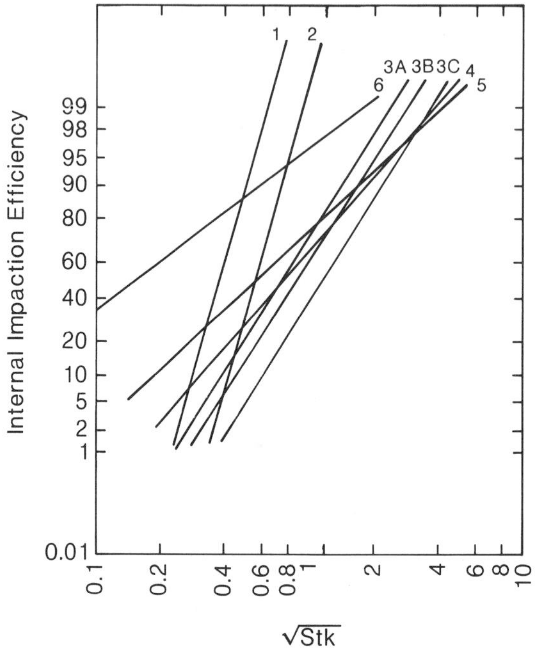

FIG. 8. *Capture efficiency by inertial impaction on various shapes. 1. Jet on round plate. 2. Jet on rectangular plate. 3. Cylinder: a. $Re_D = 150$, b. $Re_D = 10$, c. $Re_D = 0.2$. 4. Sphere. 5. Rectangular half-body. 6. Focusing away. (See Table 1)*

$$\eta_{II} = 0.00376 - 0.464 \text{ Stk} + 9.68 \text{ Stk}^2 - 16.2 \text{ Stk}^3 \tag{5.38b}$$

where $0.0416 \leq \text{Stk} \leq 0.30$, and

$$\eta_{II} = \frac{\text{Stk}^2}{(\text{Stk} + 0.25)^2} \quad \text{for Stk} > 0.3 \tag{5.38c}$$

Cheng and Wang [21] give an equation for deposition in a 90° bend, in the case of ideal flow (large Re_D) and $\text{Stk} \leq 0.1$, as

$$\eta_{II} = \left[1 + \frac{\pi}{2R_o} + \frac{2}{3R_o^2} \right] \text{Stk} \tag{5.38d}$$

where R_o = radius of the bend/radius of the tube. They have generalized this for other angles in circular pipes.

In all of the work cited above flow has been assumed to be axisymmetric and time independent. Degani and Tardos [28] have shown, however, that real flows in the higher range of Re_D, say 100 to 6000 must be regarded as pulsating and non-symmetrical. They have done the calculations for spherical targets at Re_D from 200 to 10^6. Such high values are not encountered very often in particulate collection work.

3. *Isokinetic sampling* A special case of considerable practical interest is that of the inertial effects involved in drawing a sample of particulate matter from a flowing gas stream. Usually this is done by inserting a cylindrical tube ("probe", diameter D) axially aimed upstream, suctioning a sample stream at controlled velocity through the probe, and collecting the particles from it in a sampling train. Unless the mean gas velocity through the probe (u) is the same as that in the main gas stream (u_o) at the probe location ("isokinetic sampling"), the concentration of particulate aspirated into the probe may be altered by inertial effects.

In isokinetic sampling ($V_r = u/u_o = 1$) the flow lines of gas and particle within the probe remain straight and continuous with those upstream of the probe, and no inertial effects occur. If the mean velocity in the probe is greater than that of the main stream ($V_r > 1$), streamlines of flow will curve into the tube entrance. Particles having sufficient inertia (i.e. large Stk) located beyond a certain critical radius $b > D/2$, will be unable to follow the streamlines into the tube and will remain uncollected. The concentration in the sample (c) of such particles will be less than in the main stream (c_o) and $c/c_o < 1$. On the other hand, if $V_r < 1$ streamlines of gas will curve away from the probe entrance, the critical radius $b < D/2$, and particles of large Stk will be collected in excess, making $c/c_o > 1$.

For Stokes law particles Stenhouse and Lloyd [24] solved Eqns. (5.30) numerically, assuming for the fluid velocity field that of axisymmetric potential flow of a point source in planar flow. From this they determined the critical radius b. By material balance it may be shown that $c/c_o = (2b/D)^2/V_r$. Then c/c_o may be displayed as a function of V_r and Stk_p where Stk_p refers to conditions inside the probe. Their solution shows that at $Stk_p < 0.01$ the inertial sampling error is negligible and $c/c_o \approx 1$. At $Stk_p > 10$, $c/c_o \to 1/V_r$. In the intermediate range, c/c_o is shown as a graphical function of Stk_p and V_r, but unfortunately the graph is small and difficult to read. Their results agree well with experimental values, even in the case of a thick-walled probe from the front edge of which particles might be expected to bounce.

For practical computation of c/c_o in the intermediate range of Stk_p, the empirical formula of Belyaev and Levin [77] may be used. It may be expressed as

$$c/c_o = 1 + (1/V_r - 1)\frac{(2 + 0.62\ V_r)}{(1/Stk_p + 2 + 0.62\ V_r)} \tag{5.39}$$

In the limiting cases as $Stk \to 0$ or as $Stk \to \infty$, this agrees with the theory just cited. It also seems to agree pretty well with the Stenhouse and Lloyd graph, insofar as this can be read.

There are, of course, effects other than inertial which may also interfere with obtaining a truly representative sample in the train. Among these are the design of the probe (it should be thin-walled and sharp-edged), its orientation in the stream (truly axial), the turbulence of the stream, and deposition of particles on the walls inside the probe. Even if the inertial effect is eliminated by isokinetic sampling, some of these may be present to some degree.

In an attempt to include some of these effects, Selden [23] presents an empirical correlation of experimental data in the form of

$$N = \frac{\frac{c}{c_o} V_r - 1}{V_r \left(1 - \frac{c}{c_o}\right)} = \frac{Stk_p^{-0.918} \, Re_p^{0.355}}{1.53 + 0.617 \, V_r} \tag{5.40}$$

where N is defined as shown. Thus

$$\frac{c}{c_o} = \frac{1/N + V_r}{V_r \left(\frac{N+1}{N}\right)} \tag{5.40a}$$

In the limiting cases, as $Stk_p \to 0$, $N \to \infty$, and as $Stk_p \to \infty$, $N \to 0$, which again agrees with the Stenhouse and Lloyd solution. However note the presence of Re_p in N. This should in some degree take into account the effect of turbulence in the main stream.

Thus if isokinetic sampling is not achieved, the resulting inertial errors may be estimated. Seldon points out, however, that if c is measured at any two convenient non-isokinetic velocities the ratio $N_1/N_2 = (1.53 + 0.617\, V_{r2})/(1.53 + 0.617\, V_{r1})$ may easily be computed, and from that c_o determined by solving

$$\frac{N_1}{N_2} = \frac{\frac{c_1}{c_o} V_{r_1} - 1}{V_{R_1} \left(1 - \frac{c_1}{c_o}\right)} \times \frac{V_{R_2} \left(1 - \frac{c_2}{c_o}\right)}{\frac{c_2}{c_o} V_{R_2} - 1} \tag{5.41}$$

This would eliminate the necessity for obtaining precise isokinetic conditions, which in any case is difficult to accomplish experimentally.

C. *THE INTERCEPTION EFFECT*

The calculation of limiting trajectories used to obtain values for η_{II}, makes no allowance for the size of the particle. When the particle is relatively large compared to the size of the collector, collection may be enhanced by direct interception. Fuchs [1, pg. 164] has shown how the increase may be estimated for two extreme cases: (a) the inertia of the particle is so large that it moves in a straight line, i.e., $Stk_p \to \infty$, (b) the particle possesses no mass, and moves in a streamline, i.e., $Stk_p = 0$.

In the first case, all particles within a stream tube of width D will collide with the collector. In addition particles passing within a distance $d_p/2$ will touch the collector, thus adding an increment of collection efficiency equal to η_{DI}: $\eta_{DI} = (d_p/D)$ for a cylinder, and $\eta_{DI} = (d_p/D)^2$ for a sphere.

In the second case, if the stream line on which the particle center is travelling approaches closer than $d_p/2$ to the collector, the particle will touch the collector. The increment of collection efficiency will then be (letting $R = d_p/D$ represent the basic parameter): for potential flow around a cylinder (using Eqn. 5.31):

$$\eta_{DI} = 1 + R - 1/(1 + R) \approx 2R \quad (\text{for } R < 0.1) \tag{5.42}$$

for potential flow round a sphere (using Eqns. 5.33):

$$\eta_{DI} = (1 + R)^2 - 1/(1 + R) \approx 3R \quad (\text{for } R < 0.1) \tag{5.43}$$

for viscous flow round a cylinder (using Eqns. 5.32):

$$\eta_{DI} = \frac{1}{2.002 - \ln Re_D}\left[(1+R)\ \ln\ (1+R) - \frac{R(2 + R)}{2(1 + R)}\right] \tag{5.44}$$

$$\approx \frac{R^2}{2.002 - \ln Re_D} \qquad (\text{for } R < 0.07, \text{ and } Re_D << 0.5)$$

for viscous flow round a sphere:

$$\eta_{DI} = (1+R)^2 - 3(1+R)/2 + 1/2(1+R) \approx 3R^2/2 \tag{5.45}$$

(for $R < 0.1$)

D. *IMPACTION AND PARTICLE BOUNCE*

In all of the above treatment, we are dealing only with the prediction of collisions of particle with target. Whether the particle is actually collected, i.e. adheres to the target after collision, is another question. A particle striking a target at sufficiently high speed will be likely to rebound from it. Theoretical treatment of this question [30,31] indicates that bounce will occur whenever the kinetic energy of the particle after rebound is greater than the potential energy of attraction. This attraction is that due to van der Waals forces, surface tension, electrostatic, or other forces existing between particle and target surface.

The theory provides procedures for calculating critical conditions necessary for the initiation of bounce, in terms of the kinetic energy based upon the velocity component of the particle normal to the collecting surface. Although such calculations cannot be carried out for polydisperse dusts of irregular particles, they do indicate that actual capture efficiency should correlate with the kinetic energy of the impacting particle. That is, the actual capture efficiency can only become equal to the predicted collision efficiency provided the kinetic energy at impact is below a cetain critical value.

In particulate collection work, this means that aerodynamic capture may be expected to decrease with increasing Stokes number above a certain level. Ellenbecker et al [32] have found experimentally, for example, that η_{II} for particles of fly-ash impacting on the stainless steel fibres of a mat filter begins to decrease at Stk > 1, corresponding to a particle kinetic energy of about 10^{-15} J on impact. Above a kinetic energy of 10^{-13} J, particle bounce dominated the filter collection characteristics. Likewise for the impaction of particles of pollen and spores on cylindrical stems, Aylor and Ferandino [33] found a critical kinetic energy of about the same magnitude.

The critical kinetic energy for initiation of bounce will obviously depend upon the physical and chemical nature of the particles as well as that of the target surface. Experiments by Gutfinger and Friedlander [29], for instance, show that deposition on a fibrous surface may be enhanced 10 to 1000 times over that on a comparable smooth surface. It is not to be expected, therefore, that there would be a universally applicable value of the critical kinetic energy.

E. *CAPTURE BY FLUX FORCES*

In Chapter 4 the movement of particles due to various forces of bombardment by gas molecules was discussed. This includes Brownian diffusion, thermophoresis, and diffusiophoresis. Under appropriate conditions each of these may lead to capture of particles on targets. Collectively these kinds of capture are said to occur by "flux forces" $\vec{F}_F$. They are only significant for very small particles, $d_p \ll 1$ micron, for which inertial effects are negligible.

Capture by flux forces may be taken into account by adding an external force term, for each flux force acting, to the righthand side of Newton's Law. The flux forces may be converted into corresponding dimensionless flux velocities $\vec{u}_F$, to be added to Eqns. (5.29) or (5.30). Since flux forces are only significant in the range of Stokes law particles, the relationship for conversion is:

$$\vec{F}_F = 3\pi\mu d_p u_F/C \tag{5.46}$$

and the dimensionless velocity term is defined as:

$$N_F = \frac{u_F}{v_o} = \frac{u_B + u_T + (u_D + u_{St})}{v_o} \tag{5.47}$$

which may be called "flux deposition number". Equations (5.30), for very small values of Stk, become

$$\frac{d\tilde{x}}{d\theta} - \tilde{v}_x - \tilde{u}_{F_x} = 0$$
$$\frac{d\tilde{y}}{d\theta} - \tilde{v}_y - \tilde{u}_{F_y} = 0 \qquad (5.48)$$

Aerodynamic capture efficiency by flux forces may be calculated by solving these equations in a manner similar to that used for inertial impaction, incorporating an appropriate fluid velocity pattern. This has only been done for several special cases of interest in particle collection. Brownian diffusion (u_B) may arise in any system, but thermophoresis (u_T) will occur only when there is a large temperature gradient maintained between gas and target, and diffusiophoresis (u_D) only when the target is a liquid surface, or where condensation of vapor may occur on a target surface.

1. *Brownian Diffusion* A small particle, having little mass and therefore little inertia, will tend to follow the fluid streamlines in the neighborhood of a target. Bombardment by gas molecules, setting the particle into Brownian motion, will cause it to wobble randomly about a streamline. If it passes close enough to a target surface it may thus collide with that surface.

The analogy between Brownian diffusion and true diffusion in gases has been presented in terms of Eqn. (4.63). This equation may be placed in dimensionless form by using the same set of reference variables as for Eqns. (5.26), yielding:

$$\frac{d\tilde{n}}{d\theta} = \frac{\mathcal{D}}{v_o D}\left(\frac{\partial^2\tilde{n}}{\partial x^2} + \frac{\partial^2\tilde{n}}{\partial y^2} + \frac{\partial^2\tilde{n}}{\partial x^2}\right)$$

where $\tilde{n} = n/n_o$. A dimensionless parameter automatically arises, which is called the Peclet number:

$$Pe = \frac{v_o D}{\mathcal{D}}$$

This is a measure of the amount of convective transport relative to the diffusive transport of the particles. The larger the value of Pe, the less important is the diffusional process. A related dimensionless group, called the Schmidt number, may be formed by

$$Sc = Pe/Re_D = v_o D\mu/\mathcal{D} \;\; v_o \rho_f D = \nu_f/\mathcal{D} \tag{5.49}$$

In calculating these numbers, the value of $\mathcal{D}$ is obtained by the methods discussed in Chapter 4, II.A. A Brownian flux velocity u_B has been defined as

$$u_B = \mathcal{D}/\delta_B \tag{5.50}$$

where δ_B is the boundary layer thickness of the fluid over the target surface.

Using these concepts, methods of calculating the single target collection efficiency by Brownian diffusion, η_{BD}, have been developed by several authors. The basic assumption is made that during the time the gas stream flows past the target, the particles from a certain layer are able to diffuse to the target surface. The thickness of this layer is proportional to $(\mathcal{D}\, t)^{1/2}$. The appropriate length of time to use depends upon the fluid velocity distribution. Thus in the viscous flow regime $\eta_{BD} = f(Pe, Re_D)$ and in the potential flow regime $\eta_{BD} = f(Pe)$. Good summaries of the results of various methods of calculation are given by Strauss [6] and by Pich [27]. The theories are also well presented by Friedlander [34].

a. Cylindrical Targets. For viscous flow, the original method is due to Langmuir [35] using the velocity field distribution given by Lamb, Eqns. (5.32). His results may be expressed as

$$\eta_{BD} = \frac{1.71}{(2 - \ln Re_D)^{1/3}} Pe^{-2/3} \tag{5.51}$$

Similar equations have been derived by Natanson [36], Friedlander [37], and Crawford [38] in which the constant factors are 2.92, 2.22 and 2.854 respectively. Likewise an equation by Torgeson [39], differs but slightly in numerical values:

$$\eta_{BD} = \frac{2.06}{(2 - \ln Re)^{0.4}} Pe^{-0.6} \tag{5.52}$$

Ranz [40], using the analogy between heat and mass transfer, gives

$$\eta_{BD} = \frac{\pi}{Pe} \left(\frac{1}{\pi} + 0.55\ Re_D^{1/2}\ Sc^{1/3} \right) \tag{5.53}$$

$$\eta_{BD} = \frac{1}{Pe} + 1.727 \frac{Re_D^{1/6}}{Pe^{2/3}}$$

which is recommended for $0.1 < Re_D < 10^4$, and $Sc < 100$ $(10 < Pe < 10^6)$.

For potential flow, where the velocity field is independent of Re_D and corresponds to conditions of high Re_D, Stairmand [41] and Natanson [36] have presented similar results

$$\eta_{BD} = K/Pe^{1/2} \tag{5.54}$$

where $K = \sqrt{8} = 2.83$ (Stairmand), or $K = \sqrt{32/\pi} = 3.19$ (Natanson).

From an inspection of these formulas, it is clear that collection efficiency due to Brownian diffusion will be very small unless the Peclet number is small, say less than 10. It is larger at higher Reynold's numbers (based upon target size). Note that it is theoretically possible to have a value of $\eta_{BD} > 1$ because Brownian diffusion may cause particles from beyond the distance D to collide with the target.

Example 2. Compare the relative importance of aerodynamic capture by inertial impaction, direct interception and Brownian diffusion for particles of unit density, having d_p ranging from 0.001 μm to

20 μm, upon a cylinder of 100 μm diameter from dry air at 20°C and 1 atm flowing at 10 cm/s.

$$\text{The value of } Re_D = \frac{10 \times (100 \times 10^{-4}) \times 1.205 \times 10^{-3}}{1.81 \times 10^{-4}} = 0.66$$

so the viscous flow velocity field must be used. Selecting the appropriate equations and charts, as listed, the various parameters and corresponding capture efficiencies may be determined as tabulated below.

d_{p_i} -μm	Stk	η_{II}-%	R	η_{DI}-%	Pe	η_{BD}-%
0.001	-	-	-	-	1.28	108
0.01	-	-	-	-	1.28×10^2	5.0
0.2	-	-	-	-	5.88×10^4	0.08
1	3.54×10^{-3}	0	0.01	0	3.70×10^5	0.02
10	3.08×10^{-1}	10(9.1)	0.1	0.4	-	-
20	1.23	50(57.3)	0.2	1.5	-	-

η_{II} estimated from Fig. 8; values in ( ) from Eqn. (5.38a); η_{DI} calculated from Eqn. (5.44); η_{BD} calculated from Eqn. (5.51).

It is evident from this example that diffusion contributes nothing as a collection mechanism unless the particles are very small. But as the particle size decreases, collection by diffusion increases while that due to inertial impaction decreases. At some intermediate particle size (e.g., 1 micron, in the example) there is a minimum collection where neither inertia nor diffusion is effective.

A similar effect is obtained with respect to velocity v_o. At low velocities diffusion will predominate, at high velocities inertia will be predominant. There will be a minimum collection efficiency determined primarily by the size ratio R (direct interception), at some intermediate velocity.

b. Other Target Shapes. Except for spheres very little has been done to estimate collection by Brownian diffusion on shapes other than cylindrical.

For isolated spherical targets the following are available. Johnstone and Roberts [42] suggest (D = diameter of sphere):

$$\eta_{BD} = \frac{4}{Pe}\left(2 + 0.557\, Re_D^{1/2}\, Sc^{3/8}\right) \qquad (5.55)$$

or

$$\eta_{BD} = \frac{8}{Pe} + 2.23\, Re_D^{1/8}\, Pe^{-5/8}$$

Crawford [38, pg. 380] derives a formula:

$$\eta_{BD} = \frac{4.18}{Sc^{2/3} Re_D^{1/2}} = 4.18\, Re_D^{1/6}\, Pe^{-2/3} \qquad (5.56)$$

while Pfeffer [48] gives:

$$\eta_{BD} = 3.96\, Pe^{-2/3} \qquad (5.57)$$

Strauss [6, pg. 291] cites:

$$\eta_{BD} = 8\sqrt{2/Pe}/3\pi = 1.20\, Pe^{-1/2} \qquad (5.58)$$

Strauss [6] also gives, for a side of a strip of width W:

$$\eta_{BD} = \sqrt{8\mathcal{D}/\pi v_o W} \qquad (5.59)$$

Friedlander [34] has discussed diffusional deposition in ducts.

2. *Thermophoresis* Whenever a target surface is sufficiently colder than the gas, so that a large temperature gradient exists across the fluid boundary layer on the target, capture will occur due to thermophoresis. The thermal flux force F_T is given by Eqn. (4.62), and the thermal velocity u_T by Eqn. (4.61). These will always act normal to the target and the velocity components $\tilde{u}_{T_x}$ and $\tilde{u}_{T_y}$ must be reckoned accordingly. A thermal flux deposition

number may be calculated using (5.44). This will obviously be a function of the dimensionless group Th given by Eqn. (4.61).

The flux force deposition number for very small particles (Th≈0.50) may be estimated

$$N_{FT} = \frac{\tilde{u}_T}{v_o} = - \frac{0.50}{v_o} \frac{\nu_f}{T} \frac{dT}{dx} \tag{5.60}$$

This may be appreciable even at small values of ΔT between target surface and gas because the gradient will be across a very thin thermal boundary layer, say of the order of 10^{-3} cm thick. In dry air at 1 atmos., 300°C, with ΔT = 10°C, and v_o = 10 cm/sec, u_T = 4.2×10^{-4} (10×10^3) = 4.2 cm/sec, and N_{FT} = 0.42.

To make such calculations it is necessary to have a method of estimating the thermal boundary layer thickness. For spherical targets, according to Johnstone and Roberts [42] this will be

$$\Delta x = D/(2 + 0.557\ Re_D^{0.5}\ Pr^{0.375}) \tag{5.61}$$

where $Pr = C_p \mu / k_f$ (C_p = heat capacity, k_f = thermal conductivity) and all gas properties are evaluated at the mean boundary layer temperature.

Thermophoresis is essentially independent of the particle size. For very small particles Kn >> 1, Th ≈ 0.50 as taken above. However Th is a weak function of particle size, as discussed in Chapter 4. For estimates of aerodynamic capture where u_T is likely to be significant, however, Th may be taken as constant.

Example 3. No generalized calculations of capture efficiency due to thermophoresis have been found in the literature. However Pilat and Prem [44] have worked out some specific sample cases using the methods outlined above for a water drop 100 μm diameter as a spherical target moving at 30 cm/s. Holding the air temperature constant at 65°C they took the drop temperature to range from 10°C to 82°C, and calculated isolated target efficiencies for particles ranging from $d_{p_i} = 0.01$ μm. In this range the value of u_T is essentially independent of d_p. Some sample results are:

T_{drop}-°C	Δx - cm	dT/dx - °C/cm	u_T-cm/sec
10	-3.723×10^{-3}	-1.477×10^{4}	3.96
48	-3.726×10^{-3}	-4.562×10^{3}	1.29
65	-3.724×10^{-3}	0	0
82	-3.719×10^{-3}	$+4.571 \times 10^{3}$	-1.35

Δx calculated from (5.61), $dT/dx \simeq (T_{air} - T_{drop})/\Delta x$, u_T calculated from (5.60), using Th = 0.50.

Unfortunately it is impossible to obtain the values of η_{Th} from the published work, because they are incorporated into an overall efficiency. It is evident however that they have an appreciable effect at drop temperatures away from 65°C.

3. *Diffusiophoresis* Whenever a gaseous concentration gradient exists, diffusiophoresis would enter into the capture process. It will aid in capture if the gradient is toward the target, and oppose capture if the gradient is away from the target. The most likely circumstance under which this could arise would be in case the target is a liquid surface. Condensation of vapor on the liquid would contribute to capture, evaporation of liquid would oppose capture. Here the Stefan flow discussed in Chapter 4, section II.B, will come into play. The diffusion force F_D may be calculated by Eqn. (4.75), the diffusion velocity $(u_D + u_{St})$ by Eqn. (4.74).

The flux force deposition number will be

$$N_{FD} = \tilde{u}_D = -\left(\frac{\sqrt{m_B}}{n_A\sqrt{m_A} + n_B\sqrt{m_B}}\right)\frac{D_{AB}}{v_o n_A}\frac{\partial n_B}{\partial x} \qquad (5.62)$$

It is evident that N_{FD} is completely independent of particle size. In case the diffusion effect is created by a liquid surface, the values of n_A and n_B may be represented in terms of the partial pressure of the vapor in a given system and (5.62) becomes

$$N_{FD} = -\frac{K}{v_o}\frac{dp}{dx} \tag{5.63}$$

so that $N_{FD} > 0$ for condensation ($dp/dx < 0$), $N_{FD} < 0$ for evaporation ($dp/dx > 0$), and K is a constant depending upon the system, the temperature, and the total pressure. At the liquid surface p will be equal to the saturation vapor pressure at the temperature of the liquid. For the system water-vapor diffusing in air at 0°C, 1 atm. this becomes

$$N_{FD} = -\frac{1.9 \times 10^{-4}}{v_o}\frac{dp_w}{dx}$$

where dp_w is expressed in mb/cm.

A relationship similar to (5.61) is available [42] for estimating the boundary layer thickness Δx due to diffusivity in the case of a spherical target

$$\Delta x = D/(2 + 0.557\, Re_D^{0.5}\, Sc_D^{0.375}) \tag{5.64}$$

where the Schmidt number is based upon the diffusivity of the target vapor, instead of the particles as in (5.49).

Sparks and Pilat [45] performed calculations for the combined effect of inertia and diffusion flux force, using Eqns. (5.30) with $u_D = N_{FD}$ added, such as is illustrated in (5.48). They assumed potential flow around a water drop, using (5.33) for the velocity components, and calculated η as a function of Stk for three cases: $dp/dx = -10^5$mb/cm, $dp/dx = 0$, and $dp/dx = 10^5$mb/cm, all with $v_o = 100$ cm/s. The results are presented graphically, since the equations can only be solved numerically. At Stk = 1, for example, $\eta_{II} = 0.45$ whereas with condensation the combined efficiency rises to 0.95, and with evaporation it drops to 0.16.

Example 4. Pilat and Prem [44] also incorporated diffusiophoresis into their sample calculations mentioned above under thermophoresis, for the same set of conditions. The air was taken to be saturated with water vapor at 65°C. For drop temperature < 65°C, condensation occurs ($dp/dx < 0$), and diffusiophoresis augments thermophoresis as a combined flux force capture mechanism. For drop temperature > 65°C the reverse is true. Some samples of diffusion velocities calculated are:

T_{drop} - °C	p_v -mmHg	Δx - cm	dp/dx-mb/cm	(u_D+u_{St}) - cm/sec
10	9.21	3.77×10^{-3}	-6.30×10^4	11.9
48	83.7	3.76×10^{-3}	-3.68×10^4	7.0
65	187.5	3.72×10^{-3}	0	0
82	384.9	3.72×10^{-3}	7.06×10^4	-13.4

Δx is calculated from (5.64), P_v is vapor pressure of water at indicated temperature, $(u_D + u_{St})$ is calculated from (5.63), and is independent of d_p. Again it is not possible to separate out the diffusiophoretic effect on the capture efficiency, but it is evident that it is much greater than the thermophoretic effect. To give some idea of the combined influence it may be noted that for 1 μm particles the total isolated drop capture efficiency at 65°C is given as 0.015, while at 10°C it is 2.7 (270%!) and at 82°C it is below 0.001. Obviously flux force capture may be very significant. This may be of great importance in the case of scrubbers.

F. *COLLECTION BY ELECTROSTATIC ATTRACTION*

Some aerosol particles seem to acquire electrostatic charges rather easily either during their generation or during flow through a gas stream. Likewise target elements such as the fibers making up a filter may acquire charges due to the friction caused by the passage of a gas stream over them. Sometimes this effect can be enhanced by impregnating the target fibers with resinous material. Furthermore targets may be charged by imposing an electrostatic field on the collector. Whenever any of these situations occur, electrostatic forces are set up which will influence the particle collection process.

Unless an electrostatic field is deliberately maintained, the charge on the target elements will be unstable and will tend to decrease with time, owing to such factors as conductivity of the target material, passage of ionized gas molecules, radioactive irradiation, deposition of charged particles, and humidity (which tends to ground the target).

In the absence of an external electric field, there are three general cases to be considered: (i) Charged particles-neutral target. The charge on the particles induces an image charge, opposite in sign, on the collector which results in a force of attraction between particle and target, called F_{EM}. (ii) Charged target-neutral particles. An image charge of opposite sign is induced on the particles, generating a force of attraction, F_{EI}. (iii) Charged particles - charged target. Coulombic forces of attraction or repulsion act, depending upon whether the particles and the target have unlike or like charges. This force is called F_{EC}. In addition, wherever the particles are charged in the same sense, a repulsion force is produced among themselves. This is called the space-charge effect F_{ES}. Table 2 lists the calculation of these forces both for isolated spherical and isolated cylindrical shaped targets, provided the charge on the collector is constant.

The analysis of the effect of these forces, together with the other mechanisms, upon aerodynamic capture is complicated. In principle the appropriate electrostatic forces should be summed:

$$F_E = F_{EM} + F_{EI} + F_{EC} + F_{ES} \tag{5.65}$$

where there is a constant charge on the collector. The total electrostatic force is then added to all other forces in Newton's Law for the calculation of trajectories. As in the case of

TABLE 2 *Electrostatic forces between spherical particle and collector of constant charge*

Nature of Force	Spherical Collector	Cylindrical Collector
	D = diameter Q = charge on collector (constant)	D = diameter ε_D = dielectric constant of collector
F_{EM} - Image: charged particle-neutral collector	$\frac{q^2 D}{8\pi\varepsilon_o r^3} - \frac{2q^2 Dr}{\pi\varepsilon_o (4r^2 - D^2)^2}$	$\frac{q^2}{4\varepsilon_o (r - \frac{D^2}{2})} \frac{\varepsilon_D - 1}{\varepsilon_D + 1}$
F_{EI} - Image: neutral particle-charged collector	$-(\frac{\varepsilon - 1}{\varepsilon + 2}) \frac{d_p^3 Q^2}{16\pi\varepsilon_o r^5}$	$\frac{4Q^2}{\varepsilon_o} (\frac{\varepsilon - 1}{\varepsilon + 2}) \frac{(\frac{dp}{2})^3}{r^3}$
F_{EC} - Coulombic: charged particle-charged collector	$\frac{Qq}{4\pi\varepsilon_o r^2}$	$\frac{2Qq}{\varepsilon_o r}$
F_{ES} - Space-charge: repulsion between particles	$\frac{-q\ D\ N}{24\varepsilon_o r^2}$	

N = number of particles per unit volume
r = distance between particle and collector
q = charge on particle
ε = dielectric constant of particle
ε_o = specific inductive capacity of space = 8.85×10^{-21} coulombs^2/dyne cm^2

Source: Ref. [46] and [47]

inertial impaction, it is necessary to assume some fluid velocity field for the calculations. The resulting differential equations would be similar to (5.25) through (5.30), but with an added term including F_E in the sum of forces on the right side.

There will correspond to each such force term a dimensionless parameter which is formed for Stokes law particles (including Cunningham correction) when the equations are put into the dimensionless form of (5.30). The collection efficiency η_E for an isolated single target is then found to be a function of Re_D, Stk, and the corresponding electrostatic parameters, which are listed in Table 3. Inertialess particles may be considered by placing Stk = 0.

Kraemer and Johnstone [46] have calculated values of η_E for inertialess particles on isolated spherical targets, allowing for direct interception. Their results are given in the form of charts giving η_E as a function of K's in each of the three basic cases, calculated for potential flow and viscous flow, each at several values of R. These charts are also reproduced by Strauss [6, pg. 304]. In general they show $\log\eta$ roughly linear in log K, with higher efficiencies for potential than for viscous flow. Values of $\eta_E > 1$ are of course possible.

Nielsen and Hill [48] have extended this study considerably, using the same three basic parameters as Kraemer and Johnstone and adding two more. These deal with the presence of an external electric field, one where particles only are charged and the other where neither particles nor collector are charged. Their results agree with Kraemer and Johnstone in the duplicated cases, and their chart is similar, producing some useful approximations as tabulated. It is of special interest to note that particles may be collected on the near (downstream) side of a target under certain conditions.

Pich [27] has stated the corresponding parameters (no external field) for the case of cylindrical collectors, and has quoted the expressions derived by Natanson [47] for the collection efficiencies. These are also listed in Table 3.

TABLE 3 *Dimensionless parameters K representing electrostatic forces on Stokes Law particle*

Nature of Force	Spherical Collector Q_a = charge per unit area of collector		Cylindrical Collector Q = charge per unit length of collector	
	Parameter - K	Efficiency - η_E	Parameter - K	Efficiency - η_E
F_{EM} - charged particle: neutral	$\frac{Cq^2}{3\pi^2 \mu d_p v_o \varepsilon_o D^2}$	$\approx 1.58\ K_{EM}^{1/2}$ (b) $\approx 2.89\ K_{EM}^{0.353}$ (a) $0.002 \leq K_{EM} \leq 0.1$	$\frac{(\varepsilon_D - 1)}{(\varepsilon_D + 1)} \frac{Cq^2}{3\pi \mu d_p D^2 v_o \varepsilon_o}$	$(6\pi)^{1/3} K_{EM}^{1/3}$ (a) $\frac{2}{(2-\ln Re)^{1/2}} K_{EM}^{1/2}$ (b)
F_{EI} - neutral particle, charged collector $K = K_{EI}$	$(\frac{\varepsilon-1}{\varepsilon+2}) \frac{2Cd_p^2 Q_a^2}{3\pi\mu D v_o \varepsilon_o}$	$\approx (\frac{15\pi}{8} K_{EI})^{0.4}$	$\frac{4}{3\pi} (\frac{\varepsilon-1}{\varepsilon+2}) \frac{Cd_p^2 Q^2}{D^3 \mu v_o \varepsilon_o}$	$(\frac{3\pi}{2})^{1/3} K_{EI}^{1/3}\ \frac{2r}{D} >> 1$
F_{EC} - charged particle, charged collector $K - K_{EC}$	$\frac{CqQ_a}{3\pi \mu d_p v_o \varepsilon_o}$	$-4K_{EC}$ (a) and (b)	$\frac{4QqC}{3\pi \mu d_p D v_o \varepsilon_o}$	$-\pi K_{EC}$
F_{ES} - repulsion	$\frac{Cq^2 DN}{18\pi\varepsilon_o \mu d_p v_o}$			

C = Cunningham correction factor

η_E = isolated single-target efficiency (may be > 1)

Qa, Q,q may be expressed as integral multiples of e

e = unit electronic charge = 1.601×10^{-19} coulombs

Source: Ref. [46,47,48]

(a) potential

(b) viscous flow

The K parameters given in Table 3 may be formed, as shown by Ranz and Wong [50], by evaluating the corresponding force at an appropriate value of r and then dividing by the drag force taken as $3\pi d_p \mu v_o/C$.

Example 5. Illustrate the effect of electrostatic charges by calculating the isolated target efficiencies for particles of unit density with d_p ranging from 0.001 to 1 microns, captured on a conducting (uncharged) cylinder 100 microns in diameter from a stream at 20°C, 1 atm, dry air flowing at 10 cm/s. Consider the particles to be charged with 10 electrons each, and with 300 electrons each.

As calculated in previous examples Re = 0.66 for the conditions, so η_E will be calculated from K_{EM} according to case (b) in Table 3. Since the value of ε_D is not given, take $(\varepsilon_D - 1)/(\varepsilon_D + 1) \approx 1$ and calculate K_{EM} for each particle size from

$$K_{EM} = \frac{C}{d_p} \frac{[n(1.601 \times 10^{-19})]^2}{3\pi(1.81\times10^{-4})(100\times10^{-4}) \times 10 \times 8.86 \times 10^{-21}}$$

$$= \frac{C}{d_p} \times 1.70 \times 10^{-12}\, n^2$$

where n is the number of electronic charges per particle. Then

$$\eta_E = \frac{2}{(2 - \ln 0.66)^{1/2}} K_{EM}^{1/2} = 1.27\, K_{EM}^{1/2}$$

The results are tabulated:

	n = 10 electrons		n = 300 electrons	
d_p - μm	K_{EM}	η_E	K_{EM}	η_E
0.001	0.367	0.77	330	23.1
0.01	3.77×10^{-3}	0.078	3.40	2.3
0.1	4.5×10^{-5}	0.008	0.041	0.26
1	$2\ \times10^{-6}$	0.002	1.77×10^{-3}	0.053

These results may be compared with those in Example 2. It is evident that the electrostatic effect may be very significant on fine particles. Unfortunately we do not often have sufficient data (especially for n) to make such calculations. Efforts are being made to utilize this kind of effect to enhance the collection of very fine particles. These will be discussed in connection with filters (Chapter 8) and scrubbers (Chapter 9).

G. *CAPTURE BY GRAVITY*

Gravity may affect aerodynamic capture if conditions are such that the terminal settling velocity of particles is appreciable compared to the capture velocities which may be operating. From a horizontal slow moving gas stream, large particles may tend to settle onto a target from above. In a vertical gas stream flowing downward toward a target, gravity will enchance the capture, while if the stream is flowing upward gravity will tend to detract from the capture. This effect has been demonstrated conclusively by experiments such as those of Thomas and Yoder [51]. The direction of vertical flow may thus be significant in a filter bed or filter mat.

In Eqn. (4.47), a dimensionless parameter G•Stk was developed and given as

$$G \cdot Stk = u_s/v_o$$

Ranz and Wong [50] have shown that this would represent the isolated single target efficiency of capture by settling:

$$\eta_G = G \cdot Stk = Cgd_p^2\rho_p/18\mu v_o \tag{5.66}$$

It is evident that for η_G to be significant d_p must be relatively large and v_o small. This situation does not arise frequently in the problems of capture of fine particles. In most collecting situations G•Stk will not be important. Note that in the case of a horizontal stream, there will be a non-axisymmetric situation. This case has been dealt with by Beizaie and Tien [63].

H. *CAPTURE BY COMBINATIONS OF FORCES*

When aerodynamic capture is occuring due to the simultaneous action of more than one significant mechanism, the overall isolated target efficiency may be estimated by taking the net penetration to be the product of the penetrations resulting from each mechanism acting alone, e.g.:

$$P_T = P_{II} \cdot P_{DI} \cdot P_{BD} \cdot P_E \cdot \cdot \cdot \tag{5.67}$$

This is equivalent to

$$\eta_T = 1 - (1-\eta_{II})(1-\eta_{DI})(1-\eta_{BD})(1-\eta_E) \cdot \cdot \cdot \tag{5.68}$$

The overall efficiency is not equal to the summation of the efficiencies for each mechanism operating alone. A given particle might have been collected by the action of more than one mechanism, but its collection may be counted only once. Note the similarity between this relation and that given for collectors in series as Eqn. (3.34).

Several specific combined effects have been studied individually for cylindrical and spherical targets as listed below. Presumably these results would be preferable to (5.68) in calculations where they apply. Equations (5.67) or (5.68) can be used in the absence of a specific case study, wherein the several η values would each be estimated from appropriate formulas as given in the preceding sections.

1. *Cylindrical Targets* a. Inertial Impaction Plus Direct Interception. Davies [52] has calculated this combination for $Re_D = 0.2$ (viscous flow field) and the results are given by

$$\eta_{II+DI} = 0.16\ [R+(0.50+0.8R)\ Stk - 0.1052\ R\ Stk^2] \tag{5.69}$$

b. Brownian Diffusion Plus Direct Interception. Friedlander [53] developed an expression for this case in fiber filters, which when extrapolated to the case of 100% porosity, i.e. a single fiber, agreed well with several sets of experimental data for viscous flow with Stk < 0.5, giving

$$\eta_{BD + DI} = 1.3\ Pe^{-2/3} + 0.7R^2 \qquad (5.70)$$

He points out that for $R \to 0$ this agrees with the form of (5.51), and as $Pe \to \infty$ it agrees with the form of (5.44).

c. Inertial Impaction Plus Direct Interception Plus Brownian Diffusion. Davies [52] suggested simply adding 1/Pe to Stk in equation (5.69) in order to account for all three effects. This would be valid only for $Re_D = 0.2$.

2. *Spherical Targets* a. Inertial Impaction Plus Direct Interception Plus Gravity. George and Poehlein [54] have solved equations analogous to (5.30) in spherical polar coordinates including not only the inertial effects (Stk) but also direct interception (R), and gravity ($G_D \cdot Stk$). Potential flow and Stokes law particles were assumed. Their results are presented in a series of graphs in which the total isolated single target (sphere) collection efficiency is plotted against appropriate parameters.

One graph shows η_T vs $\sqrt{Stk}$ for R = 0 and $G_D = 0$, for R = 0 and $G_D = 8.8 \times 10^{-2}$, and for R = 0.1 and $G_D = 0$. Both target particle and collected particle were taken to be falling at their respective terminal velocities in a stationary gas. Consequently where $G_D > 0$, the total capture efficiency is reduced. Note that G_D is based upon target diameter, while Stk is based upon particle diameter. The results agree well with some previous calculations by Langmuir and Blodgett [55] and some experimental work by Jarman [56]. The work has been extended to the 3-dimensional case by Beizaie and Tien [63].

b. Inertial Impaction Plus Direct Interception Plus Electrostatic Collection. Nielsen and Hill extended their work on electrical forces to include inertial impaction [49], producing an extensive calculation of particle trajectories and capture efficiencies for both potential and viscous flows. The results reveal a very interesting, and fairly complex, interplay of electrical and inertial forces. For example, with a strong coulombic force ($K_{EC} = -10.0$) and high inertia (Stk = 2.5) particles may pass completely around to the downstream side of the collector and cross the center-line several times in reverse zig-zag fashion before finally being collected on the downstream side. Over half of the total collection may take place on the downstream half ("back side") of the collector if $K_{EC} < -1.0$ and Stk is not very large. As Stk increases from zero, collection efficiency (initially > 1 for $K_{EC} < 0$) decreases, passes through a minimum and eventually increases again. Similar effects are observed in the case of the image forces (represented by K_{EM} and K_{EI}), and for either potential or viscous flow.

It is difficult to summarize briefly the results of all these cases, and impossible to present them quantitatively. One must study the collection of some thirteen charts presented. It may be useful to quote, in part, the authors' conclusions:

> . . . under favorable conditions all of the electrical forces considered can significantly enhance particle collection over that obtainable by inertial impaction alone.
> . . . in many cases involving strong attractive electrical forces the collection efficiency is greatest for negligible particle inertia and decreases with increasing particle inertia.
> . . . the introduction of particle inertia to a given electrical force situation produces a greater change in η for potential flow than for Stokes (viscous) flow. With weak electrical forces η often passes through a minimum as Stk increases from zero. When a strong attractive radial force is present, a substantial portion of the total collection may occur on the downstream half of the collector.

Investigations of this sort have been extended both theoretically and experimentally by Wang, et al [57,58] to cover especially the range of conditions which might be encountered in using charged drops as collectors in scrubbers. This includes particles moving at such high velocities as to be out of the Stokes law range for the drag force, and accelerating collectors i.e. a non-constant relative velocity between particle and spherical target.

In their first work [57], taking a fixed collector position, using potential flow around the sphere [Eqns. (5.33)], assuming that F_{EC}, the Coulombic attraction force (see Table 2), is the only electrical force present, and allowing for non-Stokesian drag at Re_p up to as high as 100 (see Chapter 4), they calculated capture efficiencies in a manner analogous to Neilsen and Hill [48,49]. They also made parametric studies of the relative influence of Stk and K_{EC} (see Table 3) upon the results. Again results are presented primarily in the form of charts showing η_{EC} as a function of the various parameters Stk, K_{EC}, Re_p, and R. They confirm that, in the case of Stk = 0, $\eta_{EC} = -4\ K_{EC}$, as shown in Table 3. At low values of Re_p, their results agree with all those previously published.

The Coulombic parameter may be expressed in an alternate form in terms of the voltage applied on the collector (V_c) as

$$K_{EC} = \frac{2CqV_c}{3\pi\mu d_p D\nu_o}$$

Using this expression, it may easily be shown that the combined parameter $-2K_{EC}/\text{Stk}$ represents the ratio of the electrical energy to the kinetic energy of the particle. The calculations show that

> . . . At high values of $-2K_{EC}/\text{Stk}$, the electrostatic force dominates and the collision efficiency increases linearly, while the inertial force dominates at low values of $-2K_{EC}/\text{Stk}$. The

change of behavior occurs approximately at $-2K_{EC}/Stk = 2.5$, for Stk from 0.5 to 2.

Another interesting result is the existence of a minimum value of η_{EC} as Stk increases, for values of $-0.6 < K_{EC} < 0$. This is shown in Fig. 9. which also includes an indication of experimental results.

Painstaking and clever experimentation yielded data which agreed with the theoretical calculations. This led to the conclusion that "the competition between inertial and electrical forces in the collection of particles by a (fixed) spherical collector can be predicted accurately by means of a theoretical model which uses the non-Stokesian drag and assumes potential flow conditions."

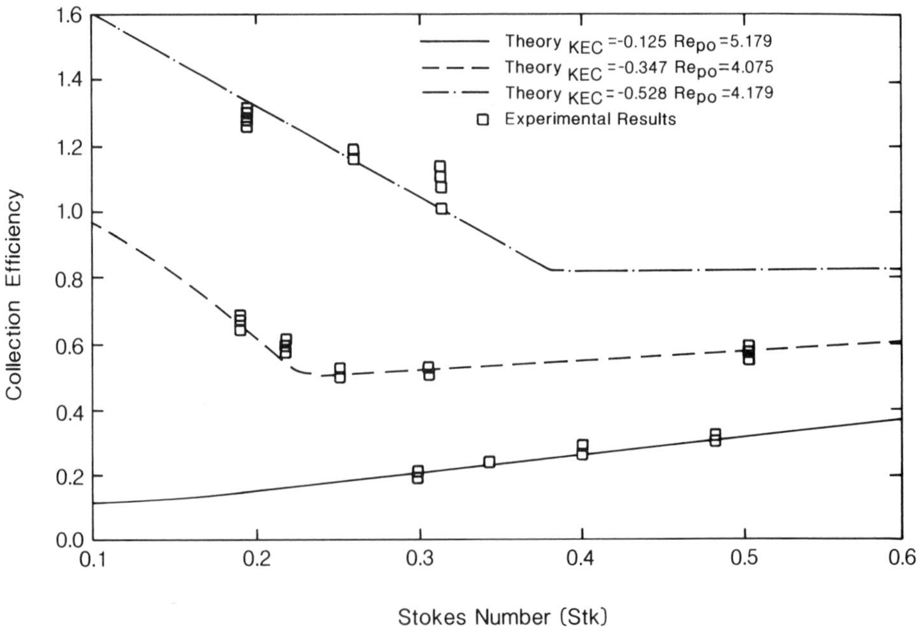

FIG. 9. *Interaction of inertial impaction and coulombic attraction for collection of charged particles on spheres. Reprinted by permission of the publisher from Aerosol Science and Technology, 5, 407. Copyright 1986 by Elsevier Science Publishing Co., Inc.*

Next [58], they set up the analog of Eqn. (4.11) for an accelerating sphere, where the acceleration is due to the drag force exerted by a surrounding stream of gas flowing at constant velocity v_o, always greater than that of the sphere. To cover the non-Stokesian range, this equation was expressed in terms of ψ, as defined in (5.37) and appears in dimensionless form as:

$$\frac{d\tilde{u}_x}{dt^*} = \frac{Re_p C_D}{24} \psi (1 - \tilde{u}_x) \text{ and } \frac{dZ^*}{dt^*} = \tilde{u}_x \qquad (5.71)$$

where $t^* = t/\tau\psi$, $\tilde{u}_x = u_x/\upsilon_o$, and $Z^* = x/\upsilon_o\tau\psi$. This was solved to give the downstream velocity as a function of distance traveled, taking the initial velocity to be zero. (A somewhat different approach to this same problem is given in Chapter 9, with a result in simpler form.)

The calculated results were then incorporated into a modified version of Eqn. (5.30), in which the coordinate system was placed on the moving sphere. In this system the flow velocity becomes $(v_o - u_x)$. Likewise in the expressions for Stk and K_{EC}, the velocity term is replaced by $(v_o - u_x)$. Solving this as above, the collection efficiency was finally obtained as a function of the distance traveled for various values of Stk and K_{EC}. Some typical results are given in terms of Z^* in Fig. 10. The most striking feature of these is that the major effect of the charge force does not become pronounced until the sphere has traveled some distance downstream. Again the calculations were reasonably well confirmed by experiment. Implications of this model for scrubber design are presented in Chapter 9.

c. Inertial Impaction Plus Flux Forces. Sparks and Pilat [45] have calculated the effect of diffusiophoresis upon inertial impaction for a target taken to be a spherical drop of water. Potential flow was assumed using equations (5.33) for the fluid velocity profile. For water-air at 0°C and 1 atm, Eqn. (4.74) becomes

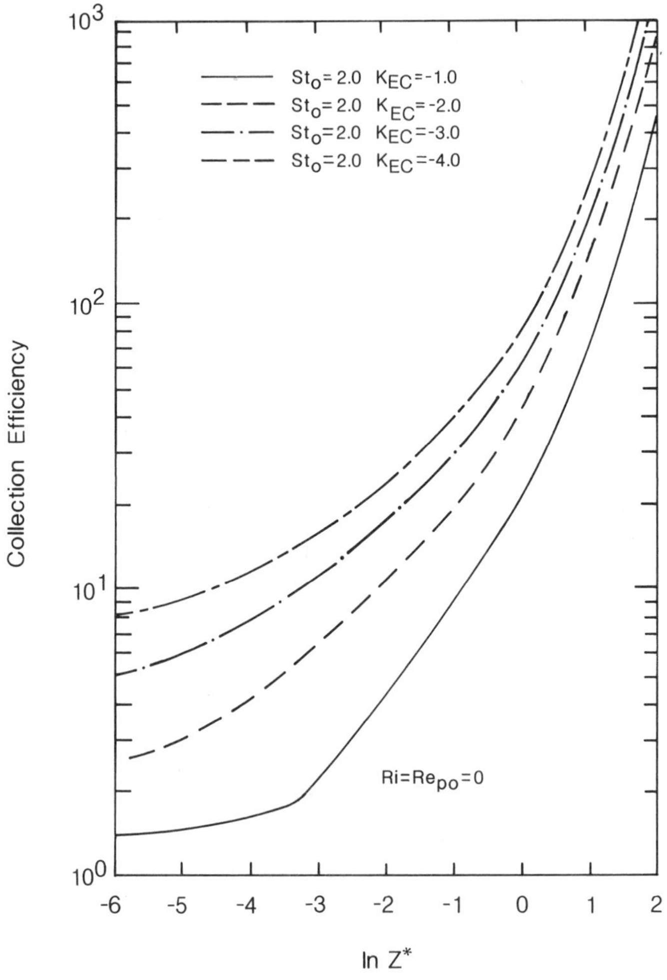

FIG. 10 *The collection efficiency of an accelerating sphere for charged particles. Reprinted by permission of the publisher from Aerosol Science and Technology, 5, 414. Copyright 1986 by Elsevier Science Publishing Co., Inc.*

$$(u_D + U_{St}) = -1.9 \times 10^{-4} \frac{dp}{dx}$$

where p is the partial pressure of the water vapor in millibars, x is the radial distance from the surface of the drop in centimeters, and v is in cm/s. Isolated single sphere aerodynamic capture efficiencies were calculated by using Eqn. (5.46) in combination with (5.30), and expressed as a function of the inertial impaction parameter K (here defined as K = 2 Stk) for three cases, all at v_o = 100 cm/s:

$\frac{dp}{dx} = 10^5$ mbar/cm	evaporation + inertial impaction $N_F = -0.19$
$\frac{dp}{dx} = 0$	inertial impaction only, $N_F = 0$
$\frac{dp}{dx} = -10^5$ mbar/cm	condensation+inertial impaction $N_F = 0.19$

Results are presented as a plot of $\eta_{II + D}$ vs K with a curve for each case. The effect of diffusiophoresis is very pronounced. For example at K = 1, η = 0.95 for condensation, η = 0.45 for impaction only, and η = 0.15 for evaporation. The calculations were continued to show the effect on the overall performance efficiency of a scrubber.

Pilat and Prem [44] extended these calculations to include the effects of Brownian diffusion and thermophoresis as well. This was for the specific case of a single water drop 100 µm diameter falling freely (u_S = 30 cm/sec) in air at 65°C, 100% relative humidity, 1 atm pressure. Although $Re_D \approx 2$, potential flow pattern was assumed. The capture of particles ranging in diameter from 0.01 to 10 microns was calculated.

In order to show the effect of the several flux forces, the temperature of the drop was considered at 10°C, 38°C, 49°C, 60°C, 65°C, and 82°C. At all drop temperatures below 65°C both thermophoresis and diffusiophoresis would aid in the capture, while above 65°C the reverse would be true. At 65°C only Brownian diffusion and inertial impaction would be acting. Brownian diffusion would contribute to particle capture at all temperatures, its effect increasing somewhat as the temperature increases. Temperature gradients ranged from -1.48×10^4 °C/cm to 4.6×10^3 °C/cm, and water partial pressure gradients from -6.4×10^4 mbar/cm to 5.3×10^5 mbar/cm. These values are so large primarily because the boundary layer around the drop is so thin, of the order of 0.00037 cm.

The methods of calculation were similar to those employed by Sparks and Pilat [45]. Flux force velocity terms were added to Eqns. (5.30). These indicate that at a given temperature the flux velocity for Brownian diffusion increases markedly as particle size decreases, while those for thermophoresis and diffusiophoresis are essentially independent of particle size.

The interplay of the influence of the several mechanisms was shown clearly by a grade efficiency plot of η_{Total} (isolated sphere) vs particle size (0.01 - 10 μm). At 65°C (drop temperature) the efficiency has a minimum value of 0.003 at $d_p \approx 0.09$ μm; Brownian diffusion causes an increase to 0.05 at $d_p = 0.01$ μm, while inertial impaction causes an increase to 1 at $d_p = 10$ μm. At 82°C, inertial impaction again brings about a high efficiency of 0.60 at $d_p = 10$ μm, but the adverse effect of thermophoresis and diffusiophoresis quickly predominates as particle size decreases, resulting in efficiency falling to 0.0001 at $d_p = 3$ μm. At all drop temperatures below 65°C efficiencies are greater the lower the temperature, reaching about 2.5 at 10°C and being rather independent of particle size below about 1 μm. Thermophoresis clearly dominates, having values of $N_{FT} \approx 0.12$ while those for diffusiophoresis are $N_{FD} \approx 0.04$.

I. *CAPTURE MECHANISMS AND PARAMETERS*

Each of the capture mechanisms discussed above has been associated with a characteristic dimensionless group or parameter. The capture efficiency due to any one mechanism is determined by the corresponding parameter through formulas or charts. In a given set of conditions e.g., particle size, target geometry and size, fluid flow rate, temperature(s), charges present, etc, etc, a calculation of the parameters will reveal which mechanisms will be predominant. It will also show what the trend of capture efficiency will be, corresponding to shifts in any of the variables. A summary of all parameters is given in Table 4.

1. *Detection of significant mechanisms and trends* Ranz and Wong [50] have stated a series of generalizations concerning mechanisms of collection, which are valuable guides. Some of them are quoted as follows:

> 1. Whenever one parameter characterizing a mechanism of collection is much greater than all other parameters, collection can be attributed entirely to the mechanism which that parameter characterizes.
>
> 2. The efficiency of collection will be negligible for any given mechanism where the parameter characterizing that mechanism is less than 10^{-2} and will be of the order of unity when that parameter is of the order of unity.
>
> 3. Although the forces of the various mechanisms are additive, the resulting individual efficiencies are not directly additive. Each mechanism will contribute to a total efficiency, however, and no combination of favorable mechanisms will cause an efficiency lower than that expected for any one of the favorable mechanisms.
>
> 4. The efficiency of collection for a given mechanism will be proportional to the size of the particle and the size of the collector (target), and to the stream velocity, in approximately the same manner that the parameter characterizing that mechanism is proportional to d_p, D, and v_o by definition.

These generalizations form the basis of practical use for the parameters in Table 4 as will be illustrated by some examples below. In addition to the trends mentioned in item 4, there are

TABLE 4 *Summary of collecting mechanisms, parameters, and trends*

Mechanism	Parameter[1]	Trends on η[2] d_p	D	v_o	Equations
All (effect of fluid velocity field)	$Re_D = \frac{v_o \rho_f D}{\mu_f}$	–	up	up	(5.31), (5.32), (5.33), (5.34)
Inertial Impaction	$Stk = \frac{C\rho_p d_p^2 v_o}{18\mu_f D}$	up	down	up	(5.38), Fig. 6, 7, 8
Direct Interception	$R = \frac{d_p}{D}$	up	down	–	(5.42), (5.43), (5.44), (5.45)
Brownian Diffusion	$Pe^{-1} = \frac{\mathcal{D}}{v_o D}$	down	down	down	(5.51) thru (5.59)
Thermophoresis	$N_{FT} = \frac{u_T}{v_o}$	–	–	–	—
Diffusiophoresis	$N_{FD} = \frac{u_D + u_{St}}{v_o}$	–	–	down	—
Gravity	$G \cdot Stk = \frac{Cg\rho_p d_p^2}{18\mu_f v_o}$	up	–	down	(5.66)
Electrostatic	K_{EM} (See Table 3)	down	down	down	Table 3
	K_{EI} (See Table 3)	up	down	down	Table 3
	K_{EC} (See Table 3)	down	down	down	Table 3
	K_{ES} (See Table 3)	down	up	down	Table 3

Notes: (1) Increase in any parameter means increase in capture efficiency due to that mechanism.
(2) Trend means effect on η as variable is increased.

the important questions of the effects of the temperature, pressure, and composition of the gas stream. These influences may be determined also from an inspection of the parameters and will be discussed in the next section.

Emi and Yoshioka [59] have illustrated a very interesting graphical method of showing the interplay of the several mechanisms (II, DI, BD, and G) over a range of conditions. This consists of making a plot with coordinates of particle size vs. stream velocity and indicating the domains in which each mechanism predominates. The exact appearance of such a plot will depend upon the sample conditions and precise methods of calculation selected.

2. *Effect of temperature, pressure, and composition of gas* It is important to consider these effects particularly because of the current interest in attempting to clean gases which arise from high temperature, high pressure operations such as pressurized fluidized bed combustion [60]. Additional comments on such applications as they relate to the specific collectors will be made in the later chapters.

The effect of temperature will be reflected by the viscosity and density of the gas, and by the Cunningham factor, diffusivity, and Knudsen number of the particles. For all gases the viscosity increases and the density decreases, as temperature increases. When the ratio $\nu_f = \mu_f/\rho_f$ appears, increasing temperature will increase this ratio. The mean free path λ of gas molecules also increases, so that Kn will increase as $\sqrt{T}$ for a fixed particle size.

The Knudsen number in turn fixes the value of C and Th. Because of the larger mean free path C will be greater at higher temperature. The net effect on C/μ will depend upon the relative changes in C and μ individually. This effect may be determined from a study of the data provided by Fig. 4 and Table 1 of Chapter 4. It is seen from Fig. 4 that the effect of temperature upon C is much greater on the finer submicron particles. For particles

$d_p > 0.30$ μm, the effect of viscosity predominates and C/μ decreases with increasing temperature. For $d_p < 0.30$, the effect of C predominates and C/μ may increase markedly with increasing temperature.

Th is independent of Kn for Kn > 1 but is a complex function usually tending to decrease as Kn decreases below 1.0. Hence as temperature increases Th may tend to increase and level off for the coarser particles, but be unaffected for the finer particles.

All diffusional processes are more rapid at higher temperatures. Values of $\mathcal{D}$ for the particles, and D_{AB} for interdiffusion of two gases A and B become larger.

The effects of increasing pressure are as follows: the viscosity is not appreciably affected unless pressure is very high, then it will be increased. The density is increased in direct proportion to pressure. Consequently ν_f is decreased, at least up to moderate pressures. The mean free path is decreased, causing a decrease in C. Since Kn will decrease, Th will tend to decrease for coarser particles and remain constant for finer particles (Kn > 1). Particle diffusivity and gas diffusivity are decreased, at least for pressure up to 10-20 atm, but data are lacking or uncertain for the effect of still higher pressures.

All of the electrostatic properties (E_O, E, E_D) are essentially independent of temperature and pressure. Hence the effects on the several K parameters are primarily those of C/μ.

In the case of air, the only significant variation in composition is that of the water vapor content, or humidity. This is limited to a range between zero and saturation which depends upon temperature. The larger the humidity the smaller is the density and smaller the viscosity but the effects are not great. For precise calculations the correct property values corresponding to the humidity should be used.

In the case of a stack gas resulting from a combustion process the oxygen content of the air is replaced to a large extent by water vapor, carbon dioxide, and perhaps some carbon monoxide. Since the resulting mixture is still largely composed of nitrogen, and since the properties of the other gases tend to offset each other, flue gas may be treated as though it were air for most calculations.

For other gases the properties must be obtained or estimated and used as they are. Some of the effects of temperature and pressure are illustrated in the sample calculations below.

Table 5 shows how all of these trends effect each of the parameters individually. The net effect of a change in conditions upon an individual capture mechanism will depend upon the interrelationships of the parameters involved. For inertial impaction, the value of Stk varies as C/μ. As explained above, this will decrease with increasing temperature for the larger particles where II is a significant capture mechanism. In the case of Brownian diffusion in viscous flow an increase in temperature will decrease Re and increase Pe^{-1}. The effects upon η_{BD} are in opposite directions ($Re_D < 1$), as may be seen from Eqns. (5.51) or (5.52); the net effect will have to be found by calculation. For direct interception, the value of R is unaffected by temperature or pressure but, in viscous flow around a cylinder there will be an effect due to Re_D as shown in Eqn. (5.44).

Where two or more mechanisms must be considered, the overall effect upon the total isolated target efficiency may be determined by calculation. Several examples of this have been published. Strauss and Thring [61] show the effect of increasing temperature upon II, DI, and BD individually for a cylindrical target and particles ranging from 0.01 μm to 5 μm. Collection by both II and DI are reduced, while that for BD is increased. The net effect is not given, but if Eqn. (5.68) is applied, it would appear that η_T decreases at higher temperatures for the larger particles and increases for the smaller ones. Strauss and Lancaster [62] have shown the combined effect of pressure and temperature upon II for 1 μm particles on 10 μm fiber in carbon dioxide. Example 6. case IV below illustrates how their results might have been obtained. These cases indicate that for II the effect of

TABLE 5 *Effect of temperature and pressure upon parameters*

Parameter	Increased T	Increased $P^{(1)}$
Re_D	ν_f up, Re_D down (Effect appears only in laminar flow)	ν_f down, Re_D up
Stk	C up, μ_f up, Stk $\propto C/\mu$	C down, Stk down
R	No effect	No effect
Pe^{-1}	$\mathcal{D}$ up, Pe^{-1} up	$\mathcal{D}$ down, Pe^{-1} down
N_{FT}	ν_f up, T^{-1} down, Th up, N_{FT} down	ν_f down, Th down, N_T down
N_{FD}	D_{AB} up, N_{FD} up	D_{AB} down, N_{FD} down
G·Stk	C up, μ up	C down, G·Stk down
K_{EM}, K_{EI}, K_{EC}, K_{ES}	C up, μ up, all K's $\propto C/\mu$	C down, all K's down

Note: (1) Effect of P increase, say up to 10 atm. only.

temperature on viscosity is greater than on the Cunningham factor, so that there is a net decrease in Stk at higher temperature.

Example 6. On the following pages are tabulated the results of sample calculations to illustrate the above procedures. For each case the set of conditions is stated at the top. Below is given the value of each dimensionless group and the corresponding isolated target efficiency. The overall efficiency is calculated by (5.68).

J. *GROWTH OF DEPOSITED PARTICLES*

The theoretical approach to the prediction of aerodynamic capture as outlined in all of the above cases is based upon the assumption that the target is clean, that is free of previously captured particles. It is evident that such a situation can be true only during a very short period after a clean target is initially exposed to flowing particles. As soon as some particles have been captured on the target surface, subsequent particles arriving no longer "see" the original target shape, but rather a more complex shape involving the already deposited particles.

The presence of a captured particle on the target surface creates what has been called a shadow zone around itself on that surface. This is an area around the particle which it is impossible for a subsequent particle to reach without first colliding with the deposited particle. Thus as more particles arrive some of them will collide and adhere in successive layers to those already deposited. This may occur due to any of the capture forces which may happen to be active at the time. Consequently particle dendrites will tend to grow outward from the surface, acting in turn as capture targets of more and more complex shapes. This also means that particle capture cannot occur uniformly over the target surface, but rather must develop at discrete locations. Figure 11 is a tracing of an electron microphotograph published by Bhutra and Payatakes [64] showing such a typical dendrite growth.

Sample cases of Example 6

		Case I			Case II		
T	P	20°C	1 atm				
	μ_f	1.81×10^{-4}	poises			Same	
	ρ_f	1.205×10^{-3}	gm/cm^3				
	ρ_p	1.0	gm/cm^3			as	
	v_o	15 cm/s.					
Shape	D	Cylinder	50 μm			Case I	
	d_p	10 μm			0.1 μm		
	C	1.02	Fig. 4.4		2.67		Fig. 4.4
	$\mathcal{D}$	2.3×10^{-8}	cm^2/s	(4.64)	6.7×10^{-6}	cm^2/s	(4.64)
	q	300 e			300 e		
	ε_D	10			10		
	Re_D	0.50	η		0.5	η	
II	Stk	0.95	0.55	(5.38a)	2.4×10^{-4}	-	
DI	R	0.20	0.013	(5.44)	0.002	-	
BD	Pe^{-1}	3.2×10^{-7}	6×10^{-5}	(5.51)	8.9×10^{-5}	2.5×10^{-3}	(5.51)
G	G·Stk	0.02	0.020	(5.66)	5.3×10^{-6}	-	(5.66)
EM	K_{EM}	3.4×10^{-4}	0.022	Tab.3	0.089	0.36	Tab.3
			$\eta_T =$ 0.57	(5.68)		$\eta_T = 0.36$	(5.68)
Comments:		II predominates although q = 300e			Electrostatic predominates on fine d_p		

		Case III			Case IV			
T	P	300°C	1 atm		300°C		10 Atm	
	μ_f	2.93×10^{-4}	poises		2.93×10^{-4}		poises	
	ρ_f	0.613×10^{-3}	gm/cm^3		0.613×10^{-2}		gm/cm^3	
	ρ_p	1.0	gm/cm^3		2.5		gm/cm^3	
	v_o	15 cm/s.			1 cm/s			
Shape	D	Cylinder 50 µm			Sphere	50 µm		
	d_p	0.1 µm			10 µm			
	C	5.56	Fig. 4.4		1.003	(4.7)		
	$\mathcal{D}$	1.7×10^{-5}	cm^2/s	(4.64)	1.85×10^{-8}	(4.64)		
	q	0			0			
	Re_D	0.16	η		0.105	η		
II	Stk	3.2×10^{-4}	–		0.095	0.033	(5.38b)	
DI	R	0.002	–		0.20	0.057	(5.55)	
BD	Pe^{-1}	2.3×10^{-4}	0.00050	(5.53)	3.7×10^{-6}	7×10^{-4}	(5.55)	(5.56)
G	G Stk	6.9×10^{-6}	–		0.47	0.47	(5.66)	
EM	K_{EM}	0	–	--	–	–		
			$\eta_T = 0.005$	(5.68)		$\eta_T = 0.49$	(5.68)	
Comments:		BD predominates; increases with with temperature			Gravity predominates at high ρ_p and low v_o			

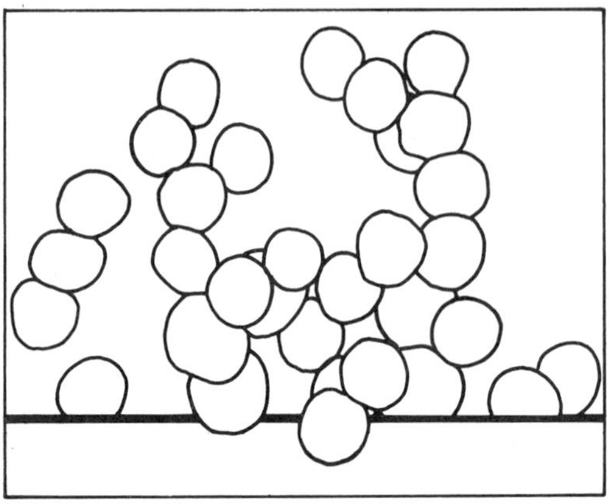

FIG. 11 *Tracing of electron microphotograph of dendrites formed by particles collected on a single fiber. Reprinted with permission from Journal of Aerosol Science* 10, *451. Copyright 1979, Pergamon Journals Ltd.*

As targets remain in service as collectors their total capture efficiency over a period of time must obviously be governed by these effects. This is especially important in something like a filter which is composed of a mat of cylindrical fibers. This subject will be dealt with in Chapter 8.

A number of theoretical studies of the dendrite growth process on single cylindrical targets have been made by computer simulation methods. Two main lines of attack have been followed, the one deterministic [65], the other stochastic [66]. Through either one it is possible to show how the dendrites grow, and how their growth is related to all the variables of the situation. Experiments, such as in [64], tend to confirm the simulation. However there is no simple way to calculate how the capture efficiency of the complex dendrite-fiber target varies as the process proceeds, and how this efficiency is related to that of the clean target, taking into account all capture forces acting.

The most useful generalization to arise from these studies is that due to Kanoaka, et al [66]. This states that the capture efficiency of the "loaded" cylinder η_M is related to that of the "clean" target η by the simple linear relationship:

$$\eta_M/\eta = 1 + \lambda m \tag{5.72}$$

where m is the accumulated mass of particles on the target, and λ is a complex function of the capture parameters. It is found that λ decreases with increasing Stk, with increasing R, and with decreasing Pe.

It is thus easy to show how m and η_M vary with time. Let M be the mass/time of particulate approaching the target. Then

$$d\eta_M = \lambda\eta \; dm \qquad \text{and} \qquad dm = \eta_M \; M \; dt$$

Combining with (5.72) and integrating gives:

$$\eta_M = \eta e^{\lambda\eta Mt} \qquad \text{and} \tag{5.72a}$$

$$m = [e^{\lambda\eta Mt} - 1]/\lambda$$

These relationships will be useful in predicting the transient behavior of a collector such as a filter.

III. AERODYNAMIC CAPTURE IN ARRAYS OF TARGETS

Collectors of the target type (fiber filters, granular bed filters, scrubbers) always function with a large number of targets operating simultaneously. The behavior of an individual target in such an array is not likely to be the same as when the target is isolated.

To develop a mathematical model for the collector performance, we begin with the study of the isolated target as presented in Section II above. Next we consider how the presence of neighboring targets in the array may modify the behavior of a single (unisolated) target. Finally the overall performance of the entire array must be examined.

The definition of aerodynamic capture efficiency (by any mechanism) is modified to apply to each single target as it functions in the array. A total single, but not isolated, target capture efficiency due to all mechanisms is obtained. The overall efficiency of the collector consists of an appropriate summation of all the single target efficiencies.

A. *MODIFIED FLUID VELOCITY PATTERNS*

The primary modification of isolated target behavior caused by the presence of neighboring targets lies in the pattern of the fluid flow field around the targets. This has an effect not only upon the aerodynamic capture efficiency but also upon the pressure drop in the gas flowing through the array of targets. (This important matter will be dealt with in the chapters devoted to collector performance theory.)

To deal with the modified flow pattern, the so-called "cell theory" was developed and applied by Happel [67], and by Kuwabara [68] in the case of cylindrical targets, and by Happel [69] and Pfeffer [70] in the case of spherical targets.

The concept of the cell theory is that in a randomly spaced (but in the case of cylinders, parallel) array of targets of diameter D, each target is surrounded by an imaginary concentric surface of similar shape and radius b, which encloses fluid and comprises a bounding free surface or cell. The value of b is determined by the porosity ε (fractional void volume) of the array such that $\varepsilon = 1-D^2/4b^2$ (cylinders), or $\varepsilon = 1-D^3/8b^3$ (spheres), and $c = 1-\varepsilon$.

Happel assumed that there was zero shearing stress at the cell boundary, Kuwabara that there was zero vorticity. Both solved the Navier-Stokes equation for viscous flow in the fluid between the target boundary and the cell boundary, assuming there was no slip on the target surface. Similar results were obtained in terms of the effect of porosity to be incorporated into the velocity component equations for viscous flow around a single target.

For flow transverse to cylinders, Kuwabara obtained

$$v_r = -\frac{v_o}{2Ku}\left[2\ \ln\frac{2r}{D} - 1+c + \frac{D^2}{4r^2}\left(1 - \frac{c}{2}\right) - \frac{2cr^2}{D^2}\right]\cos\theta$$

$$v_\theta = \frac{v_o}{2Ku}\left[2\ \ln\frac{2r}{D} + 1+c - \frac{D^2}{4r^2}\left(1 - \frac{c}{2}\right) - \frac{6cr^2}{D^2}\right]\sin\theta \qquad (5.73)$$

in which

$$K_u = -\frac{1}{2}\ \ln c - \frac{3}{4} + c - \frac{c^2}{4} \qquad (5.74)$$

and $v_o = V/\varepsilon$, where V is the superficial velocity of gas approaching the array.

Kuwabara's results for cylinders seem to agree best with experimental tests according to Davies [71, pg. 54]. As $\varepsilon \to 0$ ($c \to 1$) these equations reduce to those of Lamb (5.32) for an isolated target provided Ku is replaced by $(2.002 - \ln Re_D)$. At a value of $\varepsilon = 0.001$ the Kuwabara field is identical to the Lamb field for $Re_D = 0.495$, when Ku = 2.705. It is remarkable that in the Kuwabara viscous field there is no effect of Re_D, nor of fluid viscosity, only of the porosity of the array. Crawford [38, pg. 432] has provided a chart for Ku as a function of ε.

For spherical targets the Happel free surface model [69] has been applied by Pfeffer [70] to obtain

$$v_r = \frac{v_o \cos\theta}{W}\left(2 + 3\gamma^5 + \frac{a^3}{r^3} - (3+2\gamma^5)\frac{a}{r} - \frac{\gamma^5 r^2}{a^2}\right)$$

$$v_\theta = \frac{v_o \sin\theta}{2W}\left(4 + 6\gamma^5 - \frac{a^3}{r^3} - (3+2\gamma^5)\frac{a}{r} - \frac{4\gamma^5 r^2}{a^2}\right) \qquad (5.75)$$

in which $\gamma^3 = 1 - \varepsilon = c$ and $W = 2 - 3\gamma + 3\gamma^5 - 2\gamma^6$.

Tal and Sirignano [72] have extended the cell theory to cover higher (intermediate) Reynolds Numbers, by applying a cylindrical cell to an array of spheres.

B. *SINGLE TARGET BEHAVIOR IN ARRAYS*

In principle, single target capture efficiency should be calculated for arrays in the same manner as for isolated targets except for using the modified fluid velocity pattern. For example, Eqns. (5.73) would be substituted for v_r and v_θ in Eqns. (5.30) in order to determine inertial trajectories. In practice, the matter becomes very complex.

For arrays of cylinders Davies [71] has devoted an entire chapter to the application of the Kuwabara field to aerodynamic capture efficiency calculations. He recommends, finally, the following scheme for evaluating the overall single target efficiency incorporating inertial impaction, direct interception, and Brownian diffusion.

$$\eta_T = \eta_{DI} + \eta_{II} + \eta_{BD} \tag{5.76}$$

where

$$\eta_{DI} = \frac{1}{2Ku}\left[2(1+R)\ \ln\ (1+R) - (1+R) + (1+R)^{-1} + c\ (-2R^2 - \frac{R^4}{2} + \frac{R^5}{2} + \ldots)\right] \tag{5.77}$$

$$\eta_{II} = \frac{J}{Ku}\,Stk \qquad \text{(see Table 6)} \tag{5.78}$$

$$\eta_{BD} = \frac{2.9}{Ku^{1/3}\ Pe^{2/3}} + \frac{0.62}{Pe} + \frac{1.24\ R^{2/3}}{Ku^{1/2}\ Pe^{1/2}} \tag{5.79}$$

The last term in η_{BD} represents a combination effect of diffusion and interception. The J/Ku values in η_{II} depend only upon c and R and may be obtained from Table 6 which is recalculated from Davies.

TABLE 6 *Values of J/Ku for Equation (5.78)*

c	R = 0.01	0.02	0.05	0.1	0.2
0.005	0.00152	0.00588	0.0352	0.1276	0.434
0.01	0.00180	0.00678	0.0416	0.1510	0.514
0.02	0.00220	0.00816	0.0510	0.1852	0.626
0.05	0.00314	0.01180	0.0730	0.268	0.880
0.10	0.00460	0.01704	0.1056	0.386	1.248*

*Impossible value

The limitations on this set of equations, given by Davies [71], are as follows:

$$v_o = V/\varepsilon = V/(1-c) \qquad \text{(velocity inside array)}$$

$$Re_D = \frac{v_o D}{\nu_f} < 0.5 \qquad \text{(viscous flow thru array)}$$

$R < 0.1$

$Pe > 100$

$Stk < 0.1$ (C must be included)

Note that the calculations of the individual η's have been defined in such a way that they are merely added to get η_T and not combined as suggested by (5.68).

Example 7. Estimate the total single target efficiency under isothermal conditions for dry air at 20°C, 1 atm, flowing at 10 cm/sec through an array of cylinders each 10 microns in diameter packed with a porosity of 95% upon spherical particles of unit density and 0.01, 0.1, 1, and 2 microns in diameter respectively. Compare the result with that for an isolated cylinder under the same conditions.

For the array, $c = 0.05$; $Ku = 0.797$ (Eqn. 5.74), and $Re_D = 0.0664 < 0.5$. Parameter values are as follows:

d_p-μm	C	Stk	R	J/Ku	$\mathcal{D}\ \frac{cm^2}{sec}$	Pe
0.01			0.001	-	$5.3x10^{-4}$	18.9
0.1	2.89	0.000887	0.01	0.000314	$6.8x10^{-6}$	1470
1.0	1.166	0.0357	0.1	0.268	$2.7x10^{-7}$	$3.6x10^4$
2.0	1.083	0.1326	0.2	0.880	$1.3x10^{-7}$	$7.7x10^4$

Efficiencies:

d_p-μm	η_{DI}	η_{II}	η_{BD}	η_T
0.01	-	-	0.476	0.476
0.1	0.0002	-	0.0324	0.0326
1.0	0.0105	0.0096	0.0045	0.0245
2.0	0.0419	0.117	-	0.159

Note: η_{DI} from (5.77), η_{II} from (5.78), η_{BD} from (5.79), η_T from (5.76).

To show the effect of the array upon the single target, the above values may be compared with those calculated for an isolated cylinder. The parameters are the same except that c = 0; Ku, and J/Ku do not apply. Equation (5.69) as modified, may be used for η_T, although it is for $Re_D = 0.2$.

Efficiencies:

d_p - μm	η_T(a)	η_{DI}	η_{II}	η_{BD}	η_T(b)
0.01	0.0044	-	-	0.1438	0.144
0.1		0.0002	-	0.0079	0.008
1.0	0.0045	0.0020	0.01	0.0009	0.013
2.0	0.046	0.0075	0.02	-	0.027

Note: η_{DI} from (5.44), η_{II} from Fig. 7, η_{BD} from (5.51) and

a) $\eta_T = 0.16\ [R+(0.50+0.8R)\ (Stk+1/Pe) - 0.1052R\ (Stk+1/Pe)^2]$

b) η_T from Eqn. (5.68)

It is clear that the array has the effect of increasing η_T significantly.

To determine the effect of electrostatic charge upon arrays of cylinders, Lundgren and Whitby [73] performed experiments using charged particles and uncharged fibrous filters. They determined the single fiber efficiency caused by the charging itself using Eqn. (5.68), and correlated it with the corresponding parameter K_{EM} developed by Natanson [47] and given in Table 3. The parameter given in the table is 4π times the K_M quoted by Lundgren and Whitby. Their results for viscous flow fit well with:

$$\eta_E = 1.5\ K_M^{1/2} \tag{5.80}$$

where $K_M = K_{EM}/4\pi$ and $(\varepsilon_D - 1)/(\varepsilon_D + 1)$ was taken as 1. The value of η_E would have to be incorporated with other single target efficiencies using Eqn. (5.68) in order to get the total efficiency when charged particles are present.

For spherical targets Paretsky, et al [74] have used Pfeffer's concepts and Eqns. (5.75) to calculate single target efficiencies in the array as follows:

$$\eta_{DI} = 3R^2\ \rho(\varepsilon)^{-1} \tag{5.81}$$

$$\eta_{BD} = 5.04\ Pe^{-2/3}\ \rho(\varepsilon)^{-1/3} \tag{5.82}$$

$$\eta_G = 0.0624\ \frac{u_s}{v_o} = 0.125\ G \cdot Stk \tag{5.83}$$

in which

$$\rho(\varepsilon) = \frac{2 - 3(1-\varepsilon)^{1/3} + 3(1-\varepsilon)^{5/3} - 2(1-\varepsilon)^2}{1 - (1-\varepsilon)^{5/3}} \tag{5.84}$$

Note that $\rho(\varepsilon) \to 2$ as $\varepsilon \to 1$, and that $\eta_{DI} \to 3R^2/2$, (Eqn. 5.45), and $\eta_{BD} \to 4.0\ Pe^{-2/3}$, (Eqn. 5.57). $\rho(\varepsilon)$ appears to be valid only for values of ε near 1 because it goes through a minimum near $\varepsilon = 0.78$. Note that Tal and Sirignano's work [72] dealt only with the velocity flow pattern (and pressure drop) in an array and has not been used as a basis for estimating capture efficiencies.

η_{II} is given by a graph, similar to Fig. 8, plotted as a function of (2Stk) with curves for $\varepsilon = 0.39, 0.43, 0.49, 0.66, 1$, the last of which coincides with Herne's [13] calculation for isolated spheres. At a given value of Stk, the lower the porosity, the higher is η_{II}. These curves all predict a limiting low value of 2Stk ≈ 0.1 below which η_{II} is zero. However some experimental data show collection to occur in the range of $10^{-3} < \eta_{II} < 10^{-2}$ for Stk $\approx (1\text{-}4) \times 10^{-3}$ at $\varepsilon \approx 0.42$.

All of the above work relates only to the behavior of clean targets. There is no doubt however that dendrites will grow on the targets within the array, just as described above for isolated targets. It is likely that η_T will obey relations similar to Eqns. (5.72) and (5.72a).

C. *OVERALL BEHAVIOR OF ARRAY*

Finally the overall efficiency of an entire array of targets must be related to the single target efficiency and the number of targets present. An elementary way to do this employs the lateral-mixing model outlined in Chapter 3.

Consider an array of N_T targets per unit volume, each target having projected area of A_T in the direction of flow. In a unit cross-sectional area of the array, having a thickness of dL in the direction of flow, there will be a total of $N_T \cdot dL$ targets and $A_T N_T \cdot dL$ total projected area. Assuming all the targets to be alike and to be arranged in a regular fashion, the total single target capture efficiency η_T may be applied to every one in parallel. Then the number of particles approaching the target per unit time will be nv, and the reduction will be $v_o dn$, which must be

$$-v_o dn = nv \times \eta_T \times A_T N_T dL \tag{5.85}$$

Assuming η_T, A_T, N_T and ε to be constant, integration gives the total capture efficiency of thickness L of the array

$$-\int_{n_0}^{n} \frac{dn}{n} = \frac{\eta_T A_T N_T}{\varepsilon} \int_0^L dL \qquad (5.86)$$

$$-\ln P = \eta_T A_T N_T L/\varepsilon \quad \text{or} \quad \eta = 1 - \exp - \eta_T A_T N_T L/\varepsilon \qquad (5.87)$$

The dimensionless quantity $A_T N_T L/\varepsilon$ is sometimes called the solidarity factor S of the array.

For an array of cylindrical targets, each D by ℓ, packed with porosity ε

$$A_T = D\ell$$

$$N_T = (1-\varepsilon)/\frac{\pi D^2 \ell}{4}$$

Then

$$\eta = 1 - \exp -\eta_T(1-\varepsilon)4L/\varepsilon\pi D \quad \text{and} \quad S = 4(1-\varepsilon)L/\pi\varepsilon D \qquad (5.88)$$

When this is applied to an array of fibers, such as in a fibrous filter pad, D may be taken as the average fiber diameter and a fictitious length ℓ' may be defined to represent the total "effective" fiber length in a unit cross-section of filter. Thus $L = \pi D^2 \ell'/4(1-\varepsilon)$ and by substitution

$$\eta = 1\text{-}\exp\text{-}\eta_T D\ell'/\varepsilon = 1\text{-}\exp\text{-}S\eta_T \qquad (5.89)$$

Since the fibers in such a pad are not in a regular array, i.e., not all parallel to each other and perpendicular to the flow, ℓ' may be taken to represent the equivalent effective total length they would have if they were so arrayed.

For an array of spherical targets, each of diameter D:

$$A_T = \pi D^2/4$$

$$N_T = 6(1-\varepsilon)/\pi D^3$$

Then

$$\eta = 1\text{-exp} - 3(1-\varepsilon)\eta_T L/2\varepsilon D \quad \text{and} \quad S = 3(1-\varepsilon)L/2\varepsilon D \tag{5.90}$$

In case the elements of array are spaced far from each other, as drops in a spray scrubber might be, there is no interaction effect. If spheres in an array are as little as six or seven diameters apart, the capture efficiency of each will be relatively independent [76]. But the stream of particles may be subjected to possible capture by a number n of targets in series, each acting as though isolated. In that case the overall efficiency will simply be a repeated application of (5.68), in which η_T is an isolated target efficiency.

$$\eta = 1-(1-\eta_T)^n \tag{5.91}$$

or

$$P = P_T^n$$

If n is sufficiently large and η_T relatively small this will approach

$$\eta = 1\text{-exp-}n\eta_T \tag{5.92}$$

as may be shown by comparing the series expansions for (5.91) and (5.92) term by term.

In all cases the function of the array is to multiply the capture effect of a single target. Even though the single target efficiency be rather low, the overall efficiency of an array may be substantial.

Example 8. Using the results of Example 7 calculate the grade efficiency and number of transfer units of an array of cylinders each 10 microns in diameter, packed with a porosity of 95%, and the array either 1 mm or 2 mm in thickness

$$S = \frac{4}{\pi}\frac{(0.05)}{0.95}\frac{L}{10\times10^{-4}} = 67.0\ L$$

$$\eta_i = 1\text{-exp} - \eta_{T_i}\ (67.0)L$$

$$NTU_i = -\ln\ (1-\eta_i) = 67.0\ L\eta_{T_i}$$

where L = 0.1 cm, or 0.2 cm.

d_p - μm	η_{T_i}	Grade Efficiency η_i L=0.1 cm	Grade Efficiency η_i L=0.2 cm	$(NTU)_i$ L=0.1	$(NTU)_i$ L=0.2
0.01	0.476	0.959	0.9983	3.19	6.38
0.1	0.0326	0.196	0.354	0.218	0.436
1.0	0.0245	0.151	0.280	0.164	0.328
2.0	0.159	0.655	0.881	1.06	2.12

Note how the dominating collection mechanism shifts from Brownian diffusion to inertial impaction as the particle size increases, with an intermediate range in which direct interception is predominant, although no mechanism produces much collection. The number of transfer units is directly proportional to the thickness of the array.

Calculations such as these are only valid for the initial period of collection by a stationary array of closely packed elements. As collected particles accumulate on the target elements, the shape and nature of the targets changes. As described above particles collect upon each other and may build up layers or dendrites extending out from the original target surfaces eventually to meet. As these growths fill in the spaces between the targets the nature of the collection process may change. Thus the value of η_T will become a function both of time and location (distance downstream) in the array.

If the nature of this function becomes known it could be incorporated into the model theory beginning with a modified version of Eqn. (5.85). Thus equations could be developed showing how η and m would vary with time for a specific array, in a manner somewhat like Eqns. (5.72a). The performance of arrays is clearly a transient one, with respect both to total mass of collection and to pressure drop. As it relates to granular bed filters and to fiber mat filters, it will be discussed further in Chapter 8.

REFERENCES

1. N.A. Fuchs, The Mechanics of Aerosols, Pergamon Press, New York (1964). pg. 136.
2. J.D. Tuttle, M.S. Thesis, University of Cincinnati, Cincinnati, Ohio 1976.
3. H. Lamb, Hydrodynamics, 6th ed., Dover, New York, 1945.
4. D.O. Banks, and G.J. Kurowski, Aerosol Sci. and Tech. 3: 317 (1984).
5. W.E. Ranz, Bull. No. 65, Dept. of Eng. Res., The Pennsylvania State Univ., University Park, Penn. 1956.
6. W. Strauss, Industrial Gas Cleaning, 2nd ed. Pergamon Press, New York, 1975.
7. M.N. Golovin and A.A. Putnam, Ind. Eng. Chem. Fund., 1: 264 (1962).
8. K.R. May and R. Clifford, Ann. of Ind. Hyg, 10: 83 (1967).
9. R.J. Brun and H.W. Mergler, NACA TN 2904, March 1953.
10. C.N. Davies and M. Aylward, Proc. Phys. Soc. (London), B64: 889 (1951).
11. H.D. Landahl and R.G. Hermann, J. Colloid Sci., 4: 103 (1949).
12. R.G. Dorsch, P.G. Saper, C.F. Kadow, NACA TN 3099, March 1954.
13. H. Herne, Int. J. of Air Poll., 3: 26, (1960).
14. H.F. George and G.W. Poehlein, Environ. Sci. & Tech, 8: 46 (1974).
15. K.V. Beard and S.N. Grover, J. Atmos. Sci. 37: 543 (1974).
16. G.I. Tardos, N. Abuaf, C. Gutfinger, JAPCA 28: 354 (1978).
17. T.T. Mercer and R.G. Stafford, Ann. Occup. Hyg, 12: 41 (1969).
18. S.C. Stern, Ind. Eng. Chem. Fund., 1: 273 (1962).
19. T.T. Mercer and H.Y. Chow, J. Colloid & Interfac. Sci., 27: 75 (1963).

20. M.K. Householder and V.W. Goldschmidt, J. Colloid and Interfac. Sci., 31: 464 (1969).

21. Y. Cheng and C.S. Wang, Atmos. Environ., 15: 301 (1981).

22. R.I. Crane and R.L. Evans, J. Aerosol Sci., 8: 161 (1977).

23. M.G. Selden, Jr., JAPCA 27: 235 (1977).

24. J.I.T. Stenhouse, P.J. Lloyd, A.I.Ch.E. Symp. Ser., No. 137, 70: 307 (1974).

25. R. Israel and D.E. Rosner, Aerosol Sci. and Tech. 2: 45 (1983).

26. K.T. Whitby, ASHRAE J., 7: 56 (1965).

27. J. Pich in Aerosol Science (C.N. Davies, Ed.), Academic Press New York, 1966, Chap. 9.

28. D.D. Degani and G.I. Tardos, JAPCA 31: 981 (1981).

29. C. Gutfinger and S.K. Friedlander, Aerosol Sci. and Tech., 4: 1 (1985).

30. F. Loffler, Clean Air, 8(4): 75 (1974).

31. B. Dahneke, J. Colloid Interface Sci., 37(2): 342 (1971).

32. M.J. Ellenbecker, D. Leith, and J.M. Price, JAPCA, 30: 1224 (1980).

33. D.E. Aylor, and F.J. Ferrandino, Atmos. Environ., 19: 803 (1985).

34. S.K. Friedlander, Smoke, Dust & Haze, Wiley Interscience, New York (1977).

35. I. Langmuir, ORSD Report 865, (1942).

36. G.L. Natanson, Dokl. Akad. Nauk SSSR, 112: 100 (1957).

37. S.K. Friedlander, A.I.Ch.E.J., 3: 11 (1957).

38. M. Crawford, Air Pollution Control Theory, McGraw-Hill New York, 1976.

39. W.L. Torgeson in General Mills Rep. No. 1890 AEC Cont. No. AT-(11-1) -401, SC. Stern et al (1958).

40. W.E. Ranz, Tech. Rep. No. 3, Contract AT - (30-3)-28, University of Illinois (1951).

41. C.J. Stairmand, Trans. Inst. Chem. Engrs. (London), 28: 130 (1950).

42. H.F. Johnstone and M.H. Roberts, Ind. Eng. Chem., 41: 2417 (1949).

43. R. Pfeffer, Ind. Eng. Chem. Fund., 3: 380 (1964).

44. M.J. Pilat and A. Prem, Atmos. Environ., 10: 13 (1976).

45. L.E. Sparks and M.J. Pilat, Atmos. Environ., 4: 651 (1970).

46. H.F. Kraemer and H.F. Johnstone, Ind. Eng. Chem., 47: 2426 (1955).

47. G.L. Natanson, Dokl. Akad. Nauk SSSR, 112: 696 (1957).

48. K.A. Nielsen and J.C. Hill, Ind. Eng. Chem. Fund., 15: 149 (1976).

49. K.A. Nielsen and J.C. Hill, Ind. Eng. Chem. Fund., 15: 157 (1976).

50. W.E. Ranz and J.B. Wong, Ind. Eng. Chem., 44: 1371 (1952).

51. J.W. Thomas and R.E. Yoder, A.M.A. Arch. Ind. Health, 13: 545 (1956).

52. C.N. Davies, Proc. Inst. Mech. Engrs., 1B: 185 (1952).

53. S.K. Friedlander J. Colloid Interface Sci. 23: 157 (1967).

54. H.F. George and G.W. Poehlein, Environ. Sci. and Tech., 8: 46 (1974).

55. I. Langmuir and K. Blodgett, Army Air Forces Tech. Report: No. 5418, (1945).

56. R.T. Jarman, J. Agric. Engr. Res., 4: 139 (1959).

57. H.C. Wang, J.J. Stukel, and K.H. Leong, Aerosol Sci. Tech. 5: 391 (1986).

58. H.C. Wang, J.J. Stukel, and K.H. Leong, Aerosol Sci. Tech. 5: 409 (1986).

59. H. Emi and N. Yoshioka, Prediction of Collection Efficiencies of Aerosols by Fibrous Filters, First Pacific Chem. Eng. Congress, Oct. 10-14, 1972, Kyoto, Japan.

60. M.W. First, JAPCA, 35: 1286 (1985).

61. W. Strauss and M.W. Thring, Trans. Inst. Chem. Engrs. 41: 28 (1963).

62. W. Strauss and B.W. Lancaster, Atmos. Environ. 2:135 (1968).

63. M. Beizaie, and C. Tien, Can. J. Chem. Engr., 58: 12 (1980).

64. S. Buhtra, and A.C. Payatakes, J. Aerosol Sci., 10: 445 (1979)

65. A.C. Payatakes, and L. Gradon, Amer. Inst. Chem. Engr. J., 26: 443 (1980).

66. C. Kanaoka, H. Emi, and W. Tanthapanichakon, Amer. Inst. Chem. Engr. J., 29: 895 (1983).

67. J. Happel, Amer. Inst. Chem. Engr. J., 5: 174 (1959).

68. S. Kuwabara, J. Phys. Soc. Japan, 14: 527 (1959).

69. J. Happel, Amer. Inst. Chem. Engr. J., 4: 197 (1958).

70. R. Pfeffer, Ind. Eng. Chem. Fund., 3: 380 (1964).

71. C.N. Davies, Air Filtration, Academic Press, London (1973).

72. R.A. Tal and W.A. Sirignano, Amer. Inst. Chem. Engr. J., 28: 233 (1982).

73. D. Lundgren and K.T. Whitby, Ind. Eng. Chem. Proc. Des. Dev., 4: 345 (1965).

74. L. Paretsky, L. Theodore, R. Pfeffer, and A.M. Squires, JAPCA 21: 204 (1971).

75. P. Knettig and J.M. Beeckmans, Aerosol Sci. 5: 235 (1974).

76. D.E. Jacober and M.J. Mattesar, Aerosol Sci. and Tech., 4: 433 (1985).

77. S.M. Belyaev and L.M. Levin, J. Aerosol Sci., 5: 325 (1974).

PROBLEMS

1. Refer to the sampling collector described in Example 1. Assume that all conditions are as given there, except that the entire system is isothermal at 50°C. Estimate the grade efficiency of this device for collection by gravity only of particles ranging from 1 to 50 μm in diameter.

2. For conditions as stated in Example 1, except: consider the system to be isothermal at 30°C and particles 10 μm in diameter, determine:

(a) for a particle entering at a height 1 mm above the bottom plate, where it will strike that plate; and

(b) what fraction of the particles entering will be collected on that plate.

3. Reconsider Example 1 for an electrostatic instead of a thermal collector. Assume conditions are isothermal at 50°C, that there is a potential gradient from the top to the bottom plate, and that the particles are charged in such a way that the charge is proportional to the square of the diameter. Do not perform any calculations, but discuss and explain how the results of the example would differ from those given in the text.

4. A 10 μm particle of density 2.6 gm/cm^3 enters a spinning gas at a radial position of 4 cm. The outer boundary of the gas is at 10 cm radius. The gas is at 100°C and its outermost tangential velocity is 15 m/s. The vortex exponent may be taken as 0.6. Determine:

(a) the tangential velocity of the particle at its point of entry.

(b) the time required for the particle to reach the outer boundary;

(c) the initial radial velocity of the particle;

(d) whether the use of Stokes law is reasonable for these calculations. [If not, think about what changes would need to be made in the equations and calculations.]

5. Repeat No. 4 above, except assume that the particle carries a charge of 100 electrons, that there is a potential gradient of 1000 V/cm radially outward, and that the vortex exponent is 0.5. [Assume Stokes law.]

6. (a) Check the results given in the table for Example 2, for the case of a 1 μm particle.

(b) Repeat the calculations for a 1 μm particle, approaching a spherical target 100 μm in diameter, all other conditions being the same as in Example 2.

7. Show in which of the two cases A & B described below the aerodynamic capture efficiency due to inertial impaction will be greater. In both cases the target is the same size and shape, and the relative velocity between target and particle is the same.

	A	B
Temperature of air	300°C	20°C
Particle diameter	3 μm	4 μm
Particle density	2.2 gm/cm3	1.2 gm/cm^3

8. (a) Which of the following targets has the greater capture efficiency? Target A is a sphere for which $v_o/D = 100\ s^{-1}$. Target B is a cylinder of 4 times the diameter of the sphere, and having 1/2 the approach velocity. The particles have a relaxation time of 0.005 s for both targets. Stokes law may be used.

(b) Estimate the aerodynamic capture efficiency of a cylindrical target 10 μm in diameter for particles 0.006 μm diameter (density 1.5 gm/cm^3) in dry air at 20°C, 1 atm, flowing at 5 cm/s.

9. The aerodynamic capture efficiency is 25% for certain particles impinging from a jet onto a round plate (diameter = D) under a given set of conditions. What will the A.C.E. be for particles which are twice as big, on a plate of one-half the diameter, at a velocity one-quarter as much?

10. A drop of water 800 μm in diameter is falling through a current of air at 20°C, 1 atm, rising vertically at a velocity of 100 cm/s. Particles of dust (2.6 gm/cm^3) 10 μm in diameter are being carried upward by the air at the same velocity. Estimate the "capture efficiency" (A.C.E.) of the particles by the drop, due to (a) inertial impaction, (b) direct interception, and (c) Brownian diffusion. Repeat, for particles 0.1 μm.

11. A stream of air at 20°C, 1 atm, flowing at 30 cm/s carries aerosol particles (2.6 gm/cm^3) 0.1 μm in diameter. Estimate the capture efficiency of these particles by a human hair, diameter 100 μm, which is perpendicular to the flow. Show by calculation whether the deposition would be more or less if the particles were 0.01 μm in diameter, or 1 μm in diameter.

12. A sampling probe 1 cm inside diameter was installed axially in a duct where air at 300°C was flowing at 2.0 m/s. The probe collected a sample of particles 25 μm in diameter, density 0.8 gm/cm^3 with a velocity of 2.5 m/s through the probe. The

concentration of sample collected was found to be 0.500 gm/m^3. Estimate the true concentration of particles flowing through the duct.

13. The velocity in a sampling probe is 10% less than the velocity of the mainstream being sampled. For particles 0.1 μm diameter in this situation, the value of the Stokes Number is 0.01. What error would be made in the collection of particles 15 μm in diameter?

14. Predict the single-fiber target efficiency for aerodynamic capture of spherical particles of density 1.48 gm/cm^3, ranging in size from 0.1 μm to 10 μm diameter, upon a glass fiber 8.82 μm in diameter from an airstream at 42 ft/min, 75°F and atmospheric pressure. Repeat for a temperature of 350°F.

15. Verify the results of the calculations for Example 7. How do you account for the fact that the overall collection efficiency η_T is a minimum for particles somewhere in the range of 0.1 to 1.0 μm range, and that the minimum shifts to a lower value of d_p for the isolated fiber as compared to the array?

16. Verify the results of the calculations for Example 8. Why is the grade efficiency for any particle size not directly proportional to the thickness of the array?

17. Using the data and conditions of Examples 7 and 8, predict the effect upon (a) the single target efficiency, and (b) the grade efficiency, if the air velocity is changed to 1 cm/s or to 100 cm/s.

18. Assuming an array made up as described in Example 8, show how to determine the thickness which would be required to achieve a desired overall collection efficiency of a particulate material having a known particle size distribution.

6
Centrifugal Collectors

I. FUNDAMENTAL FEATURES OF CYCLONES

A. *BASIC CONCEPT AND APPLICABILITY*

Collectors in which the major collecting force is centrifugal in nature are of a type known as cyclones. The most common embodiment of the basic design concept is shown schematically in Figure 1. This is called a vertical reverse-flow cyclone with tangential inlet. It will be the principal subject of the calculations to be discussed here. This concept is a very old one, and the basic design which has evolved over time has become pretty well established.

With reference to the Figure, the dust-laden gas enters tangentially through the duct, shown as rectangular with height a and width b. This imparts the spinning motion to the stream which sets up a centrifugal force on the particles directing them radially outward toward the collecting surface. This is primarily the cylinder of height (h-S) and diameter D. As the gas stream spins below the level of the outlet (exit) duct, a cylinder of height S and diameter D_e (sometimes referred to as the "vortex

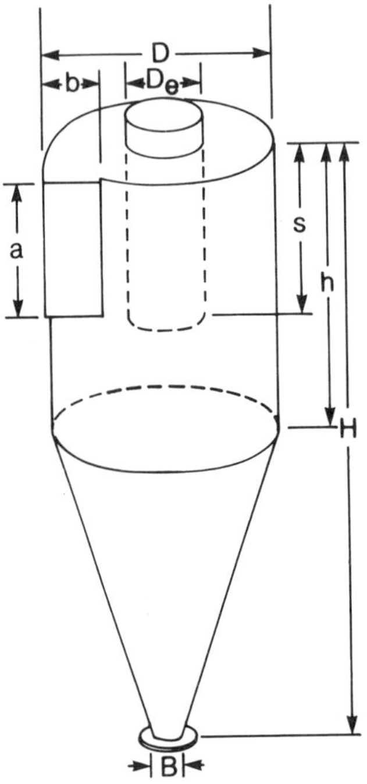

FIG. 1 *Cyclone shape and dimensions.*

finder", "exit thimble", or "tubular guard"), a radially inward motion of the gas develops. Swirling motion of the gas in a helical pattern continues into the conical base, of height (H-h) but eventually all of the gas flows inward, upward, and out through the outlet duct. This gives rise to the name "reverse-flow" for this type of design and operation. Meanwhile the dust colliding with the outer walls in the upper cylinder, as well as to some extent in the lower conical portion, moves down these walls and out through point B into a collecting hopper below, in a manner to be described below.

The collecting force developed is such as to be particularly effective on larger particles, say upwards of 10 μm in equivalent diameter. A typical experimental grade efficiency curve for solid particles [1] may be seen in Fig. 2. A cyclone is capable of

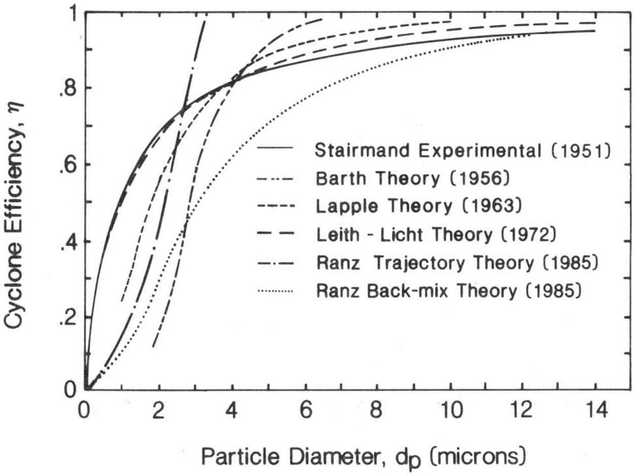

FIG. 2 *Experimental grade efficiency and model tests.*

handling very high grain loadings; values of over 1000 gr/ft^3 have been treated. Higher grain loading brings about simultaneously an increase in collection efficiency and a decrease in pressure drop for a given gas volume flow rate.

The device is relatively simple in design, construction, and operation (though not in internal flow behavior, as will be seen), having no mechanical parts except the fan required to move the gas stream. It is a relatively inexpensive collector both in initial cost and in operating cost, as the pressure drop through it is usually in the low range of 2" - 8" of water. By using suitable materials and methods of construction it may be adapted to extreme operating conditions: high temperature, high pressure, corrosive gases.

It is usually operated dry, collecting solids, but may be fitted with a spray to wet the walls, or operated to collect mist particles. Wet operation and dry operation, however, have certain intrinsically different aspects with regard to the internal collection mechanisms. It is not suitable for collecting sticky or tacky materials which would tend to cling to the interior walls, or other particles which might tend to agglomerate and clog the outlet duct.

The ordinary cyclone is incapable of collecting very fine particles to a high degree. It may well serve as a pre-cleaner to remove the coarser particles of a heavily loaded stream which will subsequently be cleaned of finer particles by one of the other collecting devices.

There are a number of variations and modifications to the basic design. Some involve a scroll ("wrap around") or helical inlet instead of the tangential duct marked a x b. Straight-through flow, without the conical portion, is also used, usually in a horizontal position. Other modifications have also been proposed in the shape of the gas outlet duct, or in that of the dust outlet hopper design, as well as the insertion of various kinds of internal vanes. All of these are intended either to improve the collection efficiency or reduce the pressure drop and may be marginally successful.

Modifications in operation include: using two or more cyclones in parallel, or in series; recycling a portion of the gas outlet stream back to the feed; the use of a secondary air stream; and electrifying the cyclone to set up a radial potential gradient to augment the collecting force. Calculations for some of these schemes will be illustrated below.

There has been a renewed interest in the application of this old concept in recent years because of the increasing interest in cleaning gases at high temperatures and pressures. This has stimulated a search for ways to improve the grade efficiency while retaining the basic features and the simplicity of operation. In this regard see reference [60] of Chapter 5.

B. *INTERNAL FLOW PATTERNS*

From extensive experimentation and observation over the years, the patterns of gas flow and of particle flow within the cyclone are well known [2-6]. It is essential to know how to describe them in order to develop useful models of cyclone performance.

1. *Gas flow* The primary pattern of gas flow is that of a double helix. Gas spirals downward from the tangential inlet in the outer zone to a certain depth near the dust hopper outlet. As it does so it "peels off" to enter an inner vortex horizontally. The inner vortex rises through the central core of the exit tube to discharge through D_e. If the body (H) is sufficiently long, at a certain axial depth known as the "natural length" the inward flow should be completed, all of the gas having flowed across an imaginary vertical cylindrical boundary formed by the extension of the outlet duct downward below S. The gas may be said to make a certain "number of turns" N_e, as it spirals down and out. N_e is reported to vary from about 0.3 to 10 for various designs, with a mean value of about 5. For a given design N_e increases as the inlet velocity increases [7]. It has been suggested that an average value of N_e may be estimated from the vertical dimensions as $N_e = [h + (H-h)/2]/a$, or roughly as $N_e = H/a$.

Alexander [8] gives an emprical formula, based upon his own experiments, for the natural length ℓ as:

$$\ell = 2.3\ D_e(D^2/ab)^{1/3} \qquad (6.1)$$

This is defined as a vertical distance measured downward from the bottom edge of the exit duct. It is noteworthy that this depends upon only four of the cyclone dimensions and is independent of the gas flow rate. This conception of the natural length may require some modification, however, due to the presence of secondary flows as described below.

The gas in the outer zone forms a free vortex such that the tangential gas velocity u_T is related to the radial position R as given by Eqn. (5.12): $u_T R^n$ = constant [2,6]. The value of n = 1 for an ideal fluid, but in the cyclone it is of the order of 0.6. It may be estimated by a method suggested by Alexander [8] which is a correlation of his own experimental measurements:

$$n = 0.67\ (D)^{0.14} \quad \text{at} \quad 10°\text{C} \quad (D \text{ in m})$$

and $$(1 - n_T)/(1 - n) = (T/283)^{0.3} \quad \text{at } T°K$$

Combining:

$$n_T = 1 - \left[1 - 0.67\ (D)^{0.14}\right] (T/283)^{0.3} \tag{6.2}$$

However, Yuu et al. [6] found n to decrease markedly with increase of grain-loading.

Tangential velocities in any horizontal plane will increase from a minimum (which may be lower than the gas inlet velocity) at the cyclone wall up to a maximum at a point identified as the radius of the central core. This radius will vary from a value equal to $D_e/2$ at the edge of the exit duct to a value of about $D_e/6$ at the bottom of the cyclone. Within the central core tangential velocities drop off to a value about the same as at the outer wall. Tangential velocities at the same radius increase markedly from top to bottom.

Superimposed upon the helical flow pattern are strong secondary flows which give rise to relatively small vertical velocities downward along the walls of the cyclone below S, and upward in the central core. The radial location at which the change from downard to upward velocities occurs is relatively closer to the cyclone wall near the top than near the bottom of the cone. At all levels, however it occurs beyond the radius defined as above for the central core.

This secondary flow may continue down along the walls of the conical section and into the dust collection hopper below B. Ranz [4] observed 5 - 7% of the gas flow to do so. Unless this gas is withdrawn in some fashion ("base purge", or "base bleeding"), it must rise again to exit through the central core thus setting up a swirling flow in the hopper itself. To the extent that this occurs, it would modify the notion of the "natural length" as described above. If the total distance (H - S) is less than ℓ, obviously the gas motion must continue in this fashion.

Secondary flow also causes some gas to move radially inward just under the top surface, thence vertically downward over the outside of the exit duct. As it reaches the level of S, this gas moves directly out, thus completely by-passing the centrifugal action. There also appears to be a rather low velocity of gas flow radially inward from the outer wall, which is constant with radial position and approximately equal at all levels below S.

The central flow in a cyclone is therefore highly turbulent and the central core should be regarded as uniformly mixed, at least with regard to small enough particle sizes. Ranz [4] suggests that this would be for

$$d_p \leq \sqrt{\mu D/\rho_p \, u_T} \tag{6.3}$$

2. *Particulate flow* Particles then apparently move toward the wall in turbulent diffusive fashion through the well-mixed central core. As they do so there is ample opportunity for numerous collisions between particles to occur, the more so the higher the concentration ("grain loading") of the particulate.

If the particles are liquid droplets they will coalesce into larger drops. When all drops reach the wall they will form a liquid film which flows as spiral streaks and bands down to the cone outlet. Ranz [4] observed that at heavy spray loadings a "rotating collar or dam of liquid was seen suspended high on the cone wall" and that "a highly agitated, continuous, spirally streaked film of liquid flowed over this collar from the cylindrical to the conical wall".

On the other hand when the particles are solid, some may rebound from collisions and some may agglomerate. Abrahamson [9] insists that this is an essential feature of collection. He attributes the increase in collection efficiency with increased grain-loading to the formation of larger aggregates by the greater amount of collisions occurring.

Upon reaching the wall some particles may bounce off, but at the wall dense fluidized layers of particles form. They do not cover the wall uniformly and slide toward the bottom, but due to the secondary gas flow arrange themselves into a regular spiral pattern of moving sand dunes which screws its way downward, making N_e turns [2,4,6]. The particles move lengthwise along this spiral path into the dust hopper. Some may be reentrained by the rising gas flow and recycled into the central core. Some particles may also be carried along the top and short-circuited out the exit tube by the secondary flow described above. All of this behavior may be seen clearly in a transparent cyclone, and the number of turns counted. It is quite evident that there are very significant differences between the collection of solid and of liquid particles.

Although the pattern of dune flow has long been observed, it is only recently that its important role in cyclone behavior has been intensively studied by Tran [10] and analyzed by Ranz [4]. He describes the spiral dune by assuming that it corresponds to the tip of a "mathematical" standing vertical wave form, having an amplitude a and a wave length λ, moving down the cyclone wall. The actual depth of the dune is represented by a portion of the mathematical wave existing a distance of (a − h) out from the wall. The value of h is a fictitious dimension, corresponding to the wall position, which has to be determined by analysis of the wave form. The vertical distance between branches of the spiral is taken to be λ, which is assumed to be equal to the height of the cyclone inlet duct. Note that a and h here are <u>not</u> the dimensions of the cyclone body.

By analyzing this standing wave form, Ranz works out the dimensions of the dune to be:

$$\text{Height } (a-h) = (a/h - 1)\, D/8\pi^2 \tag{6.4a}$$

$$\text{Width} = (\lambda/\pi) \cos^{-1}(h/a) \tag{6.4b}$$

$$\text{Cross-sectional area} = (\lambda h/\pi) \left[\sqrt{(a/h)^2 - 1)} - \cos^{-1}(h/a)\right] \tag{6.4c}$$

The theory indicates that $h = D/8\pi^2$, so that these dimensions become a function of the ratio a/h, which may also be expressed in terms of (a-h)/h. This quantity will increase with the vertical depth in the cyclone, as more particles are collected and added to the flow in the dune.

The quantity of collected material in the dune may be expressed by assuming that the particle tangential velocity along the dune is equal to the inlet tangential velocity u_T. Then the volumetric flow of particles along the dune is given by

$$u_T \, \varepsilon_{pb} \, (\lambda h/\pi) \left[\sqrt{(a/h)^2 - 1)} - \cos^{-1}(h/a) \right]$$

where ε_{pb} (≈0.3–0.5) represents the volume fraction of particles in the fluidized state of the dune. This amount of particle flow must be equal to that which has been collected from the inlet stream by the time it has reached a depth z below the inlet (by which point the spiral has made N_e turns) given by

$$u_T b \, \varepsilon^o \, \eta \, z/N_e$$

Equating these two expressions and rearranging defines a dimensionless loading parameter which determines the value of a/h at any depth in the cylone:

$$8\pi^2 \frac{\eta z \, b \, \varepsilon^o}{\lambda D N_e \varepsilon_{pb}} = \left[(a/h)^2 - 1\right]^{1/2} - \cos^{-1} (h/a) \qquad (6.5)$$

Here η represents the overall collection efficiency, and ε^o the volume fraction of particles in the feed (a measure of the loading).

The loading parameter, left side of (6.5), for a given cyclone varies mainly as ε^o. As it approaches 1, which corresponds to a high loading of say $\varepsilon^o \approx 0.01$, the value of a/h rises very fast to about 2.4. At higher loading, further increase in a/h is so rapid as to indicate that the dune flow will become overloaded, that is it will become incapable of moving the collected particles from the cyclone! At this point, for a typical operation, the dune height is predicted to be of the order of 5 mm and the width 5cm.

C. *DESIGN APPROACH*

To select or design a cylone for a specific duty requires specification of its overall size, characterized by the principal diameter D, and its shape or configuration which is determined by seven geometric ratios:

$$\frac{a}{D}, \frac{b}{D}, \frac{D_e}{D}, \frac{S}{D}, \frac{h}{D}, \frac{H}{D} \text{ and } \frac{B}{D}$$

Selection of these eight quantities will in turn depend upon the duty to be performed, as described by:

- Rate of flow, temperature, pressure of gas stream
- Composition of the gas, including especially its humidity
- Dust loading of gas
- Size-distribution of dust carried by gas
- Degree of total collection, or reduced level of emission, to be attained.

Several sets of these ratios have evolved from experience as more or less "standard" designs. Some are listed in Table 2 (p. 309). Details of the design, and slection of materials of construction, may also be influenced by the chemical and physical nature of the dust. Pressure drop must also be estimated in order to forecast energy requirements.

In order to develop an optimal design, it is necessary to have a model for the collection process which gives the grade-efficiency function, and a method of calculating the pressure drop through the device. Ideally, these models must show how the eight geometrical parameters and all the operating variables will affect the grade-efficiency and the pressure drop. They should, of course, be based upon a knowledge of the flow patterns described above.

II. MODELING CYCLONE PERFORMANCE

A. *VARIOUS APPROACHES TO EFFICIENCY*

The complexity of the flow patterns in the cyclone makes it very difficult to construct a performance model which is completely realistic in all details. Various simplifying assumptions have been introduced in different modeling approaches which lead to equations of various degrees of complexity and utility. In general it may be said that counterparts of all three of the basic modeling concepts discussed in Chapter 3 have been applied to the cyclone, as well as modifications of these. When the resulting equations are subjected to experimental verification, it is found that no one model is preeminently best suited for application in all kinds of situations. However certain modeling procedures have emerged as being valuable for practical use in design and performance prediction. These will be presented following a brief review of the several modeling approaches.

1. *Single-particle trajectory approach* This may be said to correspond to the elementary no-mix model, No. 1 of Chapter 3. It is based upon the computation of the trajectory of a particle in a spinning gas, as developed into Eqns. (5.14) or (5.15). However to apply to a cyclone, the drag force in these equations must be modified by the introduction of the radial inward gas velocity v_R in accord with the principle of Eqn. (5.1). Thus (5.15) would become:

$$\frac{d^2R}{dt^2} = \frac{u_{T_1}^2 R_1^{2n}}{R^{2n+1}} - \frac{1}{\tau}\left[\frac{dR}{dt} - v_R\right] \qquad (6.6)$$

Note: This equation is for Stokes law particles, and may not be strictly applicable to all conditions in the cyclone. Equation (5.14) may be appropriately modified to use instead.

By making various assumptions about the terms in (6.6), such as $v_R = Q/\pi D_e(H-S)$, or $v_R = 0$, or $d^2R/dt^2 = 0$, or a value for n, etc. one may derive a "critical particle diameter". This is defined in one of two ways [10,11]:

(a) Static Particle: one suspended at the edge of the central core, where the inward drag force is balanced by the maximum centrifugal force. Larger particles should spin out and be collected; smaller particles should drift into the central core and escape.

(b) Time-of-Flight Particle: one which will travel from its entry position at R_1 and just reach the cyclone wall in time t. This is found by rearranging Eqn. (5.17) to solve for d_p. R_1 and t are defined in various ways in order to find either the size of particle which is collected 100% or that which is collected 50%.

Leith [10] has given a very complete tabulated summary of the several theories which have been developed along these lines. Two of them, Barth [12] and Lapple [13], with the addition of some empirical collection data, lead to grade-efficiency curves such as are represented on Fig. 2. This illustrates that these curves do not agree well with experimental data for the collection of solid particles. Note the S-shape of these curves as contrasted with the exponential shape of the experimental curve.

Ranz [4] has suggested an elementary model of this type which may be obtained by setting n = 1, $d^2R/dt^2 = 0$, in (6.6) and calling the term in brackets U_{rel}. Then:

$$U_{rel} = \frac{\rho_p d_p^2 u_T^2}{9 \mu D} \tag{6.7}$$

where U_{rel} represents the relative velocity with which particles approach the outer wall of the cyclone. Taking into account the nature of the spiral dunes, the maximum time particles spend in the spiral may be given as the finite unwound length of the spiral divided by the inlet velocity i.e. $(N_e \pi D/u_T)$. Hence the maximum radial distance travelled will be $U_{rel}(N_e \pi D/u_T)$ and the grade-efficiency may be taken as this distance divided by the inlet width b. Thus:

$$\eta = \frac{U_{rel} N_e \pi D}{b u_T} = \frac{\pi N_e \rho_p d_p^2 u_T}{9 b \mu D} \tag{6.8}$$

A Stokes number may be defined for a cyclone as:

$$Stk_c = \frac{\rho_p d_p^2 u_T}{18\mu D} = \frac{\tau u_T}{D} \tag{6.9}$$

Then (6.7) and (6.8) may be restated in terms of Stk_c, and (6.8) will become:

$$\eta = 2\pi N_e Stk_c D/b = 2\pi N_e Stk_c'' \tag{6.10}$$

in which a modified Stk_c'' may be introduced, defining it by replacing D with b in (6.9).

Equation (6.8) is also displayed on Fig. 2 for the conditions of that experimental data. It is seen to have the parabolic shape typical of this type of model, rising to a point of 100% efficiency at a "critical" particle size. Throughout much of the range it agrees well with the Barth model.

2. *Radial back-mix approach* By analogy with the lateral-mix model, No. 2 of Chapter 3, this assumes that the uncollected dust at any level is completely remixed radially to maintain a uniform concentration in any horizontal cross-section of the cyclone. As developed by Leith and Licht [14] this approach leads to a realistic grade-efficiency function, also shown on Fig. 2. It

assumes that in a period of time dt, only those particles within a certain distance dR of the cyclone wall (indicated by radius R_2) will move to the wall and be collected. Meanwhile the particles will travel a distance $R_2 d\theta$ tangentially and dL vertically. The number of particles removed (dn') will be

$$-dn' = d\theta/2 \quad [R_2^2 - (R_2 - dR)^2] \; n \; dL \tag{6.11}$$

The total number n' of particles in the sector from which these particles are removed is

$$n' = d\theta/2 \; R_2^2 \; n \; dL \tag{6.12}$$

The fraction of particles removed in time dt is therefore

$$-\frac{dn'}{n'} = \frac{2R_2 dR - (dR)^2}{R_2^2} \approx \frac{2dR}{R_2} \tag{6.13}$$

The fraction of particles collected is related to the total average residence time through Eqn. (5.18) which was developed to show the relationship between time and the radial position of a single particle in a vortex. The rate at which the uncollected particle system moves toward the vortex wall, as a function of the time the system has spent within the vortex, is obtained [14] by setting $R_1 = 0$ in (5.18) (which assumes $v_r = 0$) solving for R and differentiating with respect to t, to obtain

$$\frac{dR}{dt} = \frac{\tau u_{T_2}^2}{R_2} \left[2(n+1) \; \tau \left(\frac{u_{T_2}}{R_2} \right)^2 t \right]^{-\frac{2n+1}{2n+2}} \tag{6.14}$$

Combining (6.13) with (6.14) and integrating over the residence time t:

$$\int_{n_o'}^{n'} \frac{dn'}{n'} = -\frac{2\tau u_{T_2}^2}{R_2^2}\left[2(n+1)\,\tau\left(\frac{u_{T_2}}{R_2}\right)^2\right]^{-\frac{2n+1}{2n+2}} \int_0^t t^{-\frac{2n+1}{2n+2}}\,dt \qquad (6.15)$$

$$\eta_i = 1 - \exp - 2\left[2(n+1)\,\tau_i\left(\frac{u_{T_2}}{R_2}\right)^2 t\right]^{\frac{1}{2n+2}} \qquad (6.16)$$

This gives the grade-efficiency of collection as a function of the residence time. Here R_2 is taken as the position of the outer wall, $R_2 = 1/2\ D$, n is found from (6.2), and u_{T_2} is replaced by the average inlet velocity u_T through the entrance duct, $u_T = Q/ab$.

There remains the question of the proper value of t to use. Different parcels of gas introduced into the cyclone at the same time may have different residence times within the unit, ranging from the shortest, corresponding to entry at level a and exit at level S, to the longest which would correspond to entry at the top and exit at depth H. For simplicity, however, it is desirable to determine an average residence time for all of the gas stream such as will account for the collection obtained.

Leith and Licht [14] found that an adequate way to do this is to set the total average residence time equal to the average time required for gas to descend from the average level of entrance to the level of the bottom of the exit pipe, plus the average time in residence below this point. Thus

$$t_{res} = (V_s + V_{nl}/2)/Q \qquad (6.17)$$

where

$$V_s = \pi(S-a/2)\ (D^2 - D_e^2)/4 \qquad (6.18)$$

$$V_{n\ell} = \frac{\pi D^2}{4}(h-S) + \frac{\pi}{4}\frac{(\ell + S-h)\ D^2}{3}\left(1 + \frac{d}{D} + \frac{d^2}{D^2}\right) - \frac{\pi D_e^2 \ell}{4} \qquad (6.19)$$

and

$$d = D - (D-B)\,(S+\ell-h)/(H-h)$$

V_s is seen to be the annular volume around the exit duct from S up to half the inlet height a/2. $V_{n\ell}$ is the annular volume around the central core from S down to the natural length ℓ, as given by (6.1).

A practical cyclone should have a physical length (H-S) near its natural length ℓ. If the cyclone body is longer than ℓ, i.e., $(H-S) > \ell$, the space at the bottom of the cyclone, below the vortex turning point, will be wasted. If the cyclone body is shorter than ℓ, i.e., $(H-S) < \ell$, the full separating potential of the cyclone will not be realized. In this case $V_{n\ell}$ should be replaced by V_H which is:

$$V_H = \frac{\pi D^2}{4}(h-S) + \frac{\pi D^2}{4}\frac{(H-h)}{3}\left(1 + \frac{B}{D} + \frac{B^2}{D^2}\right) - \frac{\pi D_e^2}{4}(H-S) \qquad (6.20)$$

Equation (6.17) may be written as

$$t_{res} = (V_s + V_{n\ell}/2)/Q = K_c D^3/Q \qquad (6.21)$$

where

$$K_c = (V_s + V_{n\ell}/2)/D^3 \qquad (6.22)$$

Here K_c is a dimensionless constant for a given cyclone design which depends only upon the relative proportions of the various dimensions. For cyclones which are geometrically similar in all respects it is independent of the size. The magnitude of K_c is an indication of the relative effective volume a given design provides, in which separation of particles from gas may take place.

Returning now to Eqn. (6.16) and replacing t by $t_{res} = K_c D^3/Q$, u_{T_2} by $u_T = Q/ab$, and R_2 by D/2 the following form is obtained:

$$\eta_i = 1 - \exp - 2\left[\frac{(n+1)\ \tau_i Q}{D^3}\left(\frac{8K_c}{K_a^2 K_b^2}\right)\right]^{\frac{1}{2n+2}} \qquad (6.23)$$

This equation may be displayed in more meaningful form by defining two dimensionless parameters.

(1) A modified Stokes number (see. Eqn. (6.9)):

$$\mathrm{Stk}'_{c_i} = (n+1)\ \tau_i Q/D^3 = (n+1)\ K_a K_b\ \mathrm{Stk}_c \qquad (6.24)$$

reflecting operating conditions within the cyclone, and size of cyclone.

(2) A cyclone design number:

$$K = 8K_c/K_a^2 K_b^2 \qquad (6.25)$$

reflecting only the configuration (shape) of the cyclone. Equation (6.23) for the grade efficiency is then written

$$\eta_i = 1 - \exp -2\left[\mathrm{Stk}'_{c_i}\ K\right]^{\frac{1}{2n+2}} \qquad (6.26)$$

Equation (6.26) is equivalent to

$$\eta_i = 1 - \exp - M d_{p_i}^N \qquad (6.27)$$

where

$$M = 2\left[\frac{KQ}{D^3}\frac{\rho_p}{18\mu}\ (n+1)\right]^{N/2} \qquad (6.28)$$

$$N = 1/(n+1) \qquad (6.29)$$

The "cut" diameter, for which $\eta_i = 0.50$, will be given by

$$d_{p_c} = \left(\frac{0.6931}{M}\right)^{(n+1)} \qquad (6.30)$$

and

$$\eta_i = 1 - \exp - 0.6931\left(\frac{d_p}{d_{p_c}}\right)^{1/n+1} \tag{6.31}$$

Figure 2 illustrates how (6.26) compares with some experimental grade-efficiency curves for the collection of solid particles, along with other models in the literature.

An extension of the radial back-mix concept has been employed by Dietz [15] to develop a more detailed analysis of the internal volume and particle residence time within the cyclone. The model divides the cyclone into three regions: the entrance region (the annular space around the outlet duct at the top of the cyclone), the downflow region (corresponding to the body of the cyclone below the exit duct but outside the central core), and the central core (defined by the imaginary extension of the outlet duct down to the base). Within each of these regions separately, it is assumed that there is a uniform radial concentration of particles at a given horizontal section. The theory takes into account the interchange of particles between the downflow and core regions.

It leads to a more complex grade-efficiency function of Stokes number and configuration ratios, which may be written as:

$$\eta = 1 - [K_o - (K_1^2 + K_2)^{0.5}] \exp -\left[\frac{\pi(2K_s - K_a)}{K_a K_b} \mathrm{Stk}_c\right] \tag{6.32}$$

Here the three constants are defined as:

$$K_o = \frac{1}{2}\left[1 + K_e^{2n}\left(1 + \frac{K_a K_b}{2\pi K_\ell \mathrm{Stk}_c}\right)\right]$$

$$K_1 = \frac{1}{2}\left[1 - K_e^{2n}\left(1 + \frac{K_a K_b}{2\pi K_\ell \mathrm{Stk}_c}\right)\right] \tag{6.33}$$

$$K_2 = (K_e)^{2n}$$

In these expressions ℓ refers to the natural length (6.1) and n to the vortex exponent (6.2). The effects of the Stokes number and the configuration ratios K_a, K_b, etc. are intertwined and cannot readily be displayed separately in a generalized fashion. However a sample of the η -function is shown on Figure 3 for comparison with the other models. Note the S-shape. Dietz developed this model initially to represent performance data taken at high temperatures and pressures.

3. *Complete back-mix approach* This corresponds exactly to basic model No. 3 of Chapter 3. As developed by Ranz [4] it assumes that the cyclone vortex is a highly turbulent region of uniform particle concentration, that is a "completely stirred tank". Thus the mass ratio Y of particles to gas in the outlet flow is the same as that in the gas flow over the wall layer. Flux of particles to the wall layer is assumed everywhere to be $\rho_f Y U_{rel}$. The wall area is taken to be the unwound length of the spiral times the height of the inlet ($\pi N_e Da$). Then a mass balance over particles of size d_p will be

$$(u_T ab)(\rho_f Y_o) = (\rho_f Y U_{rel})(\pi N_e Da) + (u_T ab)(\rho_f Y) \tag{6.34}$$

From this the grade-efficiency may be obtained as

$$\eta = \frac{Y_o - Y}{Y_o} = 1 - \frac{1}{1 + \dfrac{(2\pi N_e)\,Stk_c}{K_b}} \tag{6.35}$$

in terms of Stokes number as defined in (6.9). This curve is also shown on Figure 2.

In a modification of this model [26] N_e is replaced by $\bar{H}/a$, where

$$\bar{H} = D\left[K_h + \frac{(K_H - K_h)}{(1 - K_B)}\,\frac{(K_B^{(1-2n)} - 1)}{(2n - 1)}\right] \quad \text{if } n \neq 0.5 \tag{6.36a}$$

$$\text{or} \quad \bar{H} = D\left[K_h + \frac{(K_H - K_h)}{(1 - K_B)} \ln(1/K_B)\right] \qquad \text{if } n = 0.5 \tag{6.36b}$$

This was obtained by taking into account the vortex law, Eqn. (5.12), and integrating vertically to obtain the average particle radial velocity over the entire wall.

A generalized grade-efficiency plot may be made in terms of $\sqrt{Stk_c}$, for a given cyclone configuration, i.e. values of n, K_a, K_b, K_l etc., and N_e. Figure 3 shows a comparison of models (6.10), (6.26), (6.32), (6.35), and (6.35) as modified by (6.36), in this way for the so-called Stairmand design (as used also in Fig. 2) for which $N_e = 5$, $K_a = 0.5$, $K_b = 0.2$, $K_e = 0.5$, $n = 0.54$, and $K_{\bar{H}} = \bar{H}/D = 5.58$. Experimental data may be checked for conformity to any of these models by the methods of plotting described for the three basic models in Chapter 3, Eqns. (3.52), (3.53) and (3.54). If a straight line is obtained by one of these plots it will confirm the model type and also yield experimental constants for the model.

B. *SHORTCOMINGS, COMPARISON, AND TESTING OF MODELS*

In order to select a model which is suitable for design calculations, it is necessary to compare the models critically and to test them experimentally. First it may be noted that through the Stokes number parameter, which is implicit in Eqn. (6.6), all models correctly indicate that grade-efficiency tends to increase for higher flow rate (Q), denser particles (ρ_p), lower viscosity of gas (μ), and smaller cyclone size (D), for a given cyclone configuration (set of K's). Put in another way, these effects are equivalent to saying that the ratio of particle sizes for equal grade-efficiency is

$$\frac{d_{p_1}}{d_{p_2}} = \left(\frac{Q_2}{Q_1}\frac{D_1^3}{D_2^3}\frac{\rho_{p_2}}{\rho_{p_1}}\frac{\mu_1}{\mu_2}\right)^{1/2} = \left(\frac{u_{T_2}}{u_{T_1}}\frac{D_1}{D_2}\frac{\rho_{p_2}}{\rho_{p_1}}\frac{\mu_1}{\mu_2}\right)^{1/2} \tag{6.36}$$

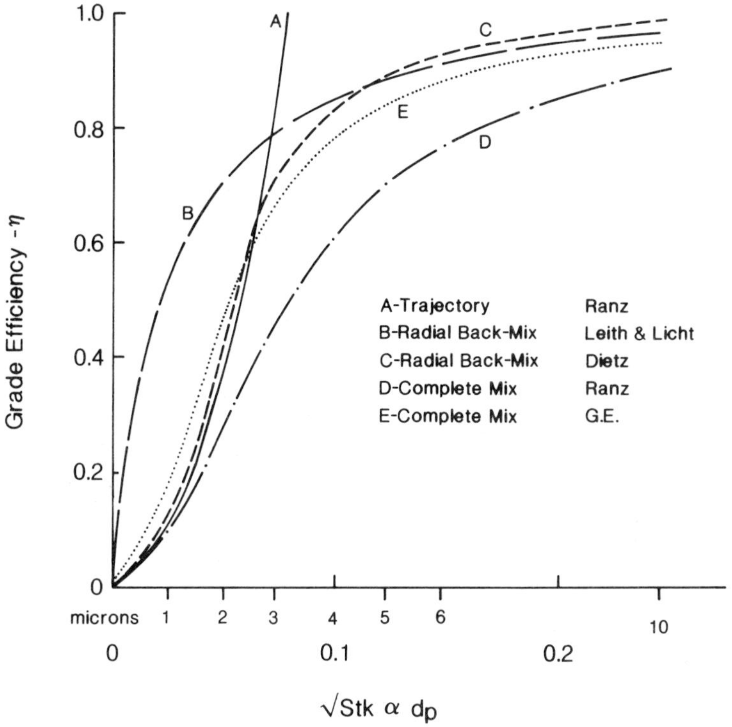

FIG. 3 *Theoretical models applied to Stairmand high efficiency cyclone configurations.*

provided the vortex exponent is essentially constant. This relationship may be used to predict how a grade-efficiency function, determined experimentally under one set of these parameters, will be transformed under another set of them in a geometrically similar cyclone.

1. *Shortcomings of models* None of the models cited takes into account the effect of grain-loading, the effect of the gas density, the degree of turbulence as may be represented by the Reynolds number, the difference in behavior of solid and liquid particles, the phenomenon of saltation, the mechanism of solids removal by the spiral dunes, nor the presence of electrostatic

effects. The known role of each of these factors will be discussed briefly, and some empirical corrections presented.

As mentioned above, it is well known that an increase in grain-loading tends to increase the collection efficiency [7]. This is probably due to to agglomeration of particles brought about by coarser particles colliding with and picking up finer ones under the more crowded conditions. Baxter [16] suggests that a correction upon overall efficiency may be made by:

$$\frac{P_1}{P_2} = \frac{100 - \eta_{M_1}}{100 - \eta_{M_2}} = \left(\frac{C_{o_2}}{C_{o_1}}\right)^{0.182} \qquad (6.37)$$

A chart showing the effect of loading, determined experimentally, is also available [7]. Masin and Koch [17] have empirically fitted this data to:

$$\ln\,(1-\eta_M) = \ln\left[(1-\eta_{M_o})/(1 + \eta_{M_o}\,[c^A-1])\right] \qquad (6.38)$$

where c is in gr/ft^3, and

$$A = 0.67 - 2.11\,\eta_{M_o} + 5.63\,\eta_{M_o}^2 - 4.00\,\eta_{M_o}^3$$

In addition an empirical correction may be made for an apparent viscosity of the stream, as first proposed by Sproull [40], given as

$$\mu_{app} = \mu\,(1 + 0.09 \log c + 0.02 \log^2 c) \qquad (6.38a)$$

In Ranz's loading parameter for the dunes, Eqn. (6.5), the volume fraction of particles in the feed ε^o appears explicitly, but is related there only to the dimensions of the dune and not to the efficiency of collection.

The density of the gas does not appear as a variable in any of the models. There is conflicting experimental evidence as to its effect. Knowlton and Bachovchin [18] found that pressure had no significant effect upon collection efficiency, and the gas density varied tenfold in their tests. On the other hand, recent experiments [19] using argon instead of air show that it has a very definite effect: increasing the gas density enhances the collection. Gas density is determined by composition (molecular weight), temperature, and pressure. There is evidence [18,19,20] that all of these factors affect efficiency. But the only quantity in the models which is directly affected by them is the gas viscosity. However the change in viscosity alone does not seem to account entirely for the effects observed [18,19,20].

Although several of the models cited refer to the importance of turbulence in the gas flow, none of them attempts to account for it in a specific way. On the other hand several studies on the use of small cyclones for sampling purposes [19,21], as well as at least one on large cyclones [20], have shown that a Reynolds number may be involved in the correlation of results. This Re is usually defined on the basis of the outlet diameter D_e. It is found that as Re increases, collection efficiency decreases (or that "cut diameter" decreases). This suggests a way in which gas density may be introduced explicitly, as well as the degree of turbulence taken into account. It has also been suggested that the vortex exponent n, Eqns. (5.12) and (6.2), might be a function of Re but no data are available. Oddly, the Stokes number is not involved in any of the studies on sampling cyclones.

There are obvious differences between the behavior of solid and liquid particles. Liquids cannot rebound from particle-to-particle collisions, cannot bounce or be reentrained from the cyclone wall, cannot be picked up from the base receiver, nor would they form fluidized spiral dunes. Since none of these effects are explicitly built into the models, experimental tests of liquid (mist) collection might provide a sensitive method of determining the "best" model on a fundamental basis. (Mists have been observed to be particularly effected by the secondary flow just under the top surface, a liquid film being carried over to the exit duct, spiralling down the outer wall of that duct and over the edge as a thin film which is readily picked up by the exit gas. This type of loss can be minimized by placing a "mist skirt" around the end of the exit tube [23].)

Such tests typically produce S-shaped grade-efficiency curves [11,22] such as those of the Barth and Lapple models shown on Fig. 2, and those of Dietz (6.32), and Ranz (6.35) shown on Fig. 3. On the other hand there are many recorded tests of solids collection (see Table 1 on p. 303) which show the exponential type of curve represented only by the Leith-Licht model, Eqn. (6.26), shown on Figs. 2 and 3. Occasionally some solids testing results in an S-shaped curve, particularly in tests at high temperatures [15,20]. It would seem that the effects of bouncing and reentrainment upon the collection efficiency of solids are somehow taken into account by the radial back-mix concepts built into (6.26).

The saltation effect places an upper limit upon the flow rate (or inlet velocity u_T) which may be used in the case of solids. This is due to the lack of deposition (or reentrainment) of particles which occurs at high velocities, and is analogous to the pick-up of solids at a high rate of bulk flow in horizontal pipelines [24]. Kalen and Zenz [25] have established a relationship for the optimum inlet velocity, and the velocity at which reentrainment becomes significant. The saltation velocity is given, empirically, by

$$u_S = 2.055 \left[\frac{4g\mu\rho_p}{3\rho_f^2}\right]^{1/3} \left[\frac{(b/D)^{0.4}}{(1 - b/D)^{1/3}}\right] D^{0.067}\, u_T^{2/3} \tag{6.39}$$

in which units of feet, pounds mass, and seconds must be used. Kalen and Zenz state that maximum collection efficiency occurs at $u_T/u_S \approx 1.25$, and significant reentrainment occurs at $u_T/u_S \approx 1.36$. For $u_T/u_S = 1.25$, (6.39) becomes:

$$u_T = 16.95 \left[\frac{4g\mu\rho_p}{3\rho_f^2}\right] \frac{(b/D)^{1.2}}{(1 - b/D)}\, D^{0.201} \tag{6.40}$$

Replacing u_T by Q/ab, and rearranging, an equation for D is obtained which may be used to select a cyclone size for a particular configuration, gas flow, and particle and gas properties.

$$D = 0.0502\left[\frac{Q^2\rho_f}{\mu\rho_p}\,\frac{(1-b/D)}{(a/D)(b/D)^{2.2}}\right]^{0.454} \quad (6.41)$$

This may become the beginning of a design procedure such as is outlined below. Since (6.39) does not account for the fact that saltation velocity depends upon particle size, it has been suggested [17] that the constant 2.055 be replaced by an empirical variable factor F, defined as:

$$F = Ga^{0.107} \qquad \text{for } Ga > 1000 \quad (6.42)$$

$$F = 4.406/Ga^{0.107} \qquad \text{otherwise}$$

where Ga is the Galileo Number as defined in Eqn. (4.25)

None of the efficiency models attempts to take into account the role of the spiral dunes. Ranz's loading factor, Eqn. (6.5), shows how to estimate the height and width of the dunes, but in order to do so the value of η must be known. This expression could lead to an estimate of the maximum rate of solids discharge at which the cyclone will choke up.

Electrostatic effects sometimes seem to occur naturally in such a way as to greatly enhance particle collection of solids by space-charge precipitation [26]. This is noted, for instance, when cyclone efficiency appears to be independent of gas flow rate, or when a cyclone operated at lower flow rates gives collection efficiencies higher than at higher flow rates, i.e. the collection efficiency exhibits a minimum value at some intermediate flow rate. Extremely high collection in a transparent cyclone constructed of plastic material has also been observed and is no doubt due to electrostatic charges. Such effects seem to be reproducible under controlled conditions, but are not entirely predictable. The effect of deliberate charging of particles has already been introduced in connection with Eqn. (5.19), and will be mentioned again in terms of the electrocyclone described below.

2. *Comparison an testing of models* Table 1 lists a summary of a number of published results of experimental tests, and analyses of them, which have been carried out on cyclones large enough to be considered representative of industrial sizes i.e. diameter from 2" to 32". (The one group on 2" diameter may be considered to be borderline in this respect.) These may be grouped into the following categories:

A. Solid particles, ambient temperatures, moderate to high grain-loading, moderate velocities; Exponential shaped grade-efficiency curves, generally well-fitted by Leith-Licht model, curve B, Fig. 3.
B. Solid particles, high temperatures and pressures, low grain-loading, high velocities; S-shaped curves, generally well-fitted by Dietz model, curve C, Fig. 3. (Perhaps would be almost as well-fitted by the General Electric back-mix model, curve E, Fig. 3.)
C. Liquid particles, ambient pressure and temperature, low grain-loading, moderate velocities; S-shaped curves, fitted only by a modified Barth model [11], such as illustrated in Fig. 2.

The radial back-mix concept is involved in the best-fit model for both groups of solids, but not for liquids. This would seem to be related primarily to the presence of rebounding and reentrainment, which may be especially prominent under the conditions of group A: lower velocities and higher grain-loading. The Leith-Licht model seems to allow reasonably well for this effect.

The role played by high temperatures and high pressures is not entirely clear, but the Dietz model works quite well for such

TABLE 1 *Summary of experimental cyclone tests*

Material	ρ_p (gm/cm^3)	Type	D (in.)	Gas	T (^{0}C)	P (atm)	Loading (gr/ft^3)	u_T (ft/s)	References
Group A: Solid particles; exponential-shaped η_i vs d_p curves									
Dust	2.0	Stairmand	8	Air	20	1	2.2–4.4	50	[1]
Coke breeze	1.55	Stairmand (near)	4	Air	20	3–55	43–870	80	[18]
	1.6	P & W	12	Air	20	1	0.3	83.5	[28]
Plastic, Clay	---	Tengbergen	11" 18"	Air	20	1	10	—	[29]
Grain Dust	---	Various	32"	Air	20	1	6.8	50	[48]
Group B: Solid particles; S-shaped η_i vs d_p curves									
PFBC Efflux	2.5	Exxon	7"	Flue gas	948	6	4.4	180	[15]
Fly ash	2.3	Stairmand	2"	N_2	20–693	2–25	0.02–4	<16	[20]
PFBC Efflux	2.5	Exxon	7"	Flue gas	838–948	4–7	low	68–180	[26]
PFBC Efflux	1.5	Exxon	6"	Flue gas	635	7	0.2	118	[27]
Group C: Liquid particles; S-shaped η_i vs d_p curves									
Mineral Oil	0.860	Stairmand	12"	Air	20	1	0.02	16–82	[11]
Mineral Oil	0.860	Spec. Exp.	12"	Air	20	1	low	16–82	[22]

conditions, at least for low grain-loadings. There is one anomalous set of data in group B, for the 2" cyclone. The grade efficiency passes through a minimum and increases for smaller particles; perhaps unobserved electrostatic effects were present.

The above comments may be used as a guide in selecting a model to use for design calculations, as will be discussed further below. An additional guide may be found in the work of Masin and Koch [17] who made an extensive study of data for high temperatures, pressures and solids loading, as might be applicable to shale retorts, coal gasifiers, and catalytic crackers. They chose to work with the Leith-Licht model over the Dietz because the applicability of the latter is "limited", and because it requires "significant additional computational effort" [30]. They found that by correcting Eqn. (6.26) on a statistical basis both for the loading effect and for saltation, a greatly improved fit to the data was obtained. The corrected equation reads:

$$\ln\left[(1-\eta_M)(1+\eta_{M_o}[c^A-1])^{1/2}\right] = -2.3\ \mathrm{SFAC}\left[\mathrm{Stk}'_c\ K/\mathrm{LFAC}\right]^{\frac{0.41}{n+1}} \qquad (6.43)$$

In this equation A, as given for (6.38), and LFAC, the polynomial factor for apparent viscosity (6.38a), correct for the loading effect. Saltation is corrected for by SFAC, defined as

$$\mathrm{SFAC} = [(u_T/u_S)/2.5]^{0.41} \qquad \text{for } u_T/u_S < 2.5 \qquad (6.44)$$

$$\mathrm{SFAC} = [(u_T/u_S)/2.5]^{-0.31} \qquad \text{otherwise.}$$

Here u_S is calculated from (6.39) as corrected by (6.39a). The use of this model involves perhaps as much computational effort as any other, but it does have the virtue of being empirically corrected to a lot of actual data, and fitted by statistical methods.

C. *PRESSURE DROP*

A number of methods have been proposed to estimate the total pressure drop in the flow of gas through a conventional tangential inlet cyclone. Unfortunately there has not been a definitive study to determine which is the best to use. The total pressure drop is measured between a point in the inlet duct just ahead of the cyclone body, and a point in the exit duct just beyond the top of the cyclone.

It is generally agreed that pressure losses or gains occur at the following places due to the phenomena listed: (a) losses due to frictional flow in the entrance duct; (b) exit losses due to the sudden expansion of the gas stream from the duct into the main cylindrical body; (c) friction losses against the surfaces bounding the gas inside the cyclone; (d) kinetic energy losses due to turbulence within the cyclone; (e) entrance losses due to the sudden contraction of the gas stream at the entrance to the exit duct; (f) static head loss due to difference in elevation between inlet and outlet duct; (g) recovery of energy in the outlet duct; and (h) losses due to frictional flow through the outlet duct. However, the relative importance of each of these items is given different weight in different methods of calculation. Item (d) is generally considered to represent a major source of loss, and some methods assume all other sources to be insignificant by comparison. Items (a), (f), and (h) will obviously be kept very small if the upstream and downstream pressure taps are taken to be very close to the cyclone body.

Most methods agree, however, in expressing total pressure drop in terms of multiple N_H of "inlet velocity heads." The inlet velocity head is defined as $\rho_f u_T^2/2g_c$. In terms of inches of water the pressure loss is as given by Eqn. (3.58):

$$\Delta P = 0.0030\ \rho_f\ u_T^2 \cdot N_H \tag{6.45}$$

where ρ_f is in lbm/ft^3 and u_T in ft/sec. The problem is then expressed in terms of finding the value of N_H.

It is well-established that grain-loading has a pronounced effect on ΔP, which may decrease by up to 50% of the ΔP where gas only is flowing [6]. Masin and Koch [17] found that this effect is well-described by:

$$\Delta P(\text{at } c) = \Delta P(\text{at } c=0)[1-0.013\ c^{0.5}] \tag{6.46}$$

where c is in gr/ft^3.

The simplest method is probably that due to Shepherd and Lapple which assumes that item (d) is the only one which needs to be taken into account. It assumes that this loss is proportional to the ratio of area of inlet duct to area of outlet duct and gives

$$N_H = 16\ ab/D_e^2 = 16\ K_a K_b/K_e^2 \tag{6.47}$$

It is noteworthy that only three of the seven dimension ratios of the cyclone affect N_H according to (6.47). Casal [31] found that an improved value of N_H could be obtained by modifying (6.47) to read

$$N_H = 11.3\ [K_a K_b/K_e^2]^2 + 3.3 \tag{6.48}$$

More elaborate equations, involving other dimensions of the cyclone, have been proposed by Barth [12], Alexander [8], First [3], ter Linden [32], Miller [33], and Stairmand [34]. Summaries of some of these are given by Strauss [35] without comparing their accuracy. Studies by Leith & Mehta [36], by Gandhi [37], and by Masin and Koch [17] concluded that one might as well use the simple Shepherd and Lapple method or the Casal modification, as any other, although in some instances it would produce large errors. Unfortunately it is not clear under what combinations of cyclone dimensions such large errors may be anticipated.

For design purposes here, N_H will be taken from (6.48) and ΔP corrected for loading by (6.46). The fact that all of the dimensions of the cyclone body itself do not affect the pressure drop so calculated, must be watched carefully. The role of N_H in the optimal design of a cyclone will be discussed further below.

III. DESIGN FOR GAS CLEANING OR PARTICULATE RECOVERY

A. *GENERAL PROCEDURE*

A cyclone may be considered for the purpose either of achieving a desired degree of gas cleaning (as in air pollution control work) or to recover a valuable particulate product (as in industrial processing). In either case there will be a specified set of working conditions to be met, usually including:

Composition, temperature, and pressure of gas;

Size-distribution, grain-loading (c_o), and nature of particulate matter;

Rate of flow of gas stream (Q_o), taken as steady at some average value, but allowing for turn-up or turn-down;

Desired collection performance, specified either in terms of a maximum allowable emission rate (E), or of a minimum desirable collection rate (C) of particulate.

The overall mass efficiency η_M, as computed from Eqns. (3.1) and (3.12), must accord with the value specified for E or C. To check on this will require a grade-efficiency function η_i, as given by one of the models described above.

A general method of design procedure may be outlined as follows, with detailed comments given below:

1. Select a design configuration; see Table 2 (p.309).
2. Select an inlet velocity (u_T) in accord with general rule-of-thumb, or with specific experience.
3. Compute cyclone diameter D from $D = (Q_o/u_T\, K_a\, K_b)^{1/2}$.
4. Compute all other cyclone dimensions from the set of K's for the selected configuration.

5. Determine ΔP from (6.45).

6. Consider whether u_T, D, or ΔP are excessively large. u_T may be checked against the saltation criterion Eqn. (6.40). If any of these are too large, consider using two or more cyclones in parallel. For P identical cyclones in parallel, repeat steps 2, 3 and 5 using Q_o/P in place of Q_o.

7. Select a grade-efficiency model from among those given above, using the general guidelines, or more specific knowledge based upon experience.

8. Using Eqn. (3.12) and the given particle size-distribution, compute η_M and check the predicted performance against the desired E or C.

9. If the desired performance is not attained, repeat the procedure using a larger value of u_T, a larger value of P or both. If the design criterion cannot be met with reasonable values a different type of collector will be needed, perhaps with the cyclone as a pre-collector.

10. Estimate the cost of the design.

B. *DETAILED SUGGESTIONS*

There are a number of specific suggestions and recommendations which may be made for carrying out the steps of the design procedure. These may be detailed as follows:

1. *Standard configurations* There are a number of cyclone configurations which have been proposed and studied sufficiently that they may be regarded as standards. The dimension ratios of these are tabulated in Table 2, along with the calculated values of ℓ (6.1), K (6.25), and N_H (6.48). Selection of one of these designs may be made on the basis of whether high efficiency or general purpose use is desired. These may also be used as standards for comparison with other proposed configurations.

On the basis of the Leith-Licht model, Eqn. (6.26) shows that K is an efficiency parameter. Equation (6.48) gives N_H as the pressure drop parameter for the cyclone configuration. Hence the ratio K/N_H may be regarded as an optimizing parameter. The Table

Table 2 *Cyclone design configurations*

		(Tangential Entry)					
		"High Efficiency"		"General Purpose"		"Experimental"	
Term	Description	Stairmand	Swift	Lapple	Swift	Dingo & Leith	Peterson & Whitby
D	body diameter	1.0	1.0	1.0	1.0	1.0	1.0
a	inlet height	0.5	0.44	0.5	0.5	0.5	0.583
b	inlet width	0.2	0.21	0.25	0.25	0.3	0.208
S	outlet length	0.5	0.5	0.625	0.6	0.558	0.583
D_e	outlet diameter	0.5	0.4	0.5	0.5	0.333	0.5
h	cylinder height	1.5	1.4	2.0	1.75	3.5	1.333
H	overall height	4.0	3.9	4.0	3.75	6.0	3.17
B	dust outlet dia.	0.375	0.4	0.25	0.4	0.375	0.5
ℓ	natural length (6.1)	2.48	2.04	2.30	2.30	1.44	2.32
K	Equation (6.25)	551.3	699.2	402.9	381.8	167.0	324.3
N_H	Equation (6.48)	5.14	7.10	6.16	6.16	23.5	6.00
K/N_H		107.2	98.5	65.4	62.0	7.10	54.1
Surf	Equation (6.49)	3.67	3.57	3.78	3.65	5.64	3.20
K/N_H Surf		29.2	29.6	17.3	17.0	1.26	16.9

Sources: Ref. [1,13,22,23,28]

shows that while the Swift design has the highest efficiency parameter, the Stairmand design ranks higher by this criterion.

Less efficient shapes may be cheaper, however, by virtue of having less total material in them. This would be indicated by their total surface ratio. A parameter reflecting this may be calculated by adding the total surface (neglecting the inlet duct) as follows:

$$\frac{\pi}{4}(D^2 - D_e^2) + \pi D_e S + \pi Dh + \frac{\pi}{2}(D + B)\sqrt{(H - h)^2 + \left(\frac{D-B}{2}\right)^2}$$

This may be put in the form πD^2 (Surf) where

$$\text{Surf} = \frac{1}{4} - \frac{1}{4}\left(\frac{De}{D}\right)^2 + \left(\frac{De}{D}\right)\left(\frac{S}{D}\right) + \frac{h}{D} + \left(\frac{1}{2} + \frac{B}{2D}\right)\sqrt{\left(\frac{H}{D} - \frac{h}{D}\right)^2 + \left(\frac{1}{2} - \frac{B}{2D}\right)^2} \tag{6.49}$$

is another dimensionless configuration parameter. Values for Surf are also tabulated.

An overall configuration optimizing parameter indicating performance (K) divided by operating cost (N_H) and by initial cost (Surf) may be constructed for comparative purposes. Values of K/N_H Surf are in the table and again rank either the Stairmand or the Swift design as first choice.

The following are some practical guidelines to consider in contemplating any proposed cyclone design. They are the results of experience and testing.

1. Dimension $a \leq S$ to prevent short circuiting of inlet dust to outlet.

2. Dimension $b \leqq (D-De)/2$ to avoid too large a ΔP.

3. Dimension $H \geqq (S+\ell)$ to keep the vortex inside the cyclone; there is no advantage in having H much greater than $(S+\ell)$.

4. A frequent constraint on the design is the head-room available. This may place a practical limit on H.

5. The angle of the conical base must be steep enough to provide for ready slippage downward of the collected dust; use approximately 7°-8°.

6. Certain trends have been observed by ter Linden [2]: η_M increases with D/D_e up to about 2.5, then levels; η_M increases with H/D_e up to about 10, then levels; η_M peaks at a value of $S/D_e \approx 1$.

2. *Inlet velocities* Since collection efficiency increases with inlet velocity (up to the saltation limit), it is desirable to use as high a value as possible. Much may depend here upon the specific character of the dust. Rule-of-thumb values are in the range of 50-90 ft/s but some of the tests cited in Table 1 have been run at as high as 150 ft/s.

As an alternative procedure, steps 2 and 3 of the above list may be reversed. The value of D may be determined first from the saltation criterion, Eqn. (6.41), and then u_T computed from $u_T = Q_o/K_aK_bD^2$. This may then be checked against experience and/or rule-of-thumb values. On occasion, especially for dusts of high specific gravity, this has given values of u_T which seem to be quite high.

3. *Pressure drop* Since an advantage of the cyclone is to have low energy consumption, a practical upper limit to ΔP is generally considered to be about 10" H_2O. For u_T = 50-90 ft/s, N_H = 5-8, the value of ΔP for air (ρ_f = 0.075 lbm/ft^3) will range from 3"-14.6" H_2O, according to Eqn. (6.45). Hence it may be desirable to couple a high value of u_T with a design having a low N_H. Recall however that this estimated value of ΔP may be significantly reduced by the grain-loading as shown by Eqn. (6.46).

In some instances of operation under high pressures, a value of ΔP as large as in the range of 50" H_2O may be used. In this case the equation given may not be reliable, and can be replaced by one given by Masin and Koch [17].

4. *Computation of overall efficiency* This computation depends upon the efficiency model selected. As indicated by the

discussion above, this will depend somewhat upon the conditions of the operation as well as upon the desire to avoid excessive computation in preliminary design work. In some cases the choice of model may not affect greatly the computed value of η_M. After reviewing the data summarized in Table 1, the author still feels that the most practical choice may well be the Leith-Licht model in the absence of compelling reasons to the contrary.

Using this model for illustration, the computation may be done as follows: Set-up Eqn. (6.27) for the grade-efficiency of each cyclone, calculating M from (6.28) and N from (6.29) using (6.2) for n. Combine the given particle-size distribution with (6.27) as set up, compute η_M by Eqn. (3.12), and so check the performance requirements as set by E or C.

If the dust happens to have a log-normal size distribution (See Chapter 2.), the calculation of η_M may be done with the aid of a chart prepared by Leith [10], reproduced as Figure 4. For this chart d_g is the MMD (mass median diameter), and σ_g the geometric standard deviation of the dust size distribution. From (6.30) the cut diameter d_{p_c} is obtained, and N is 1/(n+1). This chart represents the result of the numerical integration of (3.12) with η_i represented by (6.27) and G by Eqn. (2.37). Refer also to Eqn. (3.18). The chart may also be used where d_{p_c} and N have been obtained from an experimental η_i curve.

Occasionally it may be desired to design a cyclone to have a specified cut diameter. In this case (6.30) may be used to determine M, using an estimated value of n, say 0.7. Selecting design configuration (K), the value of Q/D^3 may be calculated from (6.28). This may be then combined with (6.41) to find both Q and D individually. The estimated value of n must then be checked against the D value through (6.1). If necessary the calculations may be repeated using the new value of n. This procedure is discussed further below in connection with the concept of cyclones in series.

Koch and Licht [38] have shown how to apply the design procedure in a quick and preliminary manner using one of the three

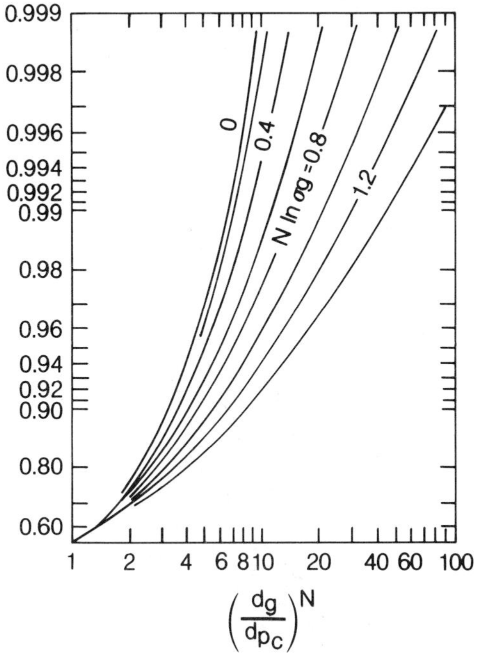

FIG. 4 *Overall efficiency for log-normally distributed particulate. Reproduced from "Handbook of Environmental Engineering - Vol. I, p.289. Copyright by Humana Press.*

standard dusts described by Stairmand (see Chapter 3, Table 1). A series of charts are presented to correct for the effect of temperature and of particle density, to select a cyclone diameter and inlet velocity, and to estimate the overall collection efficiency. All of these are in conformity with the Leith-Licht model equation given above.

The interplay of the effects of all of the operating variables and design parameters upon the cyclone performance is shown by the system analysis chart, Figure 5.

5. *Cost estimation* The purchased cost of a cyclone unit is estimated on the basis of its inlet area $A_i = ab = K_a K_b D^2$. The method recommended, quoted from Vatavuk [39], is carried out in two steps. First, the total cost of cyclone, scroll, and dust hopper is found from:

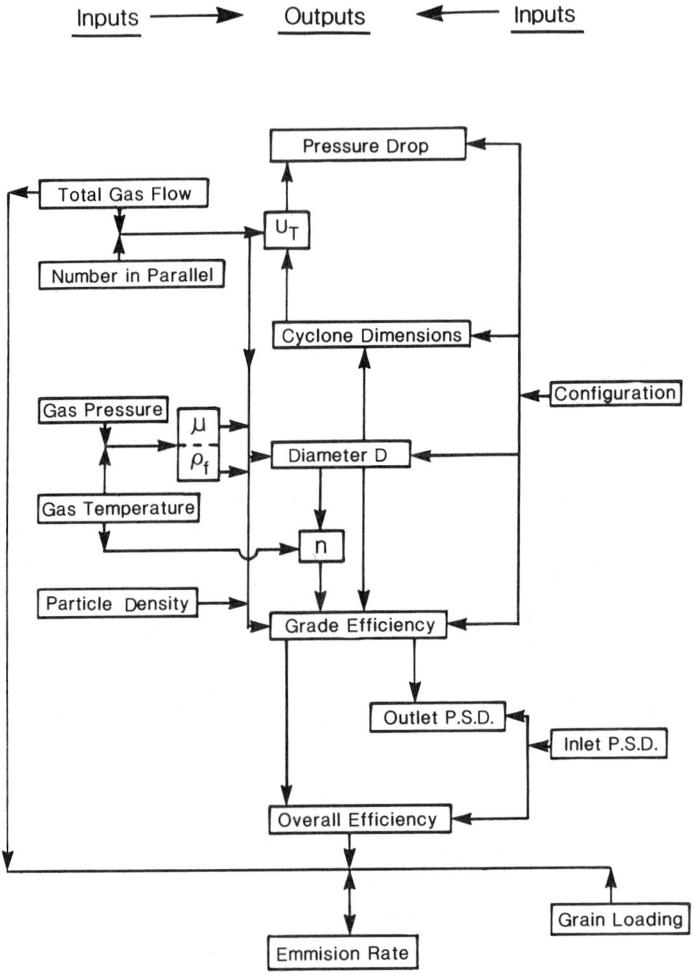

FIG. 5 *System analysis of cyclone design.*

$$C_{C+S+H} = j(t) + k(t)A + l(t)A^2 \tag{6.50}$$

where the coefficients depend upon the plate thickness and material of construction (carbon or stainless steel) according to the table, and are valid for values of A_i from 1 to 12 ft^2.

	j(t)		k(t)		l(t)	
t	CS	SS	CS	SS	CS	SS
3/16 in.	2230	3390	1210	3710	5.67	17.8
10 gauge	2030	3000	933	2380	4.15	16.6
14 gauge	1510	2000	765	1640	3.04	12.6

Second, a set of equations gives costs for supports, which apply to all plate thicknesses and materials:

$$C_{supp} = 788 + 205\,A \qquad 1 \leq A \leq 2 \tag{6.51}$$
$$= 1420 + 198\,A \qquad 2 \leq A \leq 6$$
$$= 2680 + 166\,A \qquad 6 \leq A \leq 14$$

The sum of the two items is the purchased cost for one cyclone. Where multiple units, either in series or parallel, are used add 20% for ductwork and structure to the total cost of all units. The data are based upon 1981 dollars. See Chapter 3 for further details on completing the total cost estimation for the installation.

6. *Optimized design* The interaction among all of the variables in the cyclone system, as shown in Fig. 5, is such that an unconstrained definition of an optimum design is virtually impossible. A design objective must be stated, e.g. to achieve a required collection efficiency with a minimum pressure drop, or to maximize collection efficiency without exceeding a pressure drop limit, or perhaps to achieve a specified recovery at minimum cost. No such objective can be met by a consideration of the cyclone configuration alone (set of K values); the operating conditions must also be involved.

With a set of model equations, theoretical design optimization procedures can be developed [22,36]. However, they have not as yet been verified experimentally. For example, Dirgo and Leith [22] developed four new designs each of which was supposed to be more efficient than the Stairmand (Table 2) at the same pressure drop. Only one of them actually was, on a test of collection of an oil mist. This configuration is also listed as "experimental" in Table 2 for comparison.

A useful test of the "benefit/cost" ratio of any design is to compute the value of $-\ln(1-\eta_M)/\Delta P$, called by Cooper the "effectiveness factor". For some of the designs this has been analyzed as to its sensitivity and elasticity with regard to the variables of the system as they are described by the Leith-Licht model [40]. The logarithmic form of $(1-\eta_M)$ has been used because of the exponential nature of (6.27); this causes the effectiveness factor to be expressed in a simple power law formulation. This form could also be applied to the Dietz model.

C. *EXAMPLE OF DESIGN*

The cement dust from a Portland cement kiln is log-normally distributed (as shown in Fig. 8) with an MMD of 12 microns and $\sigma_g = 3.08$, having a density of 1.5 gm/cm^3. Operating conditions are: air temperature 250°F, air pressure 1 atm., feed rate to kiln 5 tons/hour, dust rate 230 lbm per ton feed, air rate from kiln 159,600 acfm per ton feed. The Federal emission regulation (for new cement plants) is 0.30 lbm dust per ton feed. Consider whether a cyclone installation might be used to meet this emission limitation.

To meet the emission limitation will require the collection of (230-0.30) x 5 = 1148.5 lbm/hr. with an overall collection efficiency of η_M = (230-0.30)/230 x 100 = 99.78%. While it is unlikely that a cyclone could meet this efficiency, it may be worth while to consider it as a primary cleaner to be followed by a baghouse filter or an electrostatic precipitator. The grain loading of 230 x 7000/159600 = 10.1 gr/ft^3, is rather heavy to be a suitable feed for either of these collectors directly.

For purposes of illustration select the design configuration having the largest K, which is the Swift design, Table 2. Using Eqn. (6.41) calculate D for a single cyclone:

$$D = 0.0502 \left[\frac{221.7 \times (0.0561)^2 (1-0.21)}{1.512\text{x}10^{-5} \times 1.5 \times 62.4 \times 0.44 (0.21)^{2.2}}\right]^{0.454}$$

$$= \underline{5.19} \text{ ft} = \underline{1.582} \text{ m}$$

in which Q = 159600 x 5/3600 = 221.7 acfs; K_a = a/D = 0.44; K_b = b/D = 0.21. At 121 C: ρ_f = 0.0561 lbm/ft^3, μ = 2.25 x 10^{-4} x 0.0672 = 1.512 x 10^{-5} lbm/ft.s. For this design at 250 F (121 C) n = 0.685, Eqn. (6.2); a = 0.44 x 5.19 = 2.28 ft.; b = 0.21 x 5.19 = 1.09 ft; and u_T = 221.7/2.28 x 1.09 = 89.2 ft/s. This is a satisfactory value of inlet velocity.

Now set up the grade efficiency Eqn. (6.26) in which K = 699.2 (Table 2) and for which (6.24) gives:

$$\text{Stk}'_{c_i} = 1.685 \times \frac{221.7}{(5.19)^3} \frac{1.5\,(d_{p_i} \times 10^{-4})^2}{18 \times 2.25 \times 10^{-4}} = 9.89 \times 10^{-6}\, d^2_{p_i}$$

with d_p in microns. Then (6.26) reads

$$\eta_i = 1\text{-exp} -2 \left[9.89 \times 10^{-6} \times 699.2\, d^2_{p_i}\right]^{0.297}$$

or in (6.27) M = 0.4565, N = 0.594. The "cut" diameter (6.30) will be 2.02 microns. Grade efficiencies may then be calculated and are tabulated as given in Table 3 for P = 1, and plotted in Figure 6.

Table 3 *Results of design example*

Quantity		Source or Equation	Number of Cyclones in Parallel P=1	P=4	P-20
Q	acfs		221.7	55.43	11.09
D	ft.	(6.41)	5.19	2.77	1.33
u_T	ft/s		89.2	78.3	67.9
n	—	(6.2)	0.685	0.618	0.548
ΔP	"H_2O	(6.46)	9.1	7.0	5.3
$KStk'_{c_i}/d_p^2$	μm^{-2}	(6.24,6.25)	$6.92x10^{-3}$	$1.09x10^{-2}$	$1.89x10^{-2}$
M	*	(6.28)	0.4565	0.4951	0.5550
N	-	(6.29)	0.594	0.618	0.646
d_{p_c}	μm	(6.30)	2.02	1.72	1.41
η_M	%	Fig. 7	81%	84%	87%
$\frac{-\ln(1-\eta_M)}{\Delta P}$	"H_2O^{-1}	-	0.183	0.262	0.385

d_p - μm		(6.27)	Grade Efficiency - %	
1		36.7	39.1	42.6
3		58.4	62.4	67.7
5		69.5	73.8	79.2
10		83.3	87.2	91.4
20		93.3	95.7	97.8
30		96.8	98.3	99.3
40		98.3	99.2	99.8
50		99.1	99.6	99.9
60		99.4	99.8	99.9
η_M	Fig. 6	81.7	84.3	87.7

Table 3 (continued)

Results of Design Example-Outlet Particle Size Distribution

d_p-μm	G_{in}	P=1 "Area"	P=1 G_{out}	P=4 "Area"	P=4 G_{out}	P=20 "Area"	P=20 G_{out}
2.8	0.10	0.0456	0.300	0.490	0.325	0.0534	0.380
4.7	0.20	0.1078	0.505	0.1154	0.539	0.1248	0.613
6.7	0.30	0.1794	0.661	0.1909	0.695	0.1060	0.767
9.1	0.40	0.2579	0.778	0.2736	0.806	0.2934	0.870
12.0	0.50	0.3419	0.866	0.3615	0.883	0.3854	0.935
16.2	0.60	0.4306	0.928	0.4536	0.933	0.4809	0.972
22.0	0.70	0.5233	0.968	0.5462	0.980	0.5787	0.989
31.5	0.80	0.6195	0.989	0.6437	0.996	0.6777	0.998
52	0.90	0.7177	0.998	0.7431	0.999	0.7774	0.999
	1	0.8174	1	0.8431	1	0.8774	1
Pen.		0.183		0.157		0.123	

$$G_{out} = \frac{1}{P}\ G_{in} - \int_0^{G_{in}} \eta_i\, dG_{in} \qquad \text{Eqn. (3.16)}$$

$$\text{"Area"} = \int_0^{G_{in}} \eta_i\, dG \quad \text{- refers to Fig. 7}$$

$$\eta_M = \int_0^1 \eta_i\, dG_{in}$$

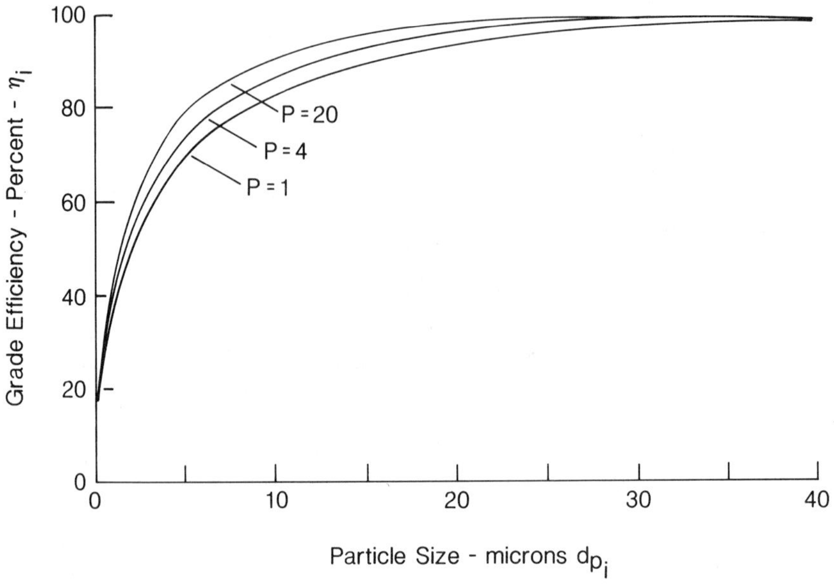

FIG. 6 *Grade efficiency for design example.*

The overall efficiency may be determined either by the graphical integration method described in Chapter 3, or by the use of Fig. 4. The graphical construction is shown in Figure 7, and gives $\eta_M = 81.7\%$. In using Fig. 4, the value of the abscissa is $(12/2.02)^{0.594} = 2.9$ and $N \ln \sigma_g = 0.594 \ln 3.08 = 0.67$. This chart gives approximately 81% for the overall efficiency.

From (6.45) with $N_H = 7.10$ (Table 2), the gas flow pressure drop is found to be

$$\Delta P = 0.0030 \times 0.0561\ (89.2)^2 \times 7.0 = 9.5\text{"}\ H_2O$$

The correction for loading, Eqn. (6.46), is 0.96, giving $\Delta P = 9.1$" H_2O, a little on the high side. Finally the "effectiveness factor" is calculated as $-\ln(1-\eta_M)/\Delta P$ and listed in Table 3.

Because ΔP is high and η_M is low, consider the use of cyclones in parallel. For illustration, the calculations are repeated with P = 4, and with P = 20. This makes Q_4 = 221.7/4 = 55.43 cfs, and Q_{20} = 221.7/20 = 11.09 cfs respectively. The results are tabulated in Table 3, along with those for P = 1 by way of comparison. Fig. 6 shows the grade efficiency curves, and Fig. 7 graphical determinations of η_M.

It is clear that the cyclone system can at best serve as a precleaner. The use of 20 small units in parallel will be desirable primarily from the point of view of the reduced pressure drop. For this case then determines what the feed to the secondary cleaner will be. The grain-loading will be reduced to 10.1 x (1-0.87) = 1.31 gr/acf.

The particle size distribution may be determined by the application of Eqn. (3.16). This requires the additional graphical integrations on Fig. 7. The inlet and outlet particle size distributions are shown in Fig. 8, and are tabulated in Table 2.

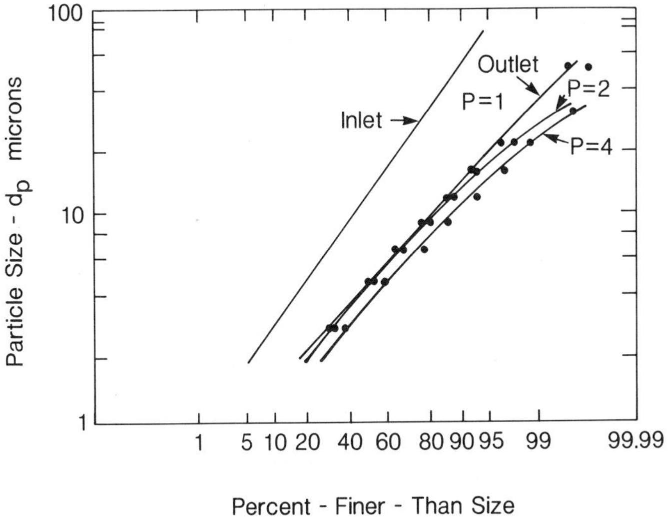

FIG. 7 *Graphical integration for overall efficiency.*

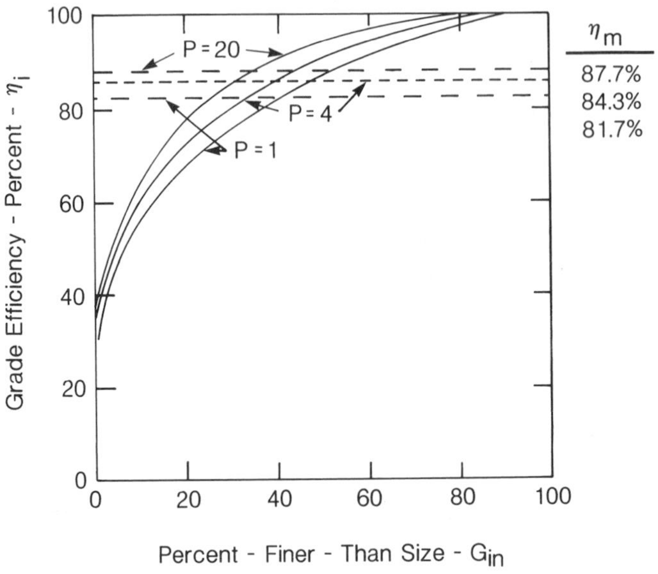

FIG. 8 *Particle size distributions for inlet and outlet of cyclone.*

Finally the dimension of each cyclone in the set of 20 are calculated from the configuration ratios given in Table 2 for the Swift "high efficiency" design.

D	= 1.33 ft	= 406 mm
a	= 0.44 x 1.33 = 0.58 ft	= 178 mm
b	= 0.21 x 1.33 = 0.27 ft	= 83 mm
S	= 0.5 x 1.33 = 0.67 ft	= 203 mm
D_e	= 0.4 x 1.33 = 0.53 ft	= 162 mm
h	= 1.4 x 1.33 = 1.82 ft	= 555 mm
H	= 3.9 x 1.33 = 5.19 ft	= 1581 mm
B	= 0.4 x 1.33 = 0.53 ft	= 162 mm

This design problem will be treated further in Chapter 7 when the design of an electrostatic precipitator as the secondary cleaner will be discussed.

IV. SPECIAL DESIGNS AND SYSTEMS

There are significant limitations upon the amount of particulate collection which may be obtained with a single reverse-flow cyclone, as has been pointed out. More complex designs and systems may be considered to achieve better performance. Among these may be listed: (a) two or more cyclones in parallel; (b) two or more cyclones in series; (c) recycling portion of emission stream; (d) skimming off portion of uncollected dust in emission stream; (e) base purge bleeding; (f) use of secondary air flow, and (g) use of electrostatic cyclones. Each of these will be discussed briefly.

As the above design example indicates, the use of P cyclones in parallel produces improved performance, and it is a customary practice. Grade efficiencies are increased and pressure drop is decreased when they are designed in the way described. The total cost is increased however, partly because of an increase in the total material and partly because of the increased complexity of duct work and installation. Furthermore, care must be taken to assure that each cyclone in the set is operating in the same manner as all the others. If one or more has a different pressure drop (due to wear, or partial blockage etc.) then they no longer operate truly in parallel and the flow rates are not the same in each. There may even be secondary circulations if all the cyclones are connected to a common dust hopper as in a multiclone. These effects can seriously affect the overall performance adversely.

Placing two or three cyclones in series downstream, so that the emission from one becomes the feed to the next, is also practiced. The principal reason for this practice is usually to recover separately two or more different size fractions of the product. The design may be based upon the alternate procedure, mentioned above, of specifying a desired cut diameter for each fraction. In this case the downstream cyclone would probably be of a more efficient configuration and/or smaller size than the upstream one. This is intended to produce increased recovery of the finer sized particles which have escaped the first unit. It is generally not possible to obtain a real sharp separation of particle sizes in the products. Note that since each unit must handle the total flow the overall pressure drop is greatly increased.

Recycling a portion of the outlet stream back into the inlet of a single unit (or set of parallel units) may also be considered as a method of increasing overall efficiency. Trasi and Licht [41] have shown that recycling a fraction r of the product gas flow will lead to a relative increase in grade efficiency of

$$\frac{E_i - \eta_i}{\eta_i} = \frac{1 + r}{1 + r\eta_i} - 1 \tag{6.52}$$

where E_i is defined as the overall grade efficiency of the recycle system. This relative increase is larger for smaller particle sizes and, for instance, will approach 50% as $\eta_i \to 0$, for r = 0.50. Unfortunately it is accompanied by a considerable increase in power consumption. For an example situation in collection of a Stairmand Fine Dust (Chapter 3), using the Leith-Licht model they found that as r increased from 0 to 1 the penetration decreased by 30% but power consumption more than doubled. However, installation of a recycle duct (and larger fan) might be a relatively inexpensive way to upgrade an existing system in some circumstances.

Another recycling design involves arranging for that portion of exit gas near the inside wall of the outlet duct to be removed ("skimmed off") through an annular scroll around the top of D_e. This is thought to remove the majority of the uncollected particles which are presumed to be exiting near the outlet tube walls. The skimmed-off stream is recycled into the inlet duct of the cyclone. Skimming of 2-10% has produced up to 30% reduction in emissions loading [42]. No theoretical analyses have been published. It would seem that the action would be like two units in series, with the added function of a very high inlet velocity due to the added power in the circulated stream.

Bleeding a small portion of the total gas flow (say 15%) from the dust outlet hopper (called "base purge") is an alternative flow scheme which has been found effective in enhancing cyclone performance. This counteracts the downward swirling flow of gas which might otherwise re-arise from the hopper carrying reentrained dust up into the outlet duct, as mentioned under the description of flow patterns above. Merely recycling this bleed back to the inlet accomplishes little [42]. Instead, the bled stream may be passed through a small auxiliary high-efficiency cleaner (say a bag filter) before rejoining the outlet gas. In this way, bleeds of 10 - 15% have been reported to reduce emissions by 30 - 40% at a cost of only one-third to one-half of that of a filter system to clean the entire flow [43].

The concept of secondary air flow is embodied in what has been variously called the "rotary flow," or "straight-through with reverse flow" cyclone [44, 45]. It consists of a cylindrical body into which two separate streams of gas are introduced. The primary flow, which is the gas to be cleaned, enters at the bottom either tangentially or through a central set of turning vanes. A secondary flow, which may be either a fresh stream or a recirculated portion of the emission gas, is introduced tangentially at the top. As particulates in the primary flow are thrown outwards toward the wall they are swept downward to the collection hopper by the secondary flow. The particles are dropped into the hopper when the secondary flow reverses and joins the primary flow. Clean gas is emitted through a central-core at the top. This design is supposed to give a much increased grade-efficiency by preventing reentrainment and offsetting the saltation effect, with cut diameters in the range of 0.5 μm. Ciliberti and Lancaster [46] studied the performance experimentally and developed a model which gives quite different results, however. They found that the grade-efficiency drops to zero at a minimum particle size in the range of 1-3 microns, with cut diameters greater than 1 micron.

Electrostatically enhanced cyclone separators have been studied both experimentally and theoretically, with very encouraging results [26, 42, 47]. But the proposed technology does not seem to have been implemented to any extent. The concept is to place an axisymmetric high-voltage electrode within the cyclone, either in the form of a rod along the centerline or by using the outlet duct itself as such. This will set up a radial potential gradient toward the grounded outer shell. Charged particles would thus be driven to the wall by electrostatic force in addition to the inertial force. This should provide for a much better collection of finer particles. An elementary illustration of a model for such a process has been presented as Eqns. (5.23, 5.24). A more complete theoretical analysis by Dietz (47) indicates that a great reduction in particle cut-size could be obtained, using a potential gradient of say 5000 V/cm. The model also indicates that a large electroyclone could be as effective as a small one in collecting fine particles. Some of the principles involved here, regarding particle charging and potential gradient, are discussed in detail in Chapter 7 dealing with electrostatic precipitators.

REFERENCES

1. C.J. Stairmand, Trans. Inst. Chem. Engrs., 29: 356 (1951).
2. A.J. ter Linden, Proc. Inst. Mech. Engin., 160: 233 (1949).
3. M.W. First, Amer. Soc. Mech. Engrs., Paper 49-A-127 (1949).
4. W. E. Ranz, Aerosol Sci. Tech., 4: 417 (1985).
5. P. Swift, Filt. and Sep., Jan/Feb: 24 (1986).
6. S. Yuu, T. Jotaki, Y. Tomita, K. Yoshida, Chem. Eng. Sci., 33: 1573 (1978).
7. "Cyclone Separators", Manual on Disposal of Refinery Wastes, Volume on Atmospheric Emission, API Publication 931, Chap. 11 (May 1975).

8. R. McK. Alexander, Proc. Austral. Inst. of Min. and Met. (N.S.), 152/3: 202 (1949).

9. J. Abrahamson, C.G. Martin, & K.K. Wong, Trans. I. Chem. E., 56: 168 (1978).

10. T.V. Tran, Ph.D Thesis, University of Minnesota (1981).

11. J. Dirgo, and D. Leith, Aerosol Sci. Tech., 4: 401 (1985).

12. W. Barth, Brennstoff-Warme-Kraft, 8: 1 (1956).

13. C.E. Lapple, in Air Pollution Engineering Manual, USEPA AP-40, 2nd ed., 1973.

14. D. Leith and W. Licht, A.I.Ch.E. Sympos. Ser., 68: No. 126, 196 (1972).

15. P.W. Dietz, A.I.Ch.E. J., 27: 888 (1981).

16. W.A. Baxter in Air Pollution-Vol. IV., 3rd Ed., A. Stern ed., Chap. 3, Academic Press, New York, 1977.

17. J.G. Masin & W.H. Koch, Environ. Prog., 5: 116 (1986).

18. T.M. Knowlton and D.B. Bachovchin, C.E.P. Tech. Man.

19. K.W. Lee, J.A. Gieseke, W.H. Piispanen, Atmos. Environ., 19: 847 (1985).

20. R. Parker, R. Jain, S. Calvert, D. Drehmel, & J. Abbott, Envir. Sci. Tech., 15: 451 (1981).

21. B.E. Saltzman, Am. Ind. Hyg. Assoc. J., 45: 671 (1984).

22. J. Dirgo, and D. Leith, Filt. and Sep., Mar/Apr.: 119 (1985).

23. P. Swift, Filt. and Sep., Jan/Feb.: 24 (1986).

24. F.A. Zenz, Ind. Eng. Chem. Fund., 3: 65 (1964).

25. B. Kalen and F.A. Zenz, A.I.Ch.E. Sympos. Ser., 70: No. 137, 388 (1974).

26. General Electric Co., FE-2357-70, Vol. A, PFB Coal Fired Combined Cycle Dev. Prog., US DOE Contract No. DE-AC21-76ET10377 (1981).

27. M. Ernst,R.C. Hoke, V.J. Siminski, J.D. McCain, R. Parker, D.C. Drehmel, Ind. Eng. Chem. Proc. Des. Dev., 21: 158 (1982).

28. C.M. Peterson and K.T. Whitby A.S.H.R.A.E.J., 42: (1965).

29. H.J. van Ebbenhorst Tengbergen, Staub, 25: 44 (1965).

30. W. Koch, Personal communcation, Jan. 27, 1986.

31. J. Casal and J.M. Martinez-Benet, Chem. Eng. 90: 99 (Jan. 24, 1983)

32. A.J. ter Linden, Tonindustrie-Zeitung, 22 (iii): 49 (1953).

33. F. Miller and M. Lissman, Amer. Soc. Mech. Engrs., Paper, New York (Dec. 1940).

34. C.J. Stairmand, Engineering, 409 (Oct. 1949).

35. W. Strauss, Industrial Gas Cleaning, 2nd ed., Pergamon Press, Oxford, (1975).

36. D. Leith and D. Mehta, Atmos. Eviron., 7: 527 (1973).

37. M. Gandhi, M.S. Project in Chem. Eng., University of Cincinnati, (1972)

38. W. Koch and W. Licht, Chem. Eng., 84: 80 (Nov. 4, 1977).

39. W.M. Vatavuk, Handbook of Air Pollution Technology, S. Calvert and H.M. Englund eds, Chap 14, Wiley Interscience, NY, 1984.

40. D.W. Cooper, Atmos. Environ., 17: 485 (1983).

41. P.R. Trasi and W. Licht, Ind. Eng. Chem. Proc. Des. Dev., 23: 479 (1984).

42. General Electric Co., FE-2357-90, Vol. B., Coal Fired Combined Cycle Dev. Prog., US DOE Contract No. EX-76-C-01-2357 (1980).

43. P.W. Sage and M.A. Wright, Filt. and Sep., Jan/Feb.: 32 (1986).

44. W. Strauss, Industrial Gas Cleaning, 2nd ed., pp. 263-4, Pergamon Press, Oxford, 1975.

45. K.R. Schmidt, Staub, 23: 491 (1963).

46. D.F. Ciliberti and B.W. Lancaster, A.I.Ch.E., 22: 394, 1150 (1976).

47. P.W. Dietz, Powder Tech., 31: 221 (1982).

48. R.V. Avant Jr., C.B. Parnell Jr., and J.W. Sorenson Jr. Paper No. 76-3543 Amer. Soc. Agric. Eng., (Dec. 1976).

PROBLEMS

1. (a) A certain cyclone installation is collecting particles of sp.gr. = 2.5 using an inlet velocity of 50 ft/s. What inlet velocity would be required to collect particles of sp.gr. = 1.5 with the same grade-efficiency ? How will the pressure drop compare with the original value ?

(b) The cut-diameter for a Swift high-efficiency design cyclone operating under a certain set of conditions is 2.0 μm and the pressure drop is 3.0" H_2O. What would be the cut-diameter and the pressure drop for a Stairmand design of the same diameter D, operating at the same flow rate, temperature, grain-loading, etc. ?

2. A cyclone designed to operate at 20 C with a flow rate of 10,800 std. cu.ft./min of air, collecting solid particles of 1.5 gm/cm^3 density, has a cut-diamter of 1.96 µm. Estimate the collection grade-efficiency of particles 1.96 µm if this same cyclone were operated at 200 C at a flow rate of 5000 scfm, collecting the same material. The cyclone is of high-efficiency Stairmand configuration and is 5 ft in body diameter.

3. Refer to Example 6 in Chapter 3, which illustrates a preliminary study survey of a gas cleaning and recovery problem. The study indicates that a system of two cyclones in series might be satisfactory. Design such a system, specifying cyclone dimensions, velocities, pressure drops, power requirements and costs. Also determine the emission rate and particle size distribution in the outlet stream.

4. Refer to the design example worked out in the text of this Chapter. For the case of the single cyclone, as worked out there, estimate the dimensions and carrying capacity of the spiral dune.

5. A system of four Stairmand high-efficiency cyclones in parallel, each 1250 mm in diameter, was proposed to collect entrained dust from a fluidized bed combustor under the following design conditions:

Gas flow	41,035 am^3/hr
Gas temperature	250 C
Particle density	1.6 gm/cm^3
Solids feed rate	220 kg/hr
Pressure drop limit	140 mm H_2O
Required efficiency	94 %

Particle size distribution:

Range - μm	Wt % in range
0 - 5	18.0
5 - 10	16.0
10 - 15	8.0
15 - 20	8.0
20 - 30	6.0
30 - 40	4.0
+ 40	40.0
Total	100.0

Determine whether the proposed design seems reasonable and is likely to meet the requirements. If it is not satisfactory, recommend changes which should be made in the design.

6. A certain dust has a particle size distribution given by two straight lines on a log-probability plot, which intersect at 5.5 μm and 20 %-less-than by weight. Above this point the one line passes through 10 μm and 50%; below, the other line passes through 2 μm and 10%. The particles have a sp.gr. = 2.5. Design a cyclone collecting system to recover at least 85% of this dust from a stream of air at 500 F, 1 atm, flowing at 8060 acfm with a grain-loading of 1.45 gr/acf. Specify number and arrangement of cyclones, their dimensions, pressure drop and power requirements.

7
Electrostatic Precipitation

I. THE ELECTROSTATIC PRECIPITATION SYSTEM

An electrostatic precipitator collects particles by means of an electrostatic force. The aerosol particles are first given a charge and then passed through an electrostatic field. The collecting force on a particle is the product of the charge (q) and the potential gradient (E) of the field: $F_E = qE$. (Refer to Eqns. 4.58-4.60). By using a very large value of E (typically, of the order of 2000 V/cm) it is possible to establish a fairly large force even upon a very small particle. The collection efficiency for very small particles may thus be made very high, and the principle finds its greatest application in the collection of them.

To implement the concept it is necessary to provide: (a) a method of giving a charge to the particles flowing in the gas stream; (b) establishment of an electrostatic field through which the gas stream may flow at right angles; (c) a method of collecting the particles after they have migrated across the stream; this involves removing the charge once the particle is essentially out of the stream.

These steps are accomplished in a device consisting essentially of two electrodes with space between them for the gas to flow. One electrode is a high-potential wire or cable, freely suspended, called the corona-discharge or "active" electrode. The other is a grounded extended surface called the collecting or "passive" electrode. One of two geometrical arrangements is usually used: the collecting electrode is a vertical hollow cylinder with the wire suspended coaxially within it; or, the collecting electrodes are vertical parallel flat plates with several wires suspended midway between the plates at uniform intervals of spacing. The device is one of the "containment zone" or "surface" collector types as described in Chapter 3.

When a sufficiently great potential difference is set up between the electrodes, a corona discharge forms about the wires. In this corona, molecules of gas (principally O_2) become ionized with charges of the same sign as the wire. They migrate across the space between the electrodes under the influence of the applied electrostatic field. In doing so they collide with and attach themselves to the aerosol particles which thus acquire charge and also migrate toward the grounded surface electrode. When the particles strike the grounded electrode they lose charge and adhere to this surface. The layer of solid particles which thus accumulates is removed by vibrating (rapping) or flushing the electrode, and drops or slides down into a collecting hopper. Thus steps (a), (b) and (c) are completed.

If the potential of the wire is made too high, there will be a breakdown and sparking over the gas stream to the grounded electrode. This, of course, destroys the collecting field. However, since it is desirable to operate as close to the limiting potential as possible, sparking is permitted to occur in controlled fashion in industrial operations. The potential of the corona wire may be either positive or negative, but in industrial cleaners it is usually made negative because a higher potential can be obtained before sparking occurs. The sparking may cause the formation of undesirable gaseous products such as ozone or oxides of nitrogen.

There is a modification of the basic design which is known as the two-stage precipitator. In this the corona-discharge particle charging is carried out in a first stage or charging zone, with the particle collection taking place in a secondary zone in which there is merely a high voltage non-discharging electrode to maintain the field.

Since the collecting surface of the flat plates may amount to as much as several hundred square feet per 1000 acfm of gas treated, the total plate surface required may be enormous. This requires that the precipitator be built up of many individual plates each of which is typically from 3 - 10 ft wide by 30 - 50 ft high. These are arranged in parallel rows bounding duct spaces between the rows of from 4" - 15" in width, in order that the gas velocity through the ducts be kept to within 4 - 8 ft/s. As many plates are placed edge to edge in a horizontal row as necessary to obtain the required total length needed to provide all of the collecting surface. The electrical supply to the wires hanging between the rows of plates must also be compartmentalized into sections. There may be up to eight sections in the total length, each section covering a number of the parallel ducts. Because of this compartmentalized construction, conditions may not be the same in all portions of the unit. This requires a special approach to the design calculations which is discussed below.

Complete consideration of the design and performance of electrostatic precipitators must of course include the purely electrical aspects of it. These involve the phenomena of corona discharge and ion-generation as well as the relationships between voltage applied, current, and field-strength produced. Since the present discussion is devoted primarily to the particle collection process itself, details of the electrical aspects will not be presented. The reader is referred to the excellent reviews by White [1], Robinson [2], or Oglesby [3] for this type of information.

The complete electrostatic precipitator system is undoubtedly one of the more complex types of collectors. There are a number of sub-systems within it, and interaction between the electrical

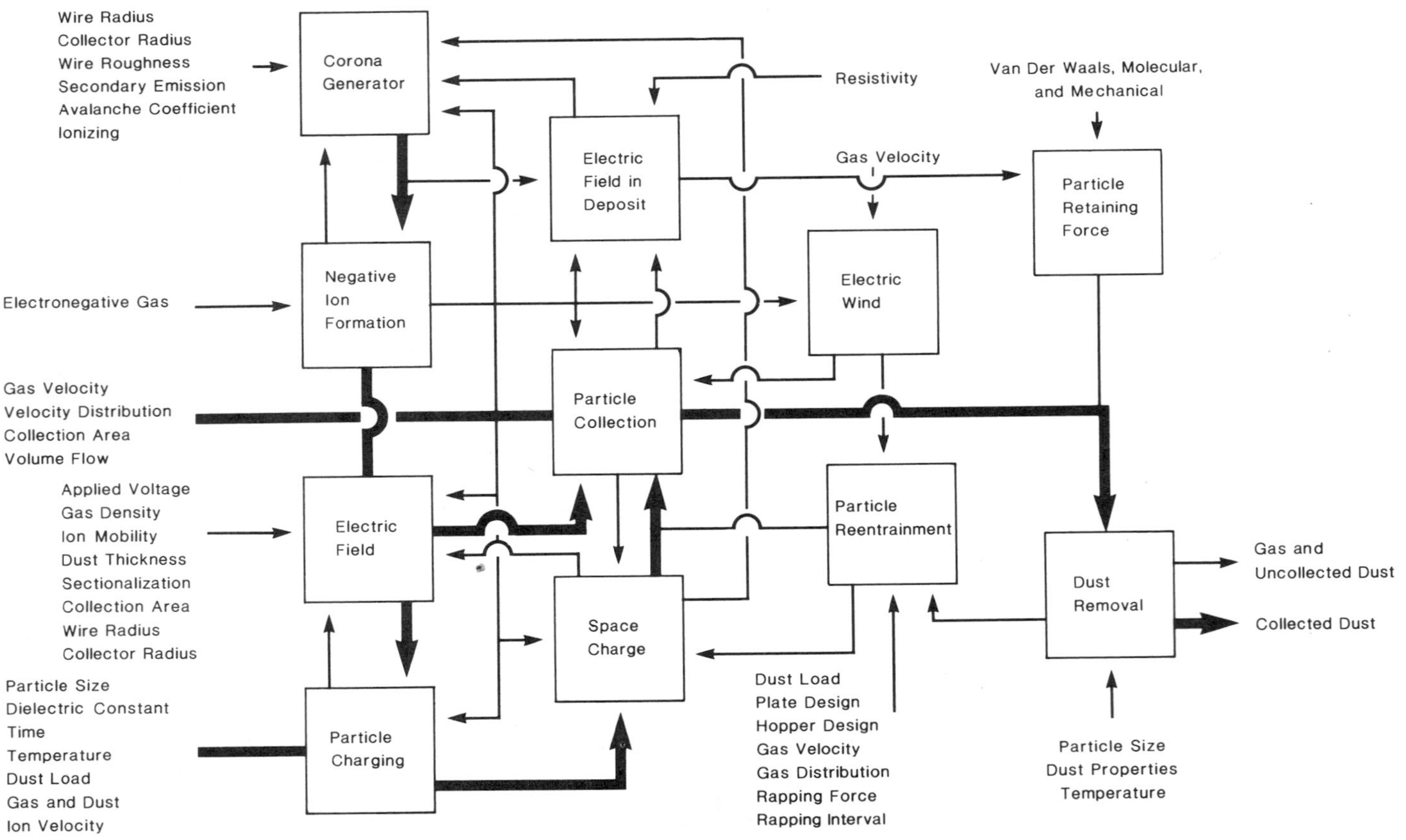

FIG. 1 *Electrostatic precipitator system model [6]. Reproduced by permission from Southern Research Institute.*

and particle aspects of it is complex. Figure 1 modified from Nichols and Oglesby [4], is an attempt to show the entire system and the interrelationships within it. The heavy lines are intended to link the principal steps involved in particle collection. The lighter lines represent the subsidiary interactions which also influence the process.

II. MODELLING THE SYSTEM

A. *GENERAL ASPECTS*

Modelling of such a complex system in its entirety will obviously be very difficult. It can only be accomplished by an overall computer program which incorporates the knowledge of all facets of the system as it has been developed by research. Such a model has been developed for the USEPA by the Southern Research Institute [5,6]. It is founded upon their exhaustive study ("Precipitator Handbook") of 1970 [3,7,8] and subsequent developments.

No attempt will be made to present this complete model here. Rather an elementary grade-efficiency equation will be developed first and then a series of important corrections to it will be described. The use of this for preliminary design calculations will be illustrated. Some of the important considerations to be taken into account are next discussed briefly.

Experimental determinations of actual grade-efficiency relationships are shown in Figs. 2, 3, and 4 [9,10]. The general shape of these curves is very significant. They all exhibit a point of minimum collection efficiency at a particle size somewhere in the range of 0.1 - 1.0 μm. The model must, of course, account for this behavior.

The gas flow through the precipitator is rather unimpeded, and the pressure drop is relatively small (typically less than 0.5" H_2O). Under usual conditions, in a parallel plate device, Re will be of the order of 10^4. A high degree of turbulence in the flow is therefore common in large scale units. Transport of fine

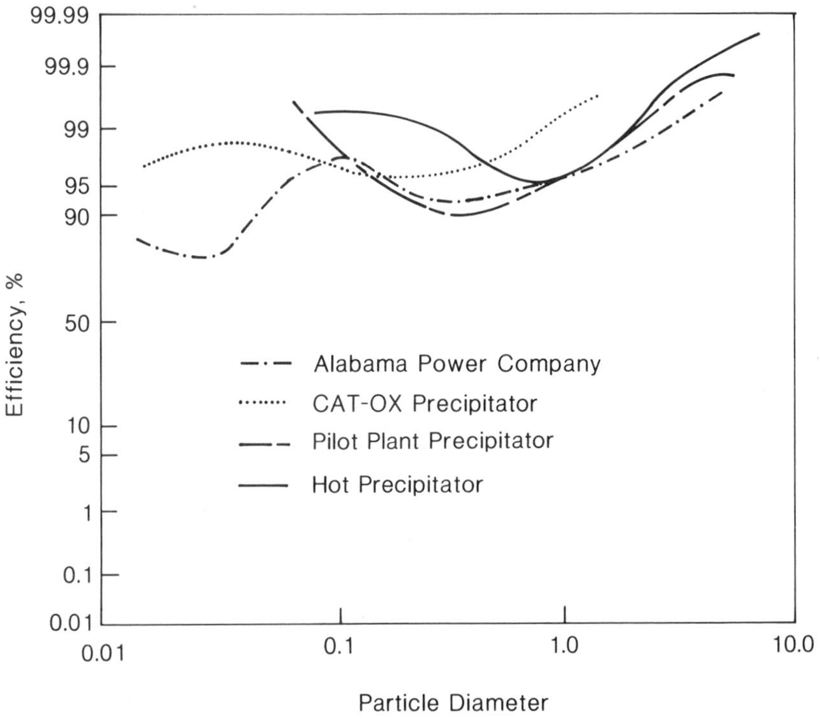

FIG. 2 *Measured grade efficiency for precipitators. Reproduced by permission of the American Institute of Chemical Engineers.*

(submicron) aerosol particles by turbulent diffusion may easily become more important than by the electrostatic force, except in the boundary layer adjacent to the collecting electrode. Coarser particles may have sufficient inertia to be relatively unaffected by the fluid turbulence. Thus a lateral-mix model (see Chapter 3) may apply to the fine particles, but not so well to the coarse ones. As collection proceeds this could lead to the central portion of the gas stream around the corona electrode becoming depleted of particles toward the downstream end of the collector. The result would be a very non-uniform distribution of dust concentration in cross-section normal to the flow.

The gas ions also stream from active toward passive electrodes under the influence of the field potential. This sets

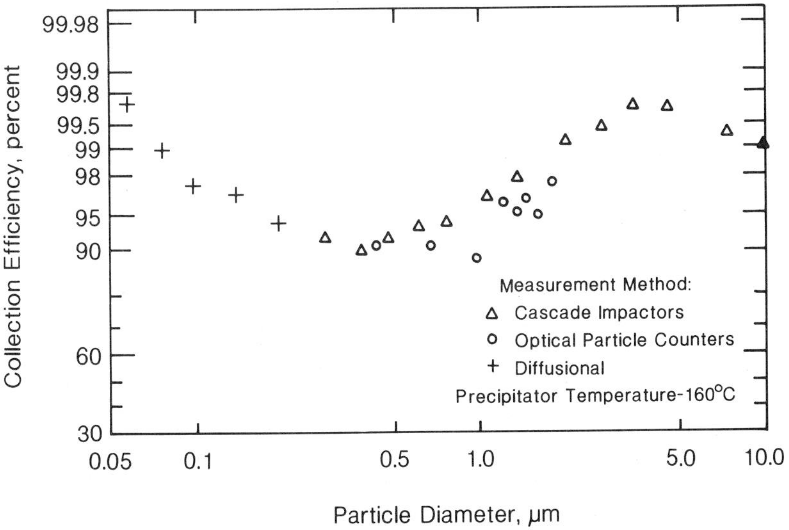

FIG. 3 *Measured fractional efficiencies for a cold side electrostatic precipitator with the operating parameters as indicated, installed on a pulverized coal boiler. Reproduced by permission from JAPCA.*

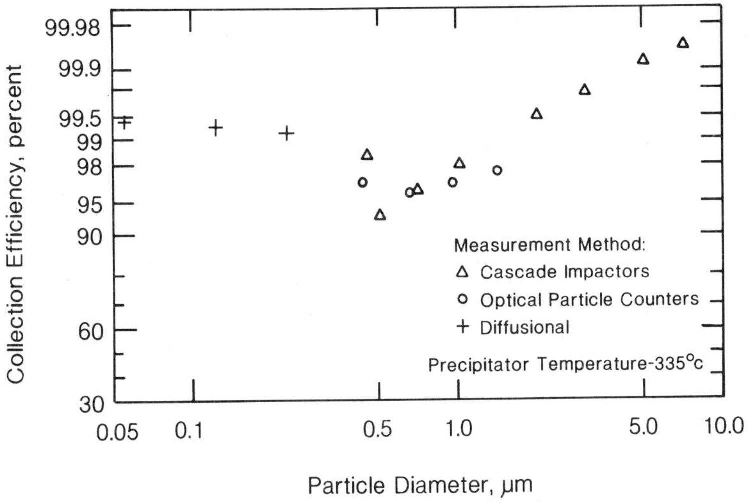

FIG. 4 *Measured fractional efficiencies for a hot side electrostatic precipitator with the operating parameters as indicated, installed on a pulverized coal boiler. Reproduced by permission from JAPCA.*

up a cross-current of gas flow referred to as the "electric wind." It obviously will have an effect upon particle distribution, collection, and reentrainment.

It cannot always be assumed that once a particle has reached the collecting electrode, it is removed and no longer influences the collecting process. If the particle material is a good conductor it may lose its charge at once and easily become re-entrained in the gas stream. If it is composed of a poorly conducting material, it may tend to cling to the electrode and build up an insulating layer upon it. This hinders the dust collection and may lead to a sparking discharge, indeed to a "reverse corona" effect. There is an optimum range, expressed in terms of particle electrical resistivity running between about 10^8 and 10^{10} ohm-cm., wherein the discharge proceeds smoothly. Electrical resistivity is affected strongly by the temperature and composition of the gas in the system. Its role in the precipitator performance is very important [11].

Nor can the importance of the rather simple geometry of the device be overlooked. Uniformity of electrical field is highly desirable and depends, among other factors, upon uniformity of electrode spacing. If the hanging wire electrode is not quite parallel to the collecting surface, then the potential gradient E becomes higher wherever the spacing is smaller. Since the vertical distance is of the order of several feet, while the electrode spacing is a matter of inches, it is easy to see how a very small departure from "plumb" can create a significant variation in E over the collecting zone. Uniformity of gas velocity distribution throughout is also very important. Regions of high velocity adversely affect the collection to a disproportionate degree, as will be shown later.

One of the features of an electrostatic precipitator is that it may be built to treat hot gas streams. In the 1970's there were a number of installations for collecting particles from hot (up to perhaps 800-900 F) stack gases, such as fly-ash arising from coal combustion, but this application seems to have ceased

[12]. Use in high-temperature high-pressure systems such as pressurized fluid-bed combustion, up to 1366 K and 35 atm, has been demonstrated [13] but has not been developed because of metallurgical problems [14]. Modelling calculations for HTHP will require an accurate knowledge of the values of all the relevant physical properties (e.g., gas viscosity, particle resistivity, ion mobility, etc.) at the temperature and pressure in question. Values in use for ordinary temperature and atmospheric pressure may be seriously in error.

B. *ELEMENTARY MODEL FOR PARTICLE COLLECTION*

An elementary, over-simplified, model of the collecting process may be developed beginning with the calculation of a collecting velocity due to the electrostatic force. This was begun, for Stokes Law particles, in Chapter 4, example 6, by setting up the equations for the trajectory of a particle having charge q in a field of potential gradient E. Based upon Eqn. (4.59), it was shown that the distance travelled in the direction of the field is:

$$z = \frac{qE\rho_p d_p}{54\pi\mu^2} \frac{t}{\tau} (1-e^{-t/\tau})$$

This indicates that after a period of time $t \approx 5\tau$, a steady state is attained with velocity $u_{E_s} = w = qE/3\pi d_p\mu$. This is called the "steady-state particle migration velocity," customarily denoted by w. It assumes that the particle has acquired and keeps a steady charge q* and that the potential gradient is constant. In case the particle is small enough to require the Cunningham factor correction to Stokes' Law it would become

$$w = q^*EC/3\pi d_p\mu \tag{7.1}$$

Sample calculations of Re_p at typical precipitator conditions indicate that Stokes' Law will be valid for particles up to 60 microns of so.

The following assumptions are now made to describe the model:

(1) The particles are considered to acquire the full charge q^* immediately on introduction into the collection zone. (The question of calculating the value of q^* will be considered later).

(2) There is a collection layer of thickness, δ, on the discharge electrode in which the flow is laminar, but in the main core of the stream it is turbulent.

(3) The thickness, δ, of the collection layer is defined as the distance away from the collection electrode to the point where the linear velocity of the laminar flow gas stream, down the precipitator, is equal to w.

(4) In the collection layer the particle always moves cross-stream at its migration velocity, which is unaffected by the velocity of the gas stream, and determined by the value of E in this layer region.

(5) In the turbulent core of the gas stream, the fluid velocity is taken as uniform at the average gas velocity v_o. (The question of non-uniform velocity will be dealt with below.)

(6) Turbulence and diffusion forces keep the particles uniformly distributed at concentration n_i across the central core in accord with (2), and the effect of migration velocity is negligible in this zone; the lateral-mix model described in Chapter 3 is assumed. (This may limit the applicability of the model to finer particles.)

(7) Dust particles are sufficiently separated that their mutual repulsion can be neglected.

(8) There are no disturbing effects such as erosion, reentrainment, uneven gas flow distribution, or back-corona present.

During a period of time Δt, defined by $\Delta t = \delta/w_i$, all particles of ith grade within the collection layer will move to the wall and, by definition, be collected. During this period of time the gas will move a distance $\Delta \ell = v_o \Delta t$. The number of particles thus removed from the volume of gas given by $(S\delta\Delta\ell)$, where S is the perimeter dimension of the collecting surface, will be

$$n_i \; S \; \delta \; \Delta\ell = n_i \; S \; \delta \; v_o \; \Delta t$$

At the same time, the total number of particles in the corresponding volume of the precipitator, having cross-sectional area A, will be $n_i \; A \cdot \Delta\ell$. Thus the fractional increment of particles removed is

$$\frac{dn_i}{\eta_i} = - \frac{S}{A} \frac{w_i}{v_o} \, d\ell$$

Then integrating, under all of the assumptions, over the length, L, of the collecting surface:

$$\int_{n_{o_i}}^{n_i} \frac{dn_i}{n} = - \frac{S}{A} \frac{w_i}{v_o} \int_0^L d\ell$$

gives the efficiency of collection as

$$\eta_i = 1 - \exp \left(-SL \, w_i / Av_o\right) \tag{7.2}$$

Note that SL equals the total surface of collecting electrode, and that $A \, v_o Q$, the volumetric rate of gas flow. The ratio SL/Q is called the "specific collecting area" SCA and often is stated in units of sq. ft/1000 acfm. It may range from 100-800 ft^2/1000 acfm.

In the case of a parallel plate precipitator A = bS, where b = half-width of spacing between plates, and in the case of a tubular precipitator $A = \pi D^2/4$ where D is the diameter of the tube. Thus (7.2) may be written

$$P_i = \exp - \frac{w_i}{v_o} \frac{L}{b} \qquad \text{(plate)} \tag{7.2a}$$

or

$$P_i = \exp - 4 \frac{w_i}{v_o} \frac{L}{D} \qquad \text{(tubular)} \tag{7.2b}$$

Since w_i depends upon d_{p_i}, equation (7.2) gives a grade-efficiency relationship. In the Stokes Law range (7.1) may be substituted for w_i, giving

$$\eta_i = 1 - \exp\left(- \frac{SL}{Q} \frac{C_i q_i^* E}{3\pi\mu d_{p_i}}\right) \tag{7.3}$$

A relationship for q_i^* is required to complete this model. Temperature and pressure will affect C_i/μ as discussed in Chapter 5, and may have some effect on q_i^*.

C. *PARTICLE CHARGING*

To express (7.3) in an explicit grade-efficiency form, a relationship between q_i^*, and d_{p_i} must be known. The development of this may be summarized as follows. For detailed derivation see Crawford [15].

Particles acquire charges by two distinct mechanisms referred to as (a) ion bombardment, and (b) ion diffusion [1]. These are also known respectively as field charging and diffusion charging. They are due (a) to motion of the ions directionally along the lines of flux, set up by the applied electric field, and (b) to random thermal motion of ions and particles (see Brownian motion, Chapter 4).

1. *Field charging* The lines of flux in the neighborhood of an uncharged particle are drawn into it. Hence ions travelling along them readily collide with and adhere to the particle primarily on the up-field surface. As charges build up, however, the flux lines are gradually repelled until finally they no longer bring ions into contact with the particles. This form of charging then

stops and the particle is said to have acquired a saturation or maximum charge from ion bombardment. The charge builds up with time according to [1]

$$q_i^* = q_{sat_i}^* \left(\frac{1}{1+t_o/t} \right) \tag{7.4}$$

where $t_o = 4\ \varepsilon_o/NeK$, a time constant independent of d_p, and

$$q_{sat_i}^* = \pi \left(\frac{3\varepsilon}{\varepsilon+2} \right) \varepsilon_o E_o d_{p_i}^2 \tag{7.5}$$

Here N = ion concentration, number/m^3; K = ion mobility, velocity per unit field strength, $m^2/s \cdot V$. The other symbols are as given in Chapter 4.

For practical purposes q_{sat}^* may be taken as attained (99%) at $t = 99\ t_o$. This is usually a very short period of time because $N \approx 10^{15}$ ions/m^3, $K \approx 10^{-4}$ m/s V, $\varepsilon_o = 8.85 \times 10^{-12}$ coulombs^2/Nm^2, and $e = 1.61 \times 10^{-15}$ coulombs. Therefore

$$t_o \approx \frac{4 \times 8.85 \times 10^{-12}}{10^{15} \times 1.61 \times 10^{-19} \times 10^{-4}} \approx 2.2 \times 10^{-4}\ \text{s}$$

and $99\ t_o \approx 0.022$ s . At a maximum gas velocity of say 8 ft/s, particles of all sizes would have acquired the saturation charge after traveling at most only about 2" into the charging zone.

The quantities in (7.4) are not directly affected by temperature and pressure. But the maximum E_o which may be used is lower at higher temperatures because spark-over occurs at lower voltage. This is offset by increased pressure, however.

2. *Diffusion charging* In the case of diffusion charging, collisions between particles and ions, resulting from the random thermal motion, continue all over the particle surface as long as the particles are in the ion field. No saturation charge exists. The diffusion charge builds up according to

$$q_i^* = \frac{2\pi\varepsilon_o kTd_{p_i}}{e} \ln\left[1 + \frac{e^2 Nd_{p_i} t}{2\varepsilon_o \sqrt{2\ nkT}}\right] \tag{7.6}$$

assuming $q_i^* = 0$ when $t = 0$. Here m = mass of an ion, and k is the Boltzman constant [1]. Higher temperature and pressure will tend to increase q_i^*.

The relative importance of the two methods of charging depends primarily upon the particle size. For particles roughly 1 μm and larger, field charging is predominant. Below about 0.4 μm diffusion charging becomes increasingly important as the values of $q_{sat_i}^*$ from (7.5) become very small. In the intermediate range, both types must be considered.

3. *Combined charging* In principle it is not satisfactory simply to add the values given by (7.5) and (7.6) in order to obtain the total charge on a particle. Rather the sum of the charging rates by field and diffusion charging should be used to give an overall rate. An overall theory of combined charging rate dq/dt, developed by Smith and McDonald [16] gives very good agreement with experimental values [1]. Unfortunately the form in which this is expressed does not lend itself to simple calculation.

Their approach is to divide the surface of the particle into three regions and to evaulate the probability that gaseous ions can impact upon each region. This is expressed as a charging rate for each region, and the total charging rate is taken as the sum of these three individual rates. The regions are defined in accord with the position of the electric field lines as they impinge on the particle.

The resulting equation for charging rate is quite complex. It can only be solved for values of q_i^* by numerical integration methods. It does however reduce to the form of (7.5) for large particles and high electric fields, and to the form of (7.6) for low fields.

An earlier expression by Cochet [17,18] for the combined effect of field charging and diffusion charging is

$$q_i^* = \left[\left(1 + \frac{2\lambda}{d_{p_i}}\right)^2 + \frac{2}{(1 + 2D/d_{p_i})} \left(\frac{\varepsilon-1}{\varepsilon+2}\right) \right] \pi \varepsilon_o E_o d_{p_i}^2 \left(\frac{t}{t+t_o}\right) \tag{7.7}$$

in which λ represents the mean-free path of the gas molecules, as discussed in Chapter 4. This lends itself to illustrative calculations, as carried out below.

It has been found, however, that if (7.5) and (7.6) are simply added together the sum fortuitously does give a reasonable value for the total charge on particles in the size range of 0.09 - 1.3 μm. The resulting equation may be written in the form

$$q_i^* = a d_{p_i}^2 + b d_{p_i} \ln(1 + c t d_{p_i}) \tag{7.8}$$

where the first term is the q_{sati}^* of (7.5). For a typical set of operating conditions: $\varepsilon = 5$, $E_o = 3 \times 10^6$ V/m, T = 300 K, N = 2 x 10^{15} ions/m^3, m = 5.3 x 10^{-26} kg (mass of an O_2 ion), Crawford [15] has shown that (7.8) would become:

$$\frac{q_i^*}{q_{sat_i}^*} = 1 + \frac{8.047 \times 10^{-9}}{d_{p_i}} \ln\left(1 + 7.79 \times 10^{10} \, t d_{p_i}\right) \tag{7.9}$$

where d_{p_i} is in meters and t in seconds.

D. *GRADE-EFFICIENCY EQUATIONS*

An explicit grade-efficiency equation may be obtained by substituting an appropriate expression for q_i^* in Eqn. (7.3). If q_i^* is taken to be as in (7.8). Equation (7.3) will become of the form

$$\eta_i = 1 - \exp - KC_i \left[ad_{p_i} + b \ln \left(1 + ctd_{p_i} \right) \right] \tag{7.10}$$

For the coarser particle size range the ln term is negligible and $C_i \approx 1$, so that

$$\eta_i = 1 - \exp - \left\{ \frac{SL}{Q} \left(\frac{\varepsilon}{\varepsilon + 2} \right) \frac{\varepsilon_o E_o E}{\mu} \right\} d_{p_i} \tag{7.11}$$

Note that this is of the form of Eqn. (3.17) with N = 1, and M representing the term in brackets which, in general, represents $SL\, w_i / Qd_{p_i}$.

In a two-stage precipitator E_o (for the charging stage) and E (for the collecting stage) are independently controlled and may very well be quite different. In a single-stage precipitator they are often assumed to be equal. However, assumption (4) in the list above requires that E be taken in the vicinity of the collecting plate; E does not necessarily equal E_o. Gooch and Francis [19] recommend taking E_o at the average value between discharge and collecting electrodes, and determining E by the numerical method of Leutert and Bohlen [20]. This is further detailed by Gooch and McDonald [21].

Equation [7.11] agrees with the form of experimental grade-efficiency determinations on spherical particles of uniform composition in which the assumptions underlying (7.2) are reasonably met. However, it cannot account for the results of tests such as are cited in Figs. 2, 3, and 4.

To account for the behavior of submicron particles either the complete form of (7.10) or another representation of q_i^*, such as (7.7), must be used. Feldman [22] incorporated (7.7) into (7.3), taking $(\varepsilon-1)/(\varepsilon+2) \approx 1$, and $t \gg t_o$, and obtained

$$\eta_i = 1 - \exp - \frac{SL}{Q} \frac{\varepsilon_o E_o E}{3\mu} (C_i A_i d_{p_i}) \tag{7.12}$$

as a grade-efficiency equation to be compared with (7.10). Here

$$A_i = \left(1 + 2\lambda/d_{p_i}\right)^2 + 2/\left(1 + 2\lambda/d_{p_i}\right)$$

and C_i is also a function of λ/d_{p_i} as given by Eqn. (4.9). The term $(C_i A_i d_{p_i})$ is size-dependent and (7.12) may be differentiated with respect to d_{p_i} in order to seek a point of minimum value of η_i. It is clear that for a given precipitator and set of operating conditions, the value of d_{p_i} for which η_i is a minimum will depend only upon λ. In turn λ is a function only of temperature, pressure, and composition of the gas, as given by Eqn. (4.7a).

Farber [23] has performed this calculation and found that

$$d_{p_{min}} = 2.855\ \lambda \tag{7.13}$$

For a typical flue gas this gives $d_{p_m} = 0.3\ \mu m$ at 160°C and $d_{p_m} = 0.4\ \mu m$ at 335°C. These values compare well with those found experimentally, shown on Fig. 3 and 4. Thus the shape of the experimental grade-efficiency curves is accounted for by the nature of the particle charging phenomena.

E. *DEUTSCH-ANDERSON EQUATION*

1. *Precipitation rate parameter* Equation (7.2) as it stands, without refinements, is known as the Deutsch-Anderson equation and has been used as the basis for much work on precipitators. While it can be taken as a grade-efficiency equation, it has been the practice to use it for overall efficiency. This is done by replacing w_i with a so-called "effective" migration velocity, or more properly "precipitation rate parameter" w_e. This quantity is taken to represent the collection behavior of an entire dust of a certain kind and under a certain set of operating conditions. It

is determined by back calculation from experimental data and so is really a performance parameter.

Quite a bit of such data is available (although much of it is proprietary) [2, 3, 7] for a variety of applications. Values of w_e are related to kind of dust, gas velocity, length of surface L, particle size, design of electrodes, rapping methods, sparking routines, etc. They may be used as a basis of empirical precipitator design usually in the form:

$$\eta_M = 1 - \exp\left(-\frac{SL}{Q} w_e\right) \tag{7.14}$$

Examination of such data indicates that the assumptions underlying (7.2) may not always be met, especially number (8). Corrections to the model for such factors as uneven gas flow distribution, particle resistivity, diffusional effects, etc. are treated briefly below.

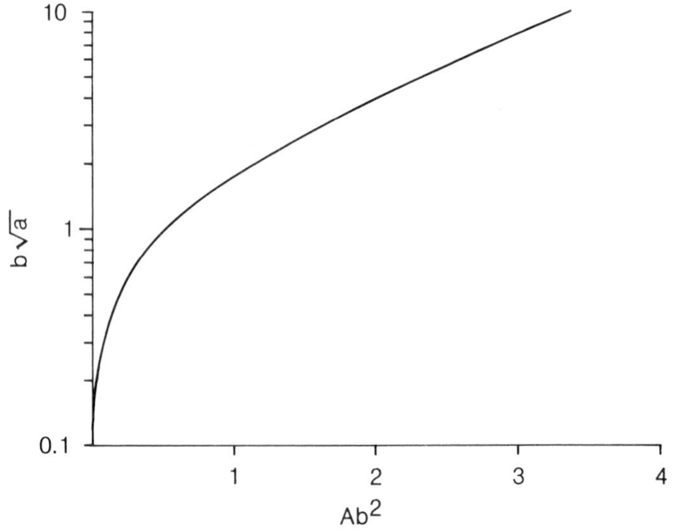

FIG. 5 *Graph for use with Equation (7.17)*

F. *OVERALL EFFICIENCY*

The calculation of overall efficiency from grade-efficiency must be done in accord with the general principles presented in Chapter 3. Thus using (7.14) or (7.2), $P = 1 - \eta_M$ is

$$P = e^{\frac{-SL}{Q} w_e} = \int_0^1 e^{\frac{-SL}{Q} w_i} dG_i = \int_0^\infty e^{\frac{-SL}{Q} w_i} q_i dd_{p_i} \tag{7.15}$$

Here $q_i = dG_i/dd_{p_i}$ as in Chapter 3, and is not to be confused with q_i^* representing charge.

Because of the complex form of expressions such as (7.7) or (7.8), as well as the possibility of complex relationships for q_i, formal evaluation of the last integral in (7.15) is likely to be impossible in the general case. However, for special cases some methods are available.

Whenever (7.11) may be taken to represent the grade efficiency, i.e., $w_i = kd_p$, and q_i is a log-normal distribution, an equation of the form of (3.18) with N = 1 applies:

$$P = \frac{1}{\sqrt{2\pi}} \int_{-\infty}^{\infty} e^{-t^2/2} \, e^{-ae^{bt}} \, dt \tag{7.16}$$

wherein

$$t = \frac{\ln d_p/\bar{d}_{PM}}{\ln \sigma_g}$$

$a = M\bar{d}_{PM}$ see Eqn. (7.11)

$b = \ln \sigma_g$

A chart is available, [2] from which $\eta_M = 1\text{-}P$ may be read as a function of a and b, but it does not permit a very precise reading. Note that a here is <u>not</u> the a in Eqn. (7.8) for q_i^*.

A unique approximation to P, developed by Kunz and Hanna [24] permits a direct calculation to be made by

$$P \approx \frac{\exp\left(-A^2b^2/2-A\right)}{\left(1+Ab^2\right)^{1/2}} \tag{7.17}$$

where A is determined from a graph, Fig. 5, of Ab^2 vs $b\sqrt{a}$. For values of $b\sqrt{a}$ off the scale of this chart, a good calculation may be made by:

$$(Ab^2) = 2\,\ell n\, b\sqrt{a} - \ell n(2\ell n\, b\sqrt{a}) + 0.3 \qquad 10 < b\sqrt{a} < 25$$

$$(Ab^2) = \frac{2ab^2\,(1 + ab^2)}{2(1 + ab^2)^2 - a^2b^4} \qquad 0 \le b\sqrt{a} \le 0.5$$

The results of the calculation agree very well with values read from the charts.

Another special case, treated by Feldman [22], uses (7.12) as the grade efficiency function combined with a log-normal inlet particle size distribution to give

$$P = \frac{1}{\sqrt{2\pi}\,\ln\sigma_g}\int_0^{\infty} \exp\left[-k(aC_{d_p}) - 0.5\left(\frac{\ln d_p/\bar{d}_{p_M}}{\ln\sigma_g}\right)^2\right] d\,\ln d_p \tag{7.18}$$

where

$$k = \frac{\varepsilon_o E_o E}{3\mu}\,\frac{SL}{Q}$$

Feldman integrated this numerically for each of several size distributions typical of fly-ash from coal combustion, to obtain P as a function of k in each case. He found that ln 1/P varied linearly with ln k in a given case, so that

$$P = c^{(10^{-4}k)^m} \tag{7.19}$$

His data are as follows:

Furnace Type	$\bar{d}_{p_m}$ -μm	σ_g	c^*	m^*
Pulverized Coal	12	3.8	0.5379	0.5000
Cyclone	2.08	9.45	0.8344	0.6765
Spreader-Stoker	68	3.54	0.1827	0.4055

*Values of c and m require that k be expressed in (m^{-1}).

Where (7.19) is combined with (7.14), an expression is obtained for the parameter

$$w_e = \frac{-(10^{-4}k)^m \ln c}{SL/Q}$$

or

$$w_e = (w_e')^m (SL/Q)^{m-1} \ln (c) \tag{7.20}$$

where $w_e' = \varepsilon_o E_o E \times 10^{-4}/3\mu$ represents a modified rate parameter which is independent of particle size. Feldman found that (7.20) accounts satisfactorily for the observed behavior of w_e in a number of respects.

A third method of computing P, due to P. Cooperman and G.D. Cooperman [25], is based upon an alternate form of (7.16), or of Eqn. (3.18) with N = 1:

$$P = \frac{1}{\sqrt{2\pi}} \int_{-\infty}^{\infty} \exp\left(-\frac{t^2}{2}\right) \exp\left(-M\bar{d}_{PM}\, \sigma_g^t\right) dt \tag{7.21}$$

They found that this integral may be evaluated rather simply and to a "satisfactory" degree of accuracy by the use of an appropriate mean value of t defined as:

$$t_{mean} = \ln m / \ln \sigma_g \tag{7.22}$$

Here m may be shown theoretically to lie between 0 and 1, and is found empirically from:

$$m = \exp\left[(0.0822 \ln P_g - 0.159)\ln \sigma_g\right] \tag{7.22a}$$

To obtain m, the value of P_g must first be found by calculating the penetration for the geometric mean particle size, i.e. by setting $d_p = \bar{d}_{PM}$ in Eqn. (7.11). Then, finally

$$P = P_g^m \tag{7.23}$$

This is equivalent to determining the overall penetration P as equal to the value of P_i calculated at a particle size corresponding to the value of t_{mean}, which is $d_{p_i} = m\bar{d}_{PM}$. The range for which (7.22) has been developed appears to be for $P_g = 10^{-4}$ to 10^{-2} and $\sigma_g = 1.5$ to 3.0.

Example 1. Consider the same Portland cement dust described in Chapter 6, for the cyclone design example. Determine whether it could be collected to meet the emission regulations by a single-stage electrostatic precipitator having about 8100 ft^2 of plate surface consisting of 9 parallel rows of plates each 15 ft. high by 34 ft. long spaced 8 in. apart. The discharge electrode potential can be as high as 32,000 V. Conditions described in Chapter 6 are: Dust: MMD = 12 μm, σ_g = 3.08 (log-normal distribution); Dielectric constant ε = 6.14; Flow: 222 acfs at 250°F, 1 atm.; Efficiency required: 99.78% by mass, overall.

Since this dust does not contain much submicron material (1% < 1 μm), Eqn. (7.11) may be used for the grade-efficiency. Since the dust is also log normally distributed (7.16) may be used to obtain the overall efficiency. Equation (7.11) becomes:

$$\eta_i = 1 - \exp - \frac{8100}{222 \times 0.3048}\left(\frac{6.14}{6.14+2}\right)\frac{8.85\times10^{-12}(3.15\times10^5)^2}{2.25 \times 10^{-5}}\left(10^{-6} d_{p_i}\right)$$

where

$$E_o = E = \frac{32{,}000}{4/12 \times 0.3048} = 315 \text{ kv/m} = 3.15 \times 10^5 \text{ N/coul.}$$

$\varepsilon_o = 8.85 \times 10^{-12}$ coul^2/Nm^2

$\mu = 2.25 \times 10^{-5}$ kg/ms @ 250^{0}F

and d_{p_i} is in microns. Thus: $\eta_i = 1 - \exp - 3.524\ d_{p_i}$. For Equation (7.16): a = 3.524 x 12 = 42.3; b = ℓn 3.08 = 1.125. Evaluating (7.16) by means of (7.17), using Fig. 5: $b\sqrt{a} = 7.32$, $Ab^2 = 2.91$, A = 2.30 and from (7.17): P = 0.0018. The overall efficiency is 99.82% which may be taken as achieving that required, considering the uncertainties involved in the calculation. This calculation is corroborated by a value of about 0.997-0.998 as read from the chart in Ref. [2]. Equation (7.22) appears not to be applicable to this case.

For this efficiency to be attained in practice there must be uniform veloctiy in the 8 ducts between the plates (v_o = 222 x 12/8 x 15 x 9 = 2.5 ft/s), uniform field potential gradient, no reentrainment, suitable particle resistivity, etc. Note that in a total L = 34 ft., there would probably be about 7 plates, each about 5 ft. wide. These factors will be investigated further in the next section.

III. CORRECTIONS TO THE MODEL

There are a number of aspects of precipitator performance which are not dealt with by the elementary model. As mentioned above, the construction of the plate area is such that operating conditions, e.g. temperature, particle concentration and size distribution, voltages and particle charges may vary along the length L. Other so-called non-ideal effects are non-uniform velocity distribution, sneakage of gas around the plates, re-entrainment of collected particles, and movement of particles due to turbulence effects. Methods of dealing with these factors will be discussed, with particular reference to the SRI/EPA computer model.

A. *INCREMENTAL LENGTH CALCULATIONS*

Changing conditions along L are dealt with by dividing the collecting surface into a number of small length increments, each short enough that conditions may be taken as constant over the increment. This divides the precipitator into a number of small precipitators in series, in each of which Eqn. (7.2) is valid for particles of a given size. For collection of particles of i-th grade in the j-th increment of length, (7.2) may be written:

$$\eta_{i,j} = 1 - \exp\,(-w_{i,j}\, S_j/Q) \tag{7.24}$$

wherein $w_{i,j}$ is the migration velocity of i-th grade particles at the conditions of the j-th increment, and S_j is the collecting surface area of the j-th increment of length.

The overall grade-efficiency for the entire precipitator is then obtained by combining the collection of all of the length increments according to the principle of collectors in series, given by Eqn. (3.34):

$$\eta_i = 1 - (1-\eta_{i,1})(1-\eta_{i,2})\ \ldots\ (1-\eta_{i,j}).. \tag{7.25}$$

The total mass efficiency of the entire precipitator is then found by summing the grade-efficienies in the usual way:

$$\eta_M = \sum_i \eta_i\, g_i \tag{7.26}$$

B. *NON-UNIFORM VELOCITY DISTRIBUTION*

The model assumes that the fluid velocity parallel to the collector surface is everywhere the same and equal to the volumetric rate of flow divided by the total cross-sectional area normal to the collecting electrode surface, i.e., $v_o = Q/A$. This condition is almost impossible to achieve in practice, but every effort should be made to do so. To this end it is essential to study the flow pattern in a proposed design by means of a scale model, usually 1/16 size. The effects of plate construction, baffles, shape of duct work, use of turning vanes and diffuser plate, etc. must all be studied in order to achieve the most uniform flow possible.

When it is possible to associate different values of v locally with corresponding portions of the collecting surface, a correction in (7.2) may be made by inserting a factor $F > 1$ (sometimes called a "quality factor" of the flow) with the value of v_o. Thus (7.2) becomes

$$P_i = \exp - \frac{SL\, w_i}{AFv_o} \tag{7.27}$$

For sake of illustration, L may be regarded here as the first length increment at the inlet. F may be evaluated if the inlet velocity distribution is known in terms of a series of local values of gas velocity, called v_k, each associated with a known cross-sectional area A_k and a corresponding perimeter S_k. Then through one of these zones, the penetration P_{ik} will be, using (7.2):

$$P_{ik} = \exp - \frac{S_k L}{A_k} \frac{w_i}{v_k} \tag{7.28}$$

The total emission will be related to these values according to

$$P_i Q = \sum_k Q_k P_{ik} = \sum_k v_k A_k P_{ik}$$

or

$$P_i = \frac{1}{v_o A} \sum_k v_k A_k P_{ik} \tag{7.29}$$

where $A = \sum A_k$. If A is divided into N equal areas, each A_k, this becomes

$$P_i = \frac{1}{Nv_o} \sum_{k=1}^{N} v_k \cdot P_{ik} \tag{7.30}$$

so that, by rearranging (7.27)

$$F = -\frac{SL}{Av_o}\frac{w_i}{\ell n P_i} \tag{7.31}$$

in which P_i is calculated from (7.29) or (7.30).

To illustrate the effect, suppose that in a plug flow situation where $v_k = v_o$ the penetration would be 0.01, but that there were actually three zones of equal areas $A_k = A/3$ in which $v_1 = v_o/2$, $v_2 = v_o$, $v_3 = 2v_o$ respectively. The actual penetration would then be obtained as follows. First from (7.2)

$$0.01 = \exp -\frac{L}{b}\frac{w_i}{v_o} \text{ or } \frac{L}{b}\frac{w_i}{v_o} = -\ \ell n\ 0.01 = 4.605$$

and from (7.28):

$$P_{i1} = \exp \frac{-4.605}{1/2} = 0.00010$$

$$P_{i2} = 0.01$$

$$P_{i3} = \exp \frac{-4.605}{2} = 0.1000$$

From (7.30)

$$P_i = \frac{1}{3v_o}\left[\frac{v_o}{2} \times 0.00010 + v_o \times 0.01 + 2v_o \times 0.1000\right] = 0.0700$$

and from (7.31)

$$F = -4.605/\ell n\ 0.0700 = 1.73$$

The use of (7.29) or (7.30) with measured velocity traverses in a working precipitator may yield values of F as high as 1.5 or more in a poorly regulated flow pattern. Where care is taken, through model studies, values of F as low as 1.1 or 1.2 should be achieved. Computations involving F are incorporated with the SRI/EPA computerized model.

B. *GAS SNEAKAGE AND DUST REENTRAINMENT*

Gooch and Francis [19] offer means of estimating corrections for these phenomena, means admittedly based upon over-simplified assumptions and requiring data which are not usually available. These methods would be useful however in estimating the magnitude of potential loss in performance due to these causes. They are also incorporated in the SRI/EPA mathematical model.

1. *Gas sneakage* This term refers to that portion of the gas flow which may by-pass the collection zone entirely. This may occur due to gas flowing through the dust collection hoppers or through the high-voltage insulation space. It is reduced or prevented by the installation of baffles which force the gas to return to the main gas passage through the collecting zone. The baffles divide the zone into a number of sections. After each section the sneakage gas is remixed into the main flow which then re-by-passes in the next section. Gooch and Francis give the following equation which may be used to correct the penetration for sneakage:

$$P_{S_i} = \left[S + (1-S)(P_i)^{1/N_s} \right]^{N_s} \tag{7.32}$$

when S = fractional amount of gas sneakage in each section; N_s = number of baffled sections; P_i = penetration of a given particle size, with no sneakage. For example, if there is 5% sneakage in each of 4 sections, an uncorrected penetration of 0.01 will be increased:

$$P_{S_i} = \left[0.05 + 0.95\ (0.01)^{1/4} \right]^4 = 0.0151$$

2. *Dust reentrainment* When the collection surface is rapped to dislodge the collected dust so that it may fall into the dust hopper, particles may become reentrained in the gas stream. They may or may not be subsequently recollected downstream. Assuming

that a fixed fraction R_i of a given dust size is reentrained and that this fraction is constant along the length of the precipitator, Gooch and Francis derive the following corrected penetration:

$$P_{R_i} = \left[R_i + (1-R_i)(P_i)^{1/N_R} \right]^{N_R} \tag{7.33}$$

where N_R = number of stages over which reentrainment occurs. This is of precisely the same form as the sneakage correction.

If, in addition to the sneakage loss cited above, there is also a 1% reentrainment, then

$$P_{R_i} = \left[0.01 + 0.99\ (0.01)^{1/4} \right]^4 = 0.0109$$

and the total penetration $P_S + P_{R_i} = 0.0260$, or a decrease in efficiency η_i from 99% to 97.4%. Since the precipitator is basically supposed to be a high-efficiency device, obviously sneakage and reentrainment must be minimized.

Because the calculations for sneakage and rapping reentrainment are of the same form, they may be lumped together by use of an overall (S+R) value in place of S in (7.32). This is done when individual values of S and R cannot be measured, and used to back-calculate a total correction to make theoretical values agree with experimental value of P_i. Values of the lumped parameter as high as 0.10 may be found in poorly operating systems.

C. *TURBULENCE EFFECTS*

The theoretical model gives no reason to expect that w_i should vary with gas velocity (v_0), the width of spacing between collection surfaces (2b), or the size (scale-up) of the precipitator. It also predicts, according to Eqn. (7.11), that w_i be directly proportional to d_{p_i}. But exception to all of these expectations are readily found to occur in practice.

P. Cooperman [26,27] and G. Cooperman [28] have developed a theory which can account for all of these phenomena through the effects of turbulence and the consequences of particle concentration gradients in the collecting zone. There are three concentration gradients: in the direction of flow, "the act of collection . . . causes the concentration of particles upstream to be greater than the concentration downstream;" perpendicular to flow, due to the electric force "concentration of particles rises from a small value near the plane of the discharge electrode to much larger values in the vicinity of the collecting electrode(s);" and vertically, because of the gravity settling of particles.

If there were no turbulence or secondary flows, the particle concentrations would not effect the particle trajectories. They would be as described in Chapter 5 for plug flow, or for laminar flow. Because of turbulence, however, there is a mixing effect which moves particles from regions of high concentration toward regions of low concentration. This effect, called turbulent diffusion, clearly opposes the collection process, first because of the lateral gradient away from the collecting electrode, and secondly because of the downstream (axial) gradient which causes the particles to move faster than the gas flow in this direction.

These concepts may be introduced into an overall unified theory of precipitator collection grade-efficiency through a mass balance which contains all of these fundamental forces. Known as the convective turbulent diffusion equation, it may be written for each particle size in terms of its concentration c_i:

$$D_{\ell_i} \frac{\partial^2 c_i}{\partial \ell^2} + D_{z_i} \frac{\partial^2 c_i}{\partial z^2} - v \frac{\partial c_i}{\partial \ell} + w_i \frac{\partial c_i}{\partial \ell} = 0 \qquad (7.34)$$

The first two terms account for turbulent diffusion in the axial (ℓ) and lateral (z) directions respectively through the mixing coefficients D_{ℓ_i} and D_{z_i}; the third term accounts for convective transport in the axial (flow) direction; and the last term the lateral transport due to the electrostatic migration. In all that follows the subscript i will be dropped, but the equations are to be understood as applying to one particle size at a time.

The solution to (7.34), subject to stated initial and boundary conditions, will give c as a function of ℓ and z, i.e. at any location in the precipitator space. From this the value of c at $\ell = L$ may be found and thence the value of η. Various solutions are available corresponding to various assumptions which have been made either about the boundary conditions or about the relative values of some of the parameters. The solution due to G. Cooperman [28] seems to be the most general, reducing to the others as special cases, and will be summarized and discussed below.

Cooperman's solution is subject to the initial conditions:

$$c(0,z) = c_o \tag{7.35}$$

$$c(\ell,z) \to 0 \text{ as } \ell \to \infty$$

where $\ell = 0$ at the inlet ($\ell = L$ at the outlet), and $z = 0$ at the surface of the collected dust layer on the collector plate. The second condition is necessary to rule out unphysical solutions.

The boundary conditions are taken as:

$$\begin{aligned} D_z \frac{\partial c}{\partial z} + wc &= 0 \qquad \text{at } z = b \text{ (center plane)} \\ D_z \frac{\partial c}{\partial z} - fwc &= 0 \qquad \text{at } z = 0 \text{ (boundary layer)} \end{aligned} \tag{7.36}$$

The first condition states that there is no net particle flux across the center plane, which is true due to symmetry. The second contains an empirical parameter f, first introduced by P. Cooperman [26], to indicate what fraction of those particles which pass from the gas stream into the boundary layer are subsequently reentrained into the gas stream: f = 0 corresponds to total collection, while f = 1 refers to total reentrainment.

This factor is meant to take into account the net effect of all the physical processes going on in the boundary layer at the plate surface, which are for the most part not well understood. In the absence of satisfactory methods of determining f, setting $f = 0$ may be a reasonable assumption.

Although the rigorous solution of (7.34) is in the form of an orthogonal series, it is satisfactory to use only the first term for the values of $\eta > 0.80$ and large L which are of interest in practical problems. This term is given [28] in dimensionless form as

$$P = (\text{const}) \exp\left\{ \left(\alpha - \sqrt{\alpha^2 + \gamma\lambda_1^2 + \beta^2/\gamma}\right) s \right\} \tag{7.37}$$

where

$$\alpha = \frac{b\,v}{2D_\ell}, \quad \beta = \frac{b\,w}{2D_\ell}, \quad \gamma = \frac{D_z}{D_\ell}, \quad s = \frac{L}{b} \tag{7.38}$$

and λ_1 is the smallest positive root of

$$\tan \lambda_1 = \frac{\beta}{\gamma} \frac{2\lambda_1 (1-f)}{\lambda_1^2 - \left(1-2f\right) \beta^2/\gamma^2} \tag{7.39}$$

The (const) varies from 1 for $\beta/\gamma \ll \pi/2$ to $\left[\sqrt{2\pi\gamma/\beta}\right]^3 \exp(\beta/\gamma)$ for $\beta/\gamma \gg \pi$. Note that the ratios α/β and β/γ reflect the relative magnitude of important parameters:

$$\alpha/\beta = v/w \quad \text{and} \quad \beta/\gamma = bw/2D_z \tag{7.40}$$

By examining the effect of the values of α, β, and γ upon P as given by (7.37), and comparing the result with that given by the elementary model in Eqn. (7.2), one may study the relative importance of turbulent mixing in affecting the grade-efficiency. This may be useful in explaining experimental results which seem to disagree with this so-called Deutsch model.

For a high degree of turbulence, i.e. very large values of the mixing coefficients, both β and β/γ are small, $\lambda_1 = \sqrt{2(1-f)\beta/\gamma}$, and const = 1. Eqn. (7.37) then reduces to

$$P = \exp \quad (\alpha - \sqrt{\alpha^2 + 2\ (1-f)\ \beta})\ s \tag{7.41}$$

This gives increasingly larger values of P than (7.2) does as D_z increases above the order of 1000 cm^2/s. The relative importance of α and β may be judged better by rewriting (7.41) so that

$$P = \exp\left\{ -\alpha \left[\left(1 + 2(1-f)\beta/\alpha^2\right)^{1/2} - 1\right] L/b \right\} \tag{7.42}$$

Two limiting special cases of this may be discerned:

(a) $\alpha^2 >> 2\beta$, so that $\left[1+2\beta(1-f)/\alpha^2\right]^{1/2} \approx 1 + \beta(1-f)/\alpha^2$

Then (7.42) becomes

$$P = \exp - (1-f)\ \frac{\beta}{\alpha}\frac{L}{b} = \exp \frac{-(1-f)w}{v}\frac{L}{b} \tag{7.43}$$

This is seen to be the same as the elementary model (7.2) with the addition of the factor (1-f). It may be called the lower turbulence case. It is approached as D_ℓ decreases below the order of 1000 cm^2/s, showing that the Deutsch Eqn. (7.2) is a special case of the general theory in which P does not depend upon D_ℓ.

(b) $\alpha^2 << 2\beta$, so that $\sqrt{1+2(1-f)\beta/\alpha^2} \approx \sqrt{2(1-f)\beta/\alpha^2} >> 1$

Then (7.42) becomes

$$P = \exp - \sqrt{2(1-f)\beta}\ \frac{L}{b} = \exp -\sqrt{(1-f)w/bD_\ell}\ L \tag{7.44}$$

This may be called the higher turbulence case. It indicates that P becomes independent of v and has reduced dependence upon w and b than in the elementary model. Thus, those conditions which produce a high degree of turbulent mixing: high electric wind, secondary gas flows promoted by very high plates or wider spacing between plates, will yield a reduced efficiency.

The other extreme case is when β/γ becomes large due to $D_z + D_\ell$ becoming small i.e. of the order 10-100 cm^2/s. Then (7.37) may be shown to become

$$P = \frac{2 \exp\left(wb/2D_\ell\right)}{\pi\left(wb/2D_\ell\right)^2} \exp - w^2L/4v\, D_\ell \tag{7.45}$$

This corresponds to the case of low turbulence such that eddy diffusion is the principal form of mixing. Equation (7.45) shows that P may be much <u>less</u> than the value obtained from (7.2).

Cooperman [27] points out that although no theoretical methods for calculating D_ℓ or f are available, values for these may be obtained by back-calculating from measured performance data. In all cases f is found to be between 0 and 1, as is to be expected. Further D_ℓ is found to be larger for finer particles, higher gas velocities, and larger spaces, again all consistent with the concepts of turbulent mixing. Sample calculations using D_ℓ from 10^2 to 10^5 cm^2/s and f in the range 0.5 to 0.7 yield values of w_e which are not only of the correct order of magnitude but also conform to the observed trends.

Although no general method is available for using it, qualitatively, at least, the Coopermans' theory appears to be quite reasonable. It confirms the existence of a phenomenon which must not be overlooked in high turbulence precipitators and indicates how precipitator performance may be improved by designs which will reduce turbulence.

IV. DESIGN EXAMPLE

The author is indebted to Dr. R. Lee Byers for suggesting the following design problem and providing some of the data for it.

In Chapter 6 the use of a cyclone was considered for the collection of a Portland cement dust. The design calculations showed that a cyclone system might be useful as a primary cleaner to be followed by a high efficiency device as a secondary cleaner needed to attain the required degree of emission control. Let us now consider the design of a single stage electrostatic precipitator for this purpose.

The particle size distribution of the cement dust entering the ESP will be taken as that leaving the cyclone system, based upon 20 cyclones in parallel. This is shown as the curve for P = 20 in Fig. 8, Chapter 6. This size distribution deviates slightly from the log-normal form above $d_p \approx 12$ μm, but the dust may be well represented by using an MMD = 3.7 μm and $\sigma_g = 2.22$.

Other conditions will be as follows: Gas flow rate, 13,300 acfm; Gas temperature 240 F; Gas pressure, 1 atm.; Gas viscosity, $\mu = 2.25 \times 10^{-4}$ poises; Dust properties: $\varepsilon = 6.14$, Inlet c_o, 1.24 gr/acf, Rate 141.5 lbm/hr.; Outlet c_o, 0.0132 gr/acf, Rate 1.5 lbm/hr.; Density, $\rho_p = 1.5$ gm/cm^3; Efficiency required: η_M = 98.94%. For data relating to similar installations consult reference [7], Tables 18.10, 18.12 and Fig. 18.10. These suggest that it would be appropriate to assume the following additional conditions. Field strength $E_o = E = 8$ kv/in. Average velocity v_o = 2-8 ft/s.

Since only a small percent of the dust is finer than 1 micron, the calculations will be carried out using Eqn. (7.11) as the basis for the grade-efficiency. The set-up is similar to that for the previous example, except that (L) is the unknown to be determined. Using the numbers presented there:

$$P_i = \exp - M d_{p_i} = \exp - 4.351 \times 10^{-4} \text{ (SL) } d_{p_i}$$

with $M = 4.351 \times 10^{-4}$ (SL) μm^{-1}, (SL) in ft^2 and d_{p_i} in μm.

The required performance to be attained is, according to Eqn. (3.12)

$$\eta_M = \int_0^1 \eta_i dG_i = 0.9894$$

or according to Eqn. (3.11)

$$P = \sum_i \Delta g_i \cdot P_i = \sum \Delta g_i \cdot \exp - Md_{p_i} = 0.0106$$

There are two methods available for solving for M. The first is by using the Kunz and Hanna approximation to the integral for η_M, given as Eqn. (7.17) above. The second is to solve the series for P by the methods of Eqns. (3.24) and (3.25). For the first, the inlet dust must be represented by a log-normal size distribution. For the second, the actual distribution may be used. Both methods will be illustrated here for purposes of comparison.

For the Kunz and Hanna method, Eqn. (7.17): MMD $\equiv \bar{d}_{p_m} = 3.7$ µm, $\sigma_g = 2.22$; $M = 4.351 \times 10^{-4}$ (SL) μm^{-1}, (SL) = ft^2; $a = M\bar{d}_{p_m} = 1.610 \times 10^{-3}$ (SL); $b = \ln \sigma_g = 0.7975$ and $P = 0.0106$. Then

$$P = \frac{e^{-0.3180A^2-A}}{(1+0.6360A)^{1/2}} = 0.0106$$

must be solved for A. This is best done by an iterative procedure, which may easily be programmed on a pocket calculator in the form:

$$A = \left(\frac{-\ln\ (0.0106)(1+0.6360A)^{1/2}-A}{0.3180} \right)^{1/2}$$

It yields A = 2.3438. Then $Ab^2 = 1.4907$ and $2.60 = b\sqrt{a}$ from Fig. 5. Finally $a = (2.60/0.7975)^2 = 10.62$ (M=2.870) and SL = $10.62/1.610 \times 10^{-3} = 6600\ ft^2$.

For the series procedure, it is necessary to choose an interval for Δg_i, then tabulate the corresponding values of d_{p_i}. Two intervals will be used: 0.05 and 0.10. The following values are read from the plot for P = 20, Fig. 8 Chapter 6.

i	$\Delta g=0.05$: Δd_p	$\Delta g=0.05$: d_{p_i}	$\Delta g=0.10$: Δd_p	$\Delta g=0.10$: d_{p_i}
1	0-0.95	0.475	0-1.28	0.64
2	0.95-1.28	1.115	1.28-1.83	1.56
3	1.28-1.58	1.43	1.83-2.38	2.11
4	1.58-1.83	1.70	2.38-2.95	2.67
5	1.83-2.10	1.97	2.95-3.65	3.30
	etc.		etc.	

A maximum of five terms in the series should be ample in order to solve for M, in each of the following equations:

$$\Delta g = 0.05: \quad \frac{0.0106}{0.05} = e^{-0.475M} + e^{-1.11M} + e^{-1.43M} + \ldots .$$

$$\Delta g = 0.10: \quad \frac{0.0106}{0.10} = e^{-0.64M} + e^{-1.56M} + e^{-2.11M} + \ldots .$$

Again, these solutions may readily be carried out by programming an iterative procedure on a pocket calculator in the form (for example):

$$M = -\frac{1}{0.64} \ln \left[0.1060 - e^{-1.56M} - e^{-2.11M} \ldots . \right]$$

The results are tabulated

Terms Used	$\Delta g=0.05$ M	$\Delta g=0.10$ M
1	3.266	3.507
2	3.481	3.564
3	3.538	3.571
4	3.559	3.572
5	3.565	3.573

Both methods appear to converge to the same value if a sufficient number of terms is used. For a quick approximation, using one term of the $\Delta g=0.10$ series will be the best.

The required plate area SL = $3.573/4.351 \times 10^{-4} = 8210 \text{ ft}^2$ by this method. Note that this result is based entirely upon the size distribution data in the fine end of the range. It uses only the lower 20% (or 40%) of the mass, according to the Δg_i selected. However, this is the critical material to be collected, in order to meet an emission limitation. The coarser material will assuredly be collected with 100% efficiency.

For the reason cited, and because it is the larger value, the design should be based upon 8200 ft^2 of collection surface rather than the value of 6600 ft^2 obtained by the Kunz and Hanna method, which assumed lognormal size distribution. The latter involves some mathematical approximation and also the reading of a chart, both points introducing an unknown degree of error.

Allowance should be made for uneven velocity distribution, for gas sneakage, and for rapping reentrainment. This may be done conveniently through use of w_e. The value of w_e, the uncorrected "effective migration velocity" or "precipitation rate parameter" will be, from

$$w_e = -Q\ell n\ P/SL = -222\ell n\ 0.0106/8210 = 0.123\ \text{ft/s.}$$

According to the methods and examples given above we may take (for all particle sizes) F = 1.2 in (7.27) and (S+R) = 0.05 in (7.32) or (7.33) assuming N = 5. For a corrected $P_{S+R} = 0.0106$, the uncorrected value from (7.32) must be

$$0.0106 = \left[\ 0.05 + (0.95)(P)^{1/5}\right]^5$$

$$P = \left[\frac{(0.0106)^{1/5} - 0.05}{0.95}\right]^5 = 0.00706$$

The corrected precipitation rate parameter will then be:

$$w_{e_{corr}} = 0.123 \times \frac{\ell n\ 0.0106}{\ell n\ 0.00706} \times \frac{1}{1.2} = 0.0941\ \text{ft/s.}$$

and the corrected surface (SL) will be

$$(SL) = -\ 222\ \ell n\ 0.0106/0.0941 = 10{,}730\ \text{ft}^2$$

The remainder of the design involves specifying plate dimensions and spacing. Assuming v_o = 2.5 ft/s, plate spacing of 8", and plate height of 15 ft, would require

$$\frac{13,300}{60} \times \frac{12}{2.5 \times 15 \times 8} = 8.87 \text{ spaces}$$

Taking 9 spaces will require the equivalent of 10 plates, each 15 ft high by

$$\frac{10,730}{15 \times 10 \times 2} = 35.8 \text{ ft. long}$$

and the actual value of

$$v_o = \frac{222}{19 \times 8/12 \times 15} = 2.47 \text{ ft/s.}$$

The assumption of N = 5 sections seem reasonable for a total plate length of nearly 36 ft. This would make the length in each section just over 7 ft (7.2 ft).

In comparing these results with the previous example it is noted that a much greater collection surface is required even though the allowable penetration is greater. This is due in part to the corrections for poor velocity distribution, sneakage and re-entrainment, and in part because the particle size distribution is finer in this case.

V. COST ESTIMATION

Once the total surface area (SL) of the plates has been determined, the purchase cost may be simply estimated from [29]

Cost = 218,000 + 6.77 (SL) for insulated unit

Cost = 158,000 + 4.50 (SL) for uninsulated unit

for a dry precipitator, when (SL) in sq. ft. ranges

$$500 \leq SL \leq 1,000,000$$

The cost of a wet precipitator may run from 2.0 to 2.5 times the cost of a dry one of the same total plate area. Costs are in 1981 dollars.

For the final example above, SL = 10,730 ft^2 and the estimated cost will be (dry, insulated) $290,640.

REFERENCES

1. H. White, in Handbook of Air Pollution Technology, S. Calvert and H.M. Englund eds., Chap. 12, Wiley Interscience, New York, 1984.
2. M. Robinson in Air Pollution Control-Part 1, W. Strauss, ed. Wiley Interscience, New York, 1971.
3. S.J. Oglesby et al., A Manual of Electrostatic Precipitator Technology - Part 1 Fundamentals, U.S. Dept. of Commerce NTIS, PB 196, 380. (Aug. 25, 1970).
4. S.J. Oglesby et al., A Manual of Electrostatic Precipitator Technology - Part 1 Fundamentals, U.S. Dept. of Commerce NTIS, PB 196, 380, page 188, (Aug. 25, 1970).
5. J.R. McDonald, A Mathematical Model of Electrostatic Precipitation: Vol. 1 Modeling and Programming, EPA 600/7-78-111a, June 1978.
6. J.R. McDonald, A Mathematical Model of Electrostatic Precipitation: Vol. 2 User Manual EPA 600/7-78-111b, June 1978.
7. S.J. Oglesby et al., A Manual of Electrostatic Precipitator Technology - Part II Application Areas, U.S. Dept. of Commerce NTIS, PB 196, 381, (Aug. 25, 1970).
8. S.J. Oglesby et al., A Manual of Electrostatic Precipitator Technology - Selected Bibliography, U.S. Dept. of Commerce NTIS PB 196, 379, (Aug. 25, 1970).
9. J.D. McCain, J.P. Gooch, W.B. Smith JAPCA, 25: 117 (1975).
10. J.H. Abbott and D.C. Drehmel, Chem. Eng. Prog., 72: 47 (1976).
11. J. Katz, JAPCA, 30: 195 (1980).
12. A.B. Walker, G. Gawreluk JAPCA, 31: 1303 (1981).
13. J.R. Bush, P.L. Feldman, M. Robinson JAPCA, 29: 365 (1979).
14. M.W. First, JAPCA, 35: 1286 (1985).
15. M. Crawford, Air Pollution Control Theory - Chap. 8, McGraw-Hill, New York, 1976.
16. W.B. Smith and J.R. McDonald, Journ. Air Poll. Contr. Assoc., 25: 168 (1975).

17. R. Cochet, Compt. Rendues 243: 243 (1956).

18. R. Cochet, Colloq. Intern. Centre Natl. Rech. Sci., (Paris) 102: 331 (1961).

19. J.P. Gooch and N.L. Francis, Journ. Air Poll. Contr. Assoc., 25: 108 (1975).

20. G. Leutert and B. Bohlen, Staub, 32: 27 (1972).

21. J.P. Gooch and J.R. Mcdonald, A.I.Ch.E. Symp. Ser. 73: 146 (1977).

22. P.L. Feldman, Paper No. 75-02-3, Annual Meeting Air Poll. Cont. Assoc., Boston, MA, 1975.

23. P.S. Farber, Paper 80-32.4, Annual Meeting Air Poll. Contr. Assoc., Montreal, Que., 1980.

24. R.G. Kunz and O.T. Hanna, Ind. Eng. Chem. Process Des. Dev., 11: 623 (1972).

25. P. Cooperman and G.D. Cooperman Atmos. Environ. 16: 307 (1982).

26. P. Coopermann, Atmos. Environ., 5: 541 (1971).

27. P. Cooperman, Proc. Fourth Int. Clean Air Cong., 835-838 (1977).

28. G. Cooperman Atmos. Environ., 18: 277 (1984).

29. W.M. Vatavuk in Handbook of Air Pollution Technology, S. Calvert and H.M. Englund, eds., Chap 14, Wiley Interscience, New York, 1984.

PROBLEMS

1. Using the conditions and values specified for Eqn. (7.9), together with appropriate values for C_i, calculate values of the size-dependent term $C_i q_i^* / d_{p_i}$ in Eqn. (7.3) after 1 sec for particle sizes in the neighborhood of 0.2 µm. Show that this term goes through a minimum value. Compare the particle size at which this minimum occurs with that obtained from Eqn. (7.13).

2. Repeat the solution to Example 1 using the Feldman approach, Eqns. (7.19), (7.20), instead of the Kunz and Hanna method. Explain any assumptions you found it necessary to make, and discuss how the results of the two methods compare with each other.

3. (a) Two solutions for required plate surface (SL) are given for the design example in Sec. IV of this chapter. Check the solutions by calculating the grade-efficiency of appropriate particle sizes and combining these grade-efficiencies into an overall efficiency. Which solution appears to be better? Discuss reasons for the result you find.

(b) Explain why the average gas velocity is not used in calculating the value of (SL) in this example. Do you think there would be any effect on the collector performance if v_o = 10 ft/s instead of 2.5 ft/s as taken? Is this related to Feldman's approach in any way?

4. Refer to Example 1 in this chapter:

(a) What is the value of the effective migration velocity (or precipitation rate parameter)?

(b) What is the "cut" diameter?

(c) What value of a "mean" particle size could be used to represent the overall performance? Does this correspond to any of the "means" defined in Chapter 2? How does it compare with Cooperman's t_{mean} as given by (7.22) and (7.23)?

(d) Estimate the value of the overall collection efficiency if the rate of gas flow were to double during operation.

5. Reconsider Example 1 in this chapter from the point of view of the Cooperman theory. Assume D_ℓ = 1000 cm2/s, and then 10000 cm2/s, with f = 0.05. What would be the collection efficiency, in each case, if all other operating conditions remain as given in the example?

6. Refer to Example 3 in Chapter 3 of this text. Assuming only the inlet particle-size distribution stated there, develop a design for an electrostatic precipitator to treat 5000 acfm at 300 f, 1 atm, to

collect this dust with 98.5% efficiency. Select any reasonable set of operating parameters, and determine the number and size of plates, spacing between plates, and any other conditions you think are important.

7. The fly-ash from a pulverized coal fired furnace has a particle-size-distribution such as given in Feldman's table just below Eqn. (7.19), and a density of 2.5 gm/cm^3. It is emitted at the rate of 170 lb/ton of coal fired in a flue gas stream (Mol. Wt. = 28.1) of 14.7×10^6 cu.ft./hr at 300 F and 1 atm. A collection system is to be designed to meet the emission regulation of 0.10 lb/million BTU. The coal used has a heating value of 12,800 BTU/lbm and is fired at the rate of 35 tons/hr. Consider the use of an electrostatic precipitator (either with or without a primary collector ahead of it) for this purpose. Estimate the collecting surface required and propose an arrangement for the plates: number in parallel, spacing, height, length and number of compartments.

8. For the Design Example in Sec. IV of this chapter, write a grade-efficiency equation in the simplest form, using the final conditions determined for the design. Compute the "cut diameter". Assuming that the ESP has been constructed in accord with this design, discuss the effect on the cut diameter if each of the following changes were to occur separately. Try to consider all aspects of the collection process, using the theoretical model as a guide wherever possible.

(a) the gas flow rate drops to 9000 acfm;
(b) the field strength drops to 6 kv/in;
(c) the temperature increases to 300 F.;
(d) the discharge wires in one of the 9 spaces between plates are disconnected.

8
Filtration

I. GENERAL CONCEPTS

Filtration may be defined as a dry method of particulate collection in which an array of many individual targets is assembled into a porous structure through which the aerosol-laden gas is passed. The aerosol particles impinge on the targets due to any of the modes of aerodynamic capture and are held there. As collection proceeds, particles impinge upon previously collected particles and a deposit is built up which in turn may itself become the principal collecting medium. The collected dust must eventually be removed somehow or it will hinder the gas flow to an intolerable degree. Filtration is therefore a repeated cyclical process of collecting and cleaning.

There are basically three types of arrays employed: (a) individual fibers loosely packed into a pad or mat; (b) granular material in either a fixed, moving, or fluidized bed; (c) fibers, woven or felted into a fabric, which is used in the form of a bag. All of them are capable of achieving very high collection efficiency with reasonable pressure drop up to the time of cleaning. Filtration is therefore used to collect very fine particulate under dry conditions. For an extensive description of types of devices and of theoretical principles three principal references may be cited: the book, *Air Filtration*, by C.N. Davies [1], the state-of-the-art study, *Handbook of Fabric Filter Technology*, by Billings and Wilder [2], and the recent *Handbook of Air Pollution Technology*, Chapter 11, by Turner and McKenna [3].

The principal limitation to its application lies in the nature of the material which may be used for the target fibers or granules. For fibrous filters or fabric filters, a wide variety of natural and synthetic fibers are available. It is necessary to select one which will have a suitably long life at reasonable cost under the conditions prevailing in a given problem. For example, a polymer (e.g., nylon, Teflon) or inorganic (e.g., fiberglass) fabric may be specified to meet the conditions of higher temperatures or contact with corrosive gas. Slag-wool or metallic mesh may be used in fibrous mats. Granular bed filters may be made of inert inorganic granules (sand, gravel, ceramics) capable of operation at very high temperatures.

The method of cleaning is also very important. Fabric used in the form of a bag or tube is cleaned mechanically by shaking, or blowing gas in the reverse direction. The material must have sufficient mechanical strength to withstand the repeated flexing involved. Fibrous pads are either discarded and replaced periodically or may be washed and/or dumped periodically, then repacked. Moving beds or fluidized beds have the possibility of being cleaned mechanically and continously.

An understanding of the initial collection process must be based upon the principles of aerodynamic capture discussed in Chapter 5. The principal mechanisms involved are inertial impaction, Brownian diffusion, and in some cases electrostatic charging. However, after the initial collecting period particles are captured upon previously collected particles in what may be a random and erratic fashion. The aerodynamic capture theory is primarily useful in predicting the performance of a clean filter.

Furthermore, in the case of fabric filters some new phenomena come into play which are not covered by the concepts of aerodynamic capture. Some particles may pass straight through without being captured at all. Collected particles are observed to seep through the layer of dust and fabric under the continued flow of gas. "Pinholes" may form which permit little plugs of particles to loosen and pass through. These are referred to as "fault processes" and they may be the principal deterrent to otherwise attaining 100% collection by use of a fabric.

Finally, the target elements themselves, in actual practice are not like the idealized targets (cylindrical fibers, spherical granules) treated in capture theory. They are irregular in shape and random in orientation in the array.

For all of these reasons the design of filtration collectors is very much more of an art, based upon prior experience, than is the design of other kinds of collectors. It is worthwhile to begin the design by calculating the magnitude of the several dimensionless groups involved in aerodynamic capture, as listed in Table 4, Chapter 5. This will serve to identify the important mechanisms and to understand, at least qualitatively, the performance trends which may be expected, as tabulated in Table 4, Chapter 5. For fibrous and granular filters the theory of arrays as given in Chapter 5 may be used to estimate collection efficiency. However, this may not be very applicable to fabric filters.

The pressure drop across a filter is also an important factor in design. It tends to increase in proportion to the amount of dust collected. The maximum pressure drop, just before cleaning or replacement, must be estimated in order to determine the power required for the operation. Usually this is the limiting factor which determines when cleaning must begin. Some theories and empirical relationships are available for estimating ΔP.

II. FIBROUS FILTERS

A. *COLLECTION EFFICIENCY*

The process of deposition of particles in a fibrous filter can be described as occuring in four stages which may be identified as follows [4]. First, when the filter is new and clean, particles deposit on the fiber surfaces by any of the mechanisms of aerodynamic capture. Second, there is a period when particles deposit on other particles, already deposited, forming dendrites. If the deposition is mainly by inertial impaction and/or interception, dendrites grow only on the front surface of fibers, but if deposition is by Brownian diffusion dendrites will grow over the entire surface [5]. Third, the dendrites grow until they intermesh, forming a coating around each fiber. Finally, the coatings bridge the gaps between fibers and form an internal cake. Thus, the entire process is intrinsically a transient one.

The treatment of arrays in Chapter 5, in terms of the single fiber efficiency η_T, deals essentially with the first stage. This forms the basis of a design procedure which will be presented first. There has been much research, largely by simulation methods (briefly summarized in Chapter 5 also), which has led to a good understanding of the dendrite growth process of the second stage. This has not yet produced a complete design method for the entire transient process. The extent of applicability of these results will be summarized briefly below.

1. *Initial capture* The overall efficiency of a clean filter may be estimated by use of Eqns. (5.88) and (5.89), for an array of fibers:

$$P_i = \exp - \eta_{T_i} \frac{(1-\varepsilon)}{\varepsilon} \frac{4L}{\pi D} = \exp - \eta_{T_i} \frac{D\ell'}{\varepsilon} = \exp - S\eta_{T_i} \tag{8.1}$$

Filters of this type are characteristically made up of fibers ranging from approximately 0.5 μm to 50 μm diameter, with porosity values usually greater than 0.95 and thickness ranging from a millimeter to several centimeters. Some typical actual values, together with the calculated solidarity factors (S) and calculated values of the total effective fiber length per unit cross-section of filter (ℓ') are:

D-μm	ε	L-cm	S	ℓ'-μm^{-1}
50	0.995	3.0	3.8	0.0756
3	0.997	1.5	19.2	6.38
11.	0.951	0.71	42.3	3.66
1.24	0.925	0.05	41.6	31.0
0.50	0.970	0.13	102.	197.9

The values of D quoted are either based upon a uniform fiber diameter or are an appropriate effective mean fiber diameter.

Values of the single fiber efficiency (η_{T_i}) in the array may be obtained by the theoretical principles outlined in Chapter 5 using the Kuwabara flow field. The most extensive study of this kind is that of Yeh and Liu [6]. It takes into account inertial impaction, direct interception, and Brownian diffusion operating simultaneously. Electrostatic and gravity effects are assumed to be negligible. The Kuwabara model is modified to allow for consideration of slip-flow at low Knudsen numbers based upon the fiber diameter: $Kn_f = 2\lambda/D$.

The velocity field, as determined by Yeh [7], reduces to that given by Eqn. (5.73), as $Kn_f \to 0$. The value of the Kuwabara number Ku, given by Eqn. (5.74), is modified to

$$\gamma = -\frac{1}{2}\ell nc - \frac{3}{4(1+Kn_f)} + \frac{c}{1+Kn_f} + \frac{Kn_f(2c-1)^2}{4(1+Kn_f)} - \frac{c^2}{4} \qquad (8.2)$$

and $\gamma \to Ku$ as $Kn_f \to 0$.

Unfortunately the only way to obtain η_{T_i} by Yeh and Liu's method is to do the numerical solution of the differential equations for particle trajectories. They present results of this in graphical form and compare them with the results of the method outlined by Davies and given by Eqns. (5.76), (5.77), (5.78), and (5.79). Within the limitations of $R < 0.1$, $Pe > 100$, stated there the agreement is good at low Kn_f. As R increases above 0.1, or as Kn_f increases above 0, there is considerable disagreement and Yeh and Liu's charts would have to be used. Their theoretical results are well substantiated by their experimental measurements [8], for the case of a uniform value of D. An alternative very similar procedure, involving somewhat different expressions for the caluclation of η's, is listed by Billiet and Stenhouse [9].

The effect of a mixture of fiber sizes in a filter has been considered by Clarenburg and Van der Wal [10]. They found that provided a mean effective fiber diameter $\bar{D}_E$ was properly defined for the mixture, the porosity of the filter affected the penetration according to

$$-\ln P = \Gamma \frac{L}{\overline{D_E^2}} \frac{(1-\varepsilon)^2}{\varepsilon} \tag{8.3}$$

where Γ was independent of the composition of the filter. The definition of $\bar{D}_E$ is given by

$$\log \overline{D_E^2} = \sum_{j=1}^{n} x_j \log \overline{D_j^2} \tag{8.4}$$

where j designates of a certain diameter present in weight fraction x_j (corrected for the shadow effect of one fibre over another). The porosity function in (8.3) is confirmed experimentally [10]. To reconcile (8.3) and (8.1) would require knowledge of how η_{T_i} is related to Γ, which must be a function of particle size. This is not available, but presumably the net effect of ε upon η_T is such as to introduce the additional $(1-\varepsilon)$ factor in (8.3). For filters of a given composition, i.e., given "mix" of fibers, the effect of varying degrees of compaction upon collection efficiency can be determined by (8.3).

The calculation of η_{T_i} by the procedure outlined above is not always as elaborate as that may seem. If the various dimensionless groups are calculated first, (see Table 4, Chapter 5) usually it will be found quickly that one mechanism predominates and subsequent calculations need only be done for that one.

The above material pertains only to the initial performance of a clean filter. However, it is observed that collection efficiency tends to increase, sometimes exponentially, as collection proceeds. Design based upon the behavior of the clean filter will therefore be conservative.

2. *Adhesion and retention* It is well recognized in all the theory of aerodynamic capture, that just because a particle collides with a target it does not necessarily adhere to the target. Adherence may be more likely if collision is due to Brownian diffusion or electrostatic forces, than if it is due to inertial impaction or direct interception. Adherence is certainly greater on fibers which have been coated with a liquid than upon dry fibers. Retention is more difficult at larger particle sizes because they are more likely to be subject to "bounce-off" or the subsequent "blow-off" (reentrainment by drag). These effects have been described in terms of a "retention" or "adherence" efficiency η_R, acting upon the initial aerodynamic capture or "collision efficiency" η_c such that the effective filter efficiency is given by

$$\eta_s = \eta_c \times \eta_R \tag{8.5}$$

A series of experiments by Freshwater and Stenhouse [11] have elucidated some of the factors affecting η_R. These experiments were done mostly on relatively coarse (d_p > 1μm) particles. They investigated the effects of (a) relative humidity of the air (% R.H.); (b) particle size (d_p); (c) air velocity (v_o) up to 600 cm/s.; (d) dust loading. The forces of adhesion increased with % R.H. and became relativley constant between 35% and 55%. According to inertial impaction theory, the single target efficiency should increase both with d_p or v_o, as they enter into Stk. But it was found that as either d_p or v_o was increased, η_R (and η_s) began to decrease above certain values as the bounce and/or drag effects increased. When fibers were oiled these effects disappeared. The same phenomena were noted with respect to dust loading. At higher loading, evidently agglomerates of captured particles build up and are more subject to removal by air drag.

The operation of filters at high velocities is very desirable because it reduces the cross-sectional area needed to handle a given volume of gas. It also may make inertial impaction a significant collection mechanism even on sub-micron particles. Consequently an entirely different line of experiments was carried out by First and Hinds [12] using sub-micron particles, both solid and liquid, and very high velocities (up to 1524 cm/s). The filter studied was a fiberglass mat composed of fibers having a median diameter of 4.1 μm, and packed to a porosity of either 99.9% (1.9 cm thick) or of 97.7% (0.40 cm thick).

The behavior of the liquid and the solid particles was essentially different as shown by the data tabulated below. These efficiencies were determined by light-scattering measurements.

	Efficiency η_s – at v_o (cm/s) –					
d_p–μm	250	500	750	1000	1250	1500
			Liquid (DOP) aerosol			
0.80	0.14	0.85	0.96	0.98	0.99	0.99+
			Solid (PSL) aerosol			
0.36	0.46	0.79	0.87(b)	0.86	0.74	0.73
0.56	0.37	0.75	0.85	0.78	0.68	0.65
0.80	0.43	0.88	0.72	0.55	0.61	0.70
1.10	0.72(a)	0.63	0.55	0.56	0.72	0.83

(a) Maximum efficiency 0.75 at v_o = 300 cm/s.

(b) Maximum efficiency 0.88 at v_o = 900 cm/s

The liquid particles are captured in almost quantitative accord with capture theory for impaction and interception, that is η_R in (5) is essentially equal to 1. The solid particles exhibit a maximum efficiency at a velocity which is higher the smaller the particle. A "blow-off" reentrainment appears to set in with η_R increasing as v_O increases. The presence of another reversal of trend at the highest velocities for the two largest particles sizes is unexplained.

Ellenbecker et al. [13] also explored this range of high velocities (in which inertial impaction is the predominant capture mechanism) by filtering both solid and liquid particles through mats of stainless steel fibers, 8 μm diameter, packed to 98.15% porosity. They found that the efficiency for collection of solid particles began to drop off in comparison with that for liquids at a value of Stk ≈ 1, this being somewhat lower at higher velocities. Put another way, the adhesion efficiency η_R began to drop below 1 when the particle kinetic energy increased above 10^{-16} J. The maximum single fiber efficiency they could obtain was about 40% for the solids, but approached 100% for liquid particles. Once again, all these experiments point up the basic difference in behavior between solids and liquids, as was emphasized in Chapter 6 on cyclone performance.

No quantitative methods have been developed by which values of η_R may be correlated or predicted.

3. *Transient behavior* The theoretical calculation of the transient behavior of fibrous filters has been summarized by Okuyama and Payatakes [14]. The methods have been developed primarily by deterministic simulation of dendrite growth and relate only to the second stage of particle deposition. They yield results for the change in both the collection efficiency and the pressure drop (see below) as the filtration proceeds.

They begin with the calculation of efficiency (and pressure drop) for a clean filter and proceed to correct this through the use of a "specific correction multiplier". This multiplier (always greater than 1) is found as a function of the "specific deposit" (volume of deposited matter per unit filter volume) integrated both throughout the depth of the filter and over the time of filtration. The calculations involve the growth of the dendrites as determined by the modes of capture, and take into account inertial impaction, interception and Brownian diffusion but no other mechanisms. Provision is made for insertion of η_R. Complete solution of the differential equations involved is only practical with the use of computer programs.

Some experimental work on the transient behavior of filters has been done by using model filters consisting of layers of wire screens collecting solid particles [15]. It was found that at lower velocities the efficiency increased monotonically with m, the mass of deposited material per unit area of filter. At higher velocities, the efficiency reached a maximum and remained there, or decreased slightly. All results could be correlated in the form:

$$\eta/\eta_o = 1 + (m/m_o)^b \tag{8.6}$$

where η_o is the initial collection efficiency and m_o is the value of m at which the increase in η over η_o is 100%. This recalls a similar result for dendrite growth on a single cylinder as expressed in Eqn. (5.72).

B. *PRESSURE DROP*

As in the case of efficiency, the calculation of pressure drop across a filter must also be considered in two cases: initial value for a clean filter, and transient value as filtration proceeds. Most of the studies available have been done on clean filters, and this case will be discussed first.

The concept of "number of inlet velocity heads" N_H, discussed in Chapter 3, is not used in relation to filters because the flow is almost always under conditions of laminar motion, i.e.,

$\Delta P \propto v_o$ (although not necessarily at the high velocities in the studies mentioned above). Provided that Kn_f is low enough that slip past the fibers is not significant, at $Re_D < 1$, Darcy's law may be taken as the basic relation for ΔP, in the form

$$- \Delta P \, D^2/L v_o \mu = f(\varepsilon) \tag{8.7}$$

The quantity $v_o L/\Delta P$ is called the "permeability" of the filter and is a unique function of the fiber size and bed porosity for geometrically similar filters in the same fluid.

Davies [16] found the form of $f(\varepsilon)$ in (8.7) to be

$$f(1-\varepsilon) = 64 \, (1-\varepsilon)^{1.5} \left[1 + 56(1-\varepsilon)^3\right] \tag{8.8}$$

by experiments with a variety of filter pads. This was also confirmed by Clarenburg and Werner [17]. Values of ε ranging from 0.70 to 0.994 were covered. For mixtures of fibers a mean square diameter D_E^2 must be used similar to that in (8.4) above. From either the Happel or the Kuwabara flow field theory (see Chapter 5), a theoretical expression for $f(\varepsilon)$ may be derived. These expressions give results which lie on either side of (8.8) depending upon whether the fibers are assumed to lie parallel or transverse to the direction of flow.

At higher flow rates (higher Re_D), and especially for relatively thick pads, Darcy's law is no longer followed both because of turbulence and also because the pads tend to be compressed. A comprehensive study by Wright, Stasny and Lapple [18] indicates that the effect of Re_D may be taken into account by a modified form of (8.7):

$$-\Delta P \cdot D^2/L v_o \mu = 8(1-\varepsilon) C_f \tag{8.9}$$

where C_f = flow coefficient, obtained from a chart, Fig. 1, as a function of Re_D and porosity ε.

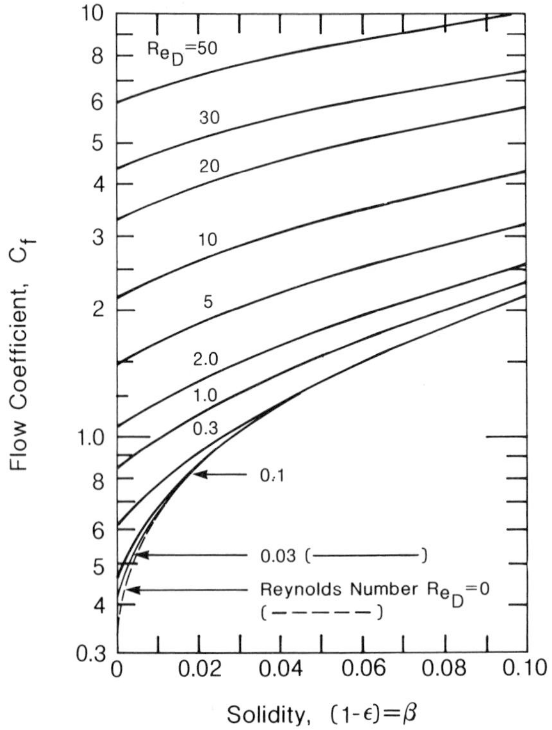

FIG. 1 *Flow coefficient for pressure drop in fibrous beds [18]. See Eqn. (8.9).*

A formula proposed by Kimura and Iinoya [19] for glass and steel fibrous filters, correlating much experimental data, may be written in the form of (8.9), giving

$$-\Delta P \cdot D^2/Lv_o\mu = (0.6Re_D + 4.7\sqrt{Re_D} + 11)\ 2(1-\varepsilon)/\pi \qquad (8.10a)$$

This would correspond to the flow coefficient C_f, given by

$$C_f = (0.6\ Re_D + 4.7\sqrt{Re_D} + 11)/4\pi \qquad (8.10b)$$

Values of C_f so calculated agree well with those read from Fig. 1, at porosities $\varepsilon \approx 0.97$ for $Re_D \leqq 5$. Above $Re_D = 5$, the calculated values correspond to larger ε's, and at $Re_D = 50$ the $C_f = 5.9$ is off the chart. Of course (8.10b) does not indicate an influence of ε upon C_f such as is given by the abscissa of Fig. 1.

A theoretical form for $f(\varepsilon)$ in (8.7), obtained from the Kuwabara flow model [20], is:

$$f(1-\varepsilon) = 16\ (1-\varepsilon)/Ku \tag{8.11}$$

where Ku is given by Eqn. (5.74). This is valid only in the range of D'Arcy's Law.

The theoretical models for transient behavior [14] also provide a method for calculating the increase in pressure drop as filtration proceeds during the second stage. As in the case of the efficiency calculation, a "specific correction factor" as a multiplier (always greater than 1) onto the ΔP value for a clean filter is determined as a function of the amount of material deposited. This is based upon the same simulation model for dendrite growth due to the three basic capture mechanisms as in the case of efficiency. Once again, computer methods are required.

The experimental work with model screen filters [15] leads to an empirical relation of the form

$$\Delta P/\Delta P_o - 1 = a\ m^b \tag{8.12}$$

No general methods for predicting a and b are available.

C. *DESIGN EXAMPLE*

A filter is to be made up of fibers having an effective diameter of 10 μm and a density of 2.34 gm/cm^3, packed with a porosity of 97.0%. It is proposed to use it to collect dust having a mass median equivalent diameter of 0.5 μm with an initial efficiency of at least 90% by weight. The gas flow is to be 500 acfm at 20 C, 1 atm. Determine the cross-sectional area and thickness of the filter required, and estimate the pressure drop across it.

For the sake of simplicity we begin this design by using only the MMD particle size, laying aside the question of particle size distribution until later. The design will be based only upon clean filter behavior.

The cross-sectional area, thickness, and pressure drop of the filter are all interrelated through the lineal velocity. If a value of v_o is specified each of these items may be determined. However, there is certainly a range of possible values for v_o, and the optimum value must be sought. We illustrate the process by selecting two values: 40 ft/min. and 100 ft/min.

For each v_o selected, all of the dimensionless parameters (Re_D, Stk, Pe^{-1}, G Stk, R, C, Ku, J) must first be calculated. Then the single target capture efficiency for each important mechanism must be found and these values combined into the overall target efficiency for the array. The procedure is as outlined in Chapter 5 and illustrated by example 7 there. To do this we need values for C = 1.33, and $\mathcal{D} = 6.26 \times 10^{-7}$ cm²/s, for the d_p = 0.5 μm .

The results of these calculations are as follows:

v_o	40 fpm	100 fpm	Eqn.
Re_D	0.135	0.338	
Stk	0.048	0.120	
R	0.05	0.05	
Pe^{-1}	3.1×10^{-5}	1.2×10^{-5}	
G Stk	1.15×10^{-4}	4.6×10^{-5}	(5.66)
c	0.03	0.03	
Ku	1.033	1.033	(5.74)
J/Ku	0.058	0.058	Table 6, Chap. 5
η_{II}	0.00269	0.00674	(5.78)
η_{DI}	0.00234	0.00234	(5.77)
η_{BD}	0.00375	0.00208	(5.79)
η_G	0.00011	0.00004	(5.66)
η_T	0.0089	0.0112	(5.76)

The values of Stk are low enough so that bouncing should be negligible and η_R = 1. Using these results, the value of L is obtained from Eqn. (5.88) and ΔP from (8.8) above. The area is found from setting $v_o = Q/A$.

v_o	40 fpm	100 fpm
S	259	206
L	6.58 cm.	5.23 cm.
ΔP	3.2 "H_2O	8.0 "H_2O
A	12.5 ft^2	5 ft^2

As the velocity increases the cross-sectional area and thickness of the filter decrease and the pressure drop increases. The optimum value of v_o will have to be determined by an economic balance over initial cost (A and L) against operating cost (power, ΔP).

The overall collection efficiency of a fibrous filter may be greatly increased if the filter is augmented by electrostatic charging. This method of filtration is finding increasing applications. Such an augmented filter may be modelled by adding the appropriate form of η_{ES} to the list involved in finding η_T. The calculation of η_{ES} would be done by the methods explained in Chapter 5. As an illustration of this, please refer to Problem 1(b) at the end of the present Chapter.

A more complete design taking into account the particle size distribution would follow the procedure illustrated by the design example for electrostatic precipitators, Chapter 7. Here, for each value of v_o, a set of parameters corresponding to several size intervals would be calculated, and a value of η_{T_i} for each interval determined. Then S would be found by solving an equation of the form:

$$P = \Delta g_i \sum_i e^{-S\eta_{T_i}}$$

Since η_{T_i} would increase rapidly with the value of d_{p_i}, the higher terms of this series would drop off rapidly. A solution could be obtained using only the first few terms as was done in the ESP example.

If calculations of η_T are performed over a range of particle sizes, it will be found that η_T goes through a minimum value at a size somewhere in the range of 0.2 - 0.8 μm. Below that range capture by Brownian diffusion predominates, and above that range inertial impaction comes to the fore. Within that range capture is due to diffusion and interception. Using their own formulation for $\eta_{II} + \eta_{BD}$, which is not unlike that of Davies, Eqns. (5.78) and (5.79), Lee and Liu [21] derived an expression for estimating the "most penetrating" particle diameter in cm as:

$$d_{p,min} = 0.885 \left\{ \left(\frac{Ku}{\varepsilon}\right)\left(\frac{\sqrt{\lambda}kT}{\mu}\right)\left(\frac{D^2}{v_o}\right)\right\}^{2/9} \tag{8.13}$$

III. GRANULAR BED FILTERS

The use of a granular bed (composed, for example, of sand or pebbles) as a filter is an old idea which has recently received renewed attention because of the possibility of using it at high temperatures and high pressures. Such conditions will be encountered in the development of gas cleaning from "advanced" energy processes such as coal gasification and fluidized bed combustion.

Granular beds may be classified as fixed, intermittently moving, continuously moving, or fluidized. The type of operation is primarily related to the method of cleaning. Cleaning requires removing the deposited dust from the granules either mechanically (shaking or moving the bed) or by back-flow of air. Fluidized beds may be continually renewed by fresh granules.

The basic model may be taken to be an array of spherical targets as presented in Chapter 5. The porosity of granular arrays is much lower than that of the fibrous filters discussed above. Fluidized or moving beds will have a higher porosity than fixed beds which usually will be of the order of ε = 0.40 - 0.50.

A. *FIXED BEDS*

This term will be taken to include all beds in which the targets are in contact with each other and not moving relative to each other during the collection process. The porosity of fixed beds is much lower than that of fibrous filters, in the range of 0.35 to 0.60 usually. Although bed filtration is obviously a transient process, almost all of the available material relates to the initial behavior of a clean bed.

1. *Collection efficiency* Because of the increasing interest in granular beds the U.S.E.P.A. commissioned a thorough review and study of them. A review report was presented by Yung, et al. [22]. After studying several models which have been proposed, as well as published experimental data and some new data of their own, they concluded that: (1) For values of $v_O < 10$ cm/s; the array behavior given by Eqn. (5.90) is valid; the total single target efficiency may be predicted by Paretsky's model, Eqns. (5.81), (5.82), (5.83), (5.84) involving II, DI, BD, and G mechanisms. (2) For values of $v_O > 10$ cm/s: inertial impaction is the primary mechanism, but none of the models available at the time were considered to be satisfactory.

Subsequently, Pendse & Tien [23] have developed a general correlation for inertial impaction and direct interception only, by using what they call the constricted tube model. In this model the bed is represented by a set of unit cells in the shape of constricted tubes, instead of the spheres in Happel's model. The result of trajectory calculations for single target efficiency is:

$$\eta_T = (1+0.04\ \text{Re})\left[\text{Stk} + 0.48\left(4-\frac{4R}{d_c^*}-\frac{R^2}{d_c^{*2}}\right)^{1/2}\left(\frac{R^{1.0412}}{d_c^*}\right)\right] \tag{8.14}$$

in which $d_c^* = d_c/D$, where d_c is the diameter of the constricted tubes; the value of d_c^* is taken as 0.35.

An extensive review of theory and experiments dealing with granular beds is also given by Tardos, et al. [24].

2. *Pressure drop* Ergun's general equation [25] for flow through packed beds is found to be applicable to the clean granular filter beds tested.

$$-\Delta P = f \frac{v_o^2 \rho_f L}{D} (1-\varepsilon)/\varepsilon^3 \tag{8.15}$$

where

$$f = 150/Re'_D + 1.75 \tag{8.16}$$

and

$$Re'_D = Dv_o\rho_f/\mu_f(1-\varepsilon) \tag{8.17}$$

It may be noted that at low values of Re'_D, the second term in (8.16) becomes negligible and (8.14) reduces to a form of Darcy's law like (8.7) in which $f(\varepsilon) = 150(1-\varepsilon)^2/\varepsilon^3$. At higher values of Re'_D, (8.14) may be expressed in terms of inlet velocity heads like Eqn. (3.58) in which

$$N_H = 2g_c \frac{fL}{D} (1-\varepsilon)/\varepsilon^3 \tag{8.18}$$

At sufficiently large values of Re'_D, $f \to 1.75$, and N_H becomes a function only of the bed target geometry (L, D, ε).

3. *Transient behavior* A method of dealing with changes in collection efficiency and pressure drop as the filtration proceeds has been proposed by Walata, et al [26]. It is similar in form to that developed for fibrous filters as given by Eqns. (8.6) and (8.12). Thus η_n/η_{M_o} and $\Delta P/\Delta P_o$ are each expressed in the form of

$$1 + am^b$$

where m represents the amount of deposition per unit volume of filter bed. Methods of evaluating a and b from experimental data are proposed, but no generalizations are available.

B. *FLUIDIZED BEDS*

A fluidized granular bed may be contrasted with a fixed one by noting that the particles are moving about and are not in contact with each other, the porosity is much higher, and gas passes through in discrete bubbles. The bubbles provide a means for particulate to pass through the bed without contacting the collection target particles of which the bed is composed. When a fixed bed is fluidized by gradually increasing the gas velocity, the overall collection efficiency is found to decrease markedly at and above the minimum v_o required for fluidization (v_{mf}). This is attributed to the onset of the bubbling process [27], [28].

Although fluidized beds can be continuously renewed without interrupting the filtering process, their use would not be attractive except for the possibility of multistage operation. Two to five stages may be superimposed vertically in a column so as to give the effect of filters in series. McCarthy, et. al. [27] have shown that this arrangement will maintain high collection efficiencies on fine particles ($d_p \approx 0.1$ μm) at velocities up to 3 v_{mf}. On any given particle size the overall efficiency of n stages will be related to that of a single stage (η_1) by Eqn. (5.91). Thus

	Overall Efficiency			
η_1	n = 2	n = 3	n = 4	n = 5
0.60	0.840	0.936	0.974	0.990
0.80	0.960	0.992	---	---
0.90	0.990	0.999	---	---
0.95	0.997	---	---	---

These calculations assume, of course, that each stage operates at the same efficiency and that there is complete mixing of uncollected particles between stages.

Some indication of the collection mechanism operating in a fluidized bed may be obtained by sample calculations assuming spherical targets. The target diameters are typically in the range of a few hundred microns. For sub-micron particulate and velocities in the neighborhood of v_{mf} (a few cm/s) these calculations show that II is insignificant and the collection is primarily due to DI for the coarser particles and BD for the finer ones. If the bed particles become charged, ES forces may increase collection enormously [28].

A comprehensive model of the collection process has been developed by Peters, et. al [29]. It incorporates all of the phenomena of fluidization mechanics and gas by-passing, as well as accounting for all of the modes of aerodynamic capture including induced electrostatic charges. The model requires computerized calculation to predict collection efficiencies, but these agree well with experimental values. It is found that the electrostatic effects, arising naturally, are an essential part of the process. This model could be used for design purposes.

Pressure drop across a bubbling bed may easily be estimated. It is equal to the total weight of bed particles divided by the cross-sectional area of the bed. The total frictional drag on all the suspended particles is just equal to the force required to suspend them, namely their weight.

IV. FABRIC FILTERS

Fabric filters differ from fibrous filters in that the fibers are woven into a fabric which has a much lower porosity than the loosely packed fibrous pads or mats. After a new fabric has gone through several cycles of use and cleaning, it retains a residual cake of dust of part of the filter medium.

The cloth fabric is usually made into cylindrical tubes, of the order of a foot in diameter by 20 or 30 feet long. A number of these tubes are suspended vertically and operate in parallel in a unit called a "baghouse." If the required number of bags is very large, they may be grouped into "compartments" of up to several hundred bags each. The bags are cleaned periodically either by mechanical shaking, by a reverse flow of air, or by a reverse pulse-jet of air. Most of the collected dust drops off into a hopper below although some of it will be resuspended in the air and refiltered. The fabric must have good strength to withstand the repeated flexing which occurs during cleaning.

The use of cloth fabric involves another important consideration, namely that the temperature of operation is limited by the properties of the fabric. In many cases this requires that the gas/dust stream be precooled before entering a fabric filter. The cooling system must become an integral part of the design.

The method of cleaning is very important and is the basis for dividing fabric filters into two classes for purposes of design and performance. The differences may be summarized briefly as follows.

The first class is that in which cleaning is performed by first stopping the flow of dusty gas, allowing dust to settle, and then either shaking the bag mechanically or passing a flow of air through in the reverse direction. This usually requires that extra filtering surface be provided to make up for that portion of the bags which is always off-stream for cleaning. Woven fabrics are used, run at relatively low filtering velocities, and the filtration medium is primarily the cake of dust which builds up on the surface of the fabric.

In the second class, called pulse-jet, cleaning is done without interrupting the flow by forcing a burst of air, lasting a few tenths of a second, down through the bag thus flexing it violently. No extra compartments are needed. Felted fabrics may be used, run at substantially higher velocities, and filtration is done largely by the composite of fabric and residual dust within it.

The nature of the capture process in either case has been viewed as analogous to that in fiber filters, but this is an over-simplified approach. The situation is more complex than that modelled by the array of cylindrical targets as explained in Chapter 5. Billings [30] has viewed it on three levels of scale:

- molecular, dealing with molecular surface interactions of individual particles and fibers; scale $< 10^{-5}$cm.
- microscopic, dealing with local flow through localized deposit of several particles and neighboring fibers; scale $\sim 10^{-2}$ cm.
- macroscopic, dealing with behavior of fabric, bag, and dust deposit as a whole; scale 10^{0} cm.

At each level he has enumerated many factors which affect the performance. Leith and First [31] have found that a significant part of the penetration is due to "fault processes" which permit particles to continue to be emitted from the deposited cake although no fresh particles are entering the filter. In spite of these considerations a system of modelling based upon the target array theory continues to be stated and will be outlined below.

However, the interaction of a particular dust with a particular fabric, under a given set of filtering and cleaning conditions seems to be highly specific. In consequence filter design must be based to a very large extent upon performance experience, more so perhaps than for any of the other collectors. This may be organized in a systematic way by the system analysis shown in Fig. 2. This will also be discussed below, along with more details of the Leith and First findings.

A. *FIBER ARRAY MODEL*

Early approaches to modelling the fabric filter followed precisely the same scheme as that given above for fibrous filters. The basic mechanisms of aerodynamic capture were assumed to account for the collision of particles with fibers, although it was

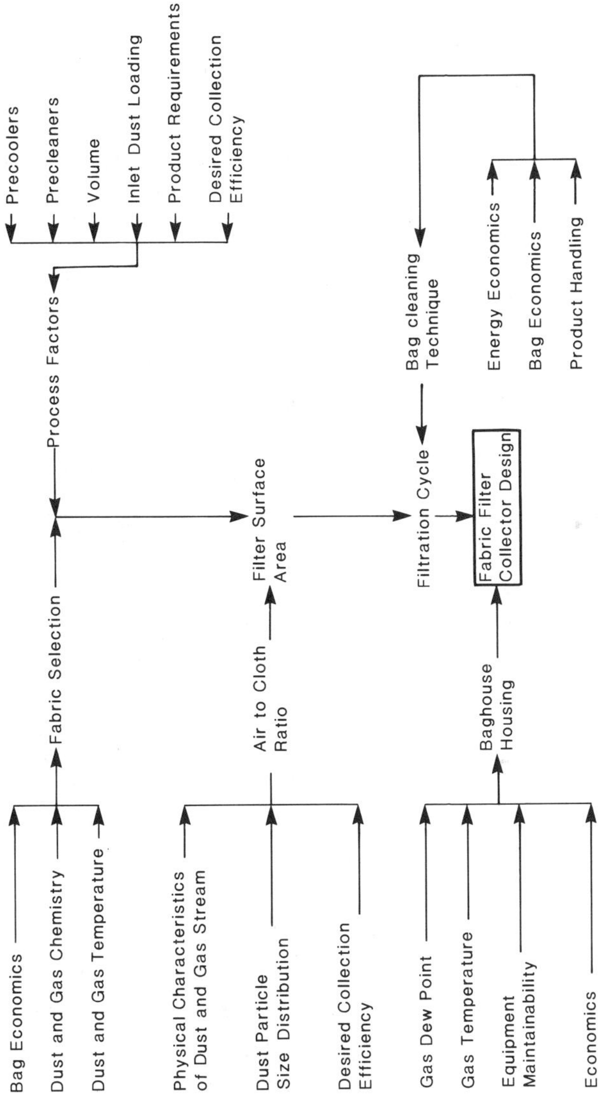

FIG. 2 *System analysis for fabric filter design.*

recognized that fibers making up a fabric are curved and intertwined, and are represented only poorly by an array of straight, parallel cylinders. These mechanisms, no doubt, are indeed involved in the initial capture in a clean fabric. But they do not take into account the subsequent behavior of the dust cake of the composite residual-dust-fabric as collecting medium, nor do they account for the "fault processes".

A representative model of this kind is that of Fraser and Foley [52]. They developed descriptions for: (1) pre-conditioning to cool the gas; (2) fallout of dust into hopper prior to reaching fabric; (3) build-up of pressure across the bags; (4) collection efficiency of the bags; (5) the effect of cleaning on the residual dust and drag. The pre-conditioning involves standard engineering calculations related to the heat exchange involved in the cooling process, which will be illustrated in the design example below. The fallout of dust pertains to the gravity settling of larger particles against the rising current of gas entering the bags. It is based upon the calculation of terminal settling velocities as presented in Chapter 4. The build-up of pressure drop across a bag is not treated by equations such as (8.7) or (8.9) above. Rather it is based upon a semi-empirical approach which is also used in the system design approach discussed below.

Collection efficiency in the Fraser and Foley model is based upon Eqn. (5.88) for the array, using modified Davies Eqn. (5.69) for the single target efficiency. This equation represents the combined effect of II, DI, and BD for an $Re_D = 0.2$, around an isolated cylinder. Note that the Kuwabara cell model, Eqns. (5.76), (5.77), (5.78), (5.79), is not used.

A test of this model against some experimental data showed that while it underestimated the grade-efficiency somewhat, it followed the trend with d_{p_i} down to about 0.3 μm. Below that particle size the model failed to predict the increase in grade-efficiency due to the increasing importance of Brownian diffusion. Further attempts to check this model against experimental performance data revealed that it could not account for the effect of the humidity of the gas nor for the behavior of different types of fabrics.

B. *SYSTEM ANALYSIS MODEL*

Because of all of the complexities mentioned above, the design of fabric filters (baghouses) is best carried out by resort to a scheme of empiricism and experience blended with theory, rather than by a mathematical model for grade-efficiency. For this purpose the System Analysis Chart [33], Fig. 2 is useful to coordinate all of the information involved.

This chart shows that three major decisions must be made: determination of an appropriate fabric, bag cleaning technique and "air-to-cloth ratio." These coupled with the process factors determine the total area of filter surface required, hence the number of bags and size of the baghouse. Subsequently the cycle of filtering and cleaning must be determined, along with the corresponding cycling of pressure drop. Depending upon which class of filter is being designed, there will be significant differences in how these steps are carried out. A comprehensive discussion is presented in the excellent book by Donovan [34], and a convenient recent summary by Turner et al. [35]. The major steps in design are given below.

1. *Air-to-cloth ratio* This term is defined as the volumetric flow rate (Q) of the gas (e.g., acfm) divided by the total filtering surface (A) of all bags (sq.ft). It is obviously the average lineal approach velocity $v_o = Q/A$. Since the value of Q is a given process factor, selection of v_o as a design parameter determines the active or net filtering surface required. It is desirable to have v_o as large as possible, consistent with attaining desired performance, in order to keep A as small as possible.

As shown in the design example for a fiber filter above, and as would be true in use of the Fraser-Foley or similar model, v_o

affects η_{T_i} and ΔP. If capture theory applies, an increase in v_O should increase η_{T_i} for larger d_{p_i} because it will increase the effectiveness of the II mechanism. On the other hand, for very fine d_{p_i} when BD is the predominant mechanism, the opposite affect would be predicted.

However, experience shows that one cannot develop a grade-efficiency relationship from classical capture theory. The mass emission rate of particles from a fabric tends to be independent of particle size, and always to increase with v_O [31]. Excessive air-to-cloth ratios must always be avoided.

One is forced to select v_O on the basis of case studies showing collection efficiency attained with a specified kind of dust using a specified kind of fabric. Extensive tables of such data are available [2,3,35,36] from which a case must be selected to correspond as nearly as possible to that for which the design is to be undertaken. Overall efficiencies of greater than 99% are customarily attainable at values of v_O usually ranging between 1-6 ft/min. for shaker/reverse-flow filters, and 5-15 ft/min. for pulse-jet units.

a. Fault Processes. Recent research by Leith and First [31], and Leith and Ellenbecker [37,38] has elucidated the collection behavior of fabric filters in terms of the concept of "fault processes" instead of capture mechanisms. Penetration is attributed primarily to the filters' inability to retain collected particles, rather than inability to capture them initially. It is postulated that essentially all particles would be captured in attempting to negotiate the tortuous passage through the composite cake/fabric layer, but that those which penetrate do so because of either "seepage" or "pinhole generation."

Seepage refers to the movement of arrested particles from their deposition sites, by a kind of flushing action of the gas. As pressure drop builds up the fabric may be stretched and particles loosened so that the drag of the gas can move them.

Particles may be loosened during the cleaning process, especially during the sudden flexing caused by a pulse-jet. This is a form of reentrainment, occurring relatively slowly. It appears to be the dominant form of penetration in pulse-jet filters and increases as the dust deposit areal density increases. The concentration of emissions is given by [37,38]

$$c_{out} = k\,W^2/t \tag{8.19}$$

where W is the areal density (mass of dust deposit per unit area of bag) at the end of the cleaning pulse, and t is the time of filtration. The constant k = 0.002 m-s/kg for all new fabrics and all dusts tested, but increases in value as the fabric ages.

Pinholes are observed to form in the surface of a filter cake, at the pores between yarns in a woven fabric, or at places where synthetic felt was needle-punched during manufacture. Agglomerates of dust collected above such a hole may break loose and slip through, producing coarser particulate in the emission than was in the feed! High localized gas velocity through a pinhole may prevent it from "healing" or becoming sealed over by additional dust. Instead anthill-shaped deposits surround the pinhole due to an inertial deposition as gas streamlines bend to pass through the hole. Particles may pass straight through the pinholes, by-passing any capture process in a manner analogous to sneakage in an electrostatic precipitator. This appears to be the primary form of penetration in shaker/reverse-air class of filters.

An empirical equation developed by Dennis and Klemm [39] may be used to predict outlet dust concentration of fly-ash from fiberglass bags:

$$c_{out} = \left[P_{ns} + (0.1-P_{ns})\exp - aW\right] c_{in} + c_R \tag{8.20}$$

$$P_{ns} = 1.5 \times 10^{-7} \exp\left[12.7(1-\exp 1.03\, v_o)\right]$$

$$a = 3.6 \times 10^{-3}\, v_o^{-4} + 0.094$$

Here: c_R = residual outlet concentration due to seepage = 5 mg/m^3 ($c_{out} = c_R$ where $c_{in} = 0$)

W = areal dust loading added during filtration g/m^2

v_o = local face velocity m/min

These Eqns. (8.19) and (8.20), are quoted not so much for use in actual computation, but to indicate the differences between seepage and pinhole emission in terms of the variables affecting each. In any case their use would require having experimentally determined values for k, W, and c_R.

The goal of reducing the size of baghouses by operating at higher air/cloth ratios can only be approached by learning how to reduce the total emission flux. If this can be done fabric filters can approach being perfect collectors even of fine particles.

2. *Pressure drop* The flow through a fabric filter (cloth plus residual or deposited dust) follows Darcys' law, (8.7) above, so that $\Delta P \propto v_o$. The proportionality factor, called the drag, includes all the properties of the dust layer and fabric, as well as the effect of temperature (primarily upon μ). Three cases must be recognized: ΔP_c: a clean, unused fabric; ΔP_r: a used fabric containing a residual dust deposit, after cleaning; ΔP_f: fabric with increasing dust layer being deposited as filtering proceeds. For a clean fabric

$$\Delta P_c = K_1 \upsilon_o \tag{8.21}$$

K_1, the "clean-fabric drag," is determined by measuring the value of v_o for which ΔP = 1/2 inch H_2O, for air flow at 20°C. This v_o is called the "permeability" of the fabric, and tables of such values are available [2]. While ΔP_c is usually a negligible part of the total ΔP in operation, relative values of K_1, which depend only upon the composition and structure of the fabric, are useful in comparing fabrics.

For the residual after cleaning

$$\Delta P_r = S_E v_o \tag{8.22}$$

when S_E is the drag characteristic of the cloth/dust system and also of the method of cleaning and cycling. It is assumed that once a cloth is "broken in" by a few cycles of use an equilibrium of residual dust in the fabric will be reached. Thereafter S_E will have the same constant value at the beginning of each cycle. $\Delta P_r \gg \Delta P_c$ and S_E automatically includes the effect of K_1.

For the filtering process, ΔP_f increases as the dust layer builds up in direct proportion to the weight of dust collected. A useful formulation for the cake filtration in shaker/reverse-air cleaning is.

$$\Delta P_f = K_2 \left(\frac{cv_o t}{7000}\right) v_o = K_2 W v_o \tag{8.23}$$

when c = grain-loading of dust, t = time of filtration and 100% collection efficiency is assumed. The quantity in parenthesis is proportional to the cake thickness L as in (8.7) above.

The total pressure drop during filtration builds up linearly with the time:

$$\Delta P_{filt} = \Delta P_r + \Delta P_f = v_o \left(S_E + \frac{K_2 c v_o}{7000} t\right) \tag{8.24}$$

provided v_o is kept constant. Values of S_E and K_2, specific for each case, may be taken from recorded case studies [3,30]. They are obtained from a plot of ΔP_{filt} vs t. The slope of the plot is proportional to K_2 while extrapolation to t =0 gives an intercept from which S_E may be calculated. The values of the K's are found to depend upon humidity of the gas, as well as temperature of operation.

The pressure drop in the non-cake filtration, characteristic of pulse-jet cleaning, is apparently better estimated by a model due to Leith and Ellenbecker [40] which involves the clean fabric drag, the system hardware, and the cleaning energy. It is given as:

$$\Delta P_{filt} = \tfrac{1}{2}\left\{P_s + \Delta P_c \sqrt{([P_s - \Delta P_c)^2 - 4\ \Delta P_f/K_3}\right\} + K_V v_o^2 \qquad (8.25)$$

Here, in addition to the terms already defined, P_S is the maximum static pressure attained in the bag during the cleaning pulse, K_3 is a coefficient relating dust removal efficiency to bag/pulse properties, and $K_V v_o^2$ is the pressure drop across the venturi nozzle through which the air pulse is introduced to the bag. To use this, values of K_1 and K_2 (as above), and K_3 must be determined by experiment.

3. *Design procedure* A simplified procedure suitable for preliminary design studies is outlined below. It includes all of the major points which need to be considered in a screening study, but does not go into great detail. More sophisticated computerized procedures are available in which allowance is made for the fact that different portions of the fabric surface, even on the same bag, may behave differently with respect to cleaning and filtering. For details see [3], [34], or [35].

a. Selection of Fabric. The first step in design is to select the fabric. This must be done with due consideration for a number of factors: ability of fabric to collect the particular dust, ability to withstand the corrosive and erosive effects of gas and dust, ability to withstand elevated temperature, suitable mechanical strength, cost, etc. These are all considered in the light of relevant case history.

The temperature and humidity of the gas are major considerations. Each type of fabric has a limitation on the maximum temperature at which it may be used. Since many filter applications are appropriate for hot gases, a precooling system must usually be designed to precede the baghouse. As cooling proceeds the dew-point of the gas must be kept in mind. Condensation of moisture in the baghouse cannot be tolerated because of the "muddy" cake which would result. It is desirable to keep the gas temperature at least 50°F above the dew-point. Sometimes it is necessary to insulate the baghouse in order to prevent further natural cooling.

Characteristics of some representative fabrics are given in Table 1.

b. Precooling System. Generally there are three methods of gas cooling to be considered when a precooling system is required:

- cooling by dilution with ambient air;
- cooling in a heat-exchanger, or waste-heat boiler;
- cooling by evaporation of a water spray into the gas.

Each method has its advantages and disadvantages.

Cooling by dilution with ambient air is simple and relatively inexpensive. However, the greater the cooling required, the more ambient air is used and consequently there is a large net increase in volume of gas to be filtered. This increases the filtering surface area and the size and cost of the baghouse.

Cooling by means of some sort of heat exchanger will require the additional equipment for the purpose. This may be as simple as an extended surface of ductwork to provide for convective and radiative cooling to the surrounding air. However, transfer rates will be low and the surface required large. If there is a great amount of heat to be removed it may be economically attractive to recover it, for example, in a waste heat boiler where there is a use for process steam.

Evaporative cooling will involve a rather simple spray system. The quantity of water required will be relatively small, hence its vapor will not increase appreciably the total volume of gas to be filtered. However, it will increase the humidity of the gas and the resulting increase in the dew-point will have to be taken into account.

Table 1 *FILTER FABRIC CHARACTERISTICS*

Fiber	Operating exposure of		Supports com-bustion	Air perme-ability cfm/ft² [a]	Composition	Abrasion	Resistance Rating to [b]	
	Steady	Surge					Mineral acids	Alkali
Cotton	180	225	yes	10–20	Cellulose	F–G	P	E
Wool	200	250	no	20–60	Protein	F–G	F	P
Nylon[d]	200	250	yes	15–30	Polyamide	E	P	E
Orlon[d]	240	260	yes	20–45	Acrylic	F–G	G	F
Dacron[d]	275	325	yes	10–60	Polyester	E	G	F
Polypropylene	190	190	yes	7–30	Olefin	G	E	E
Nomex[d]	400	425	no	25–54	Polyamide	E	F	G
Fiberglass	500	550	no	10–70	Glass	P	G	P
Teflon[d]	450	500	no	15–65	Fluorocarbon	F	E	E
Goretex[c]	450	500	no	---	Fluorocarbon film on backing	P	E	E

[a] cfm/ft^2 at 0.5 in . W.G.

[b] P = Poor, F = Fair, G = Good, E = Excellent

[c] W.R. Gore Co. Registered Trademark

[d] Dupont registered trademark.

Source: Ref. [2,3,34]

The temperature and humidity of gas at conditions expected to obtain in the baghouse must also be checked against the dew-point of the gas. The baghouse may need to be insulated to prevent condensation from occuring. Calculations for each of the methods of cooling will be illustrated in the design examples below.

c. Total Filter Surface. As soon as the fabric is selected, the appropriate air/cloth ratio may be determined. The actual volume flow rate (Q) of the gas to the filter may be calculated when the temperature of operation and method of cooling is fixed. Then A is calculated from $A = Q/v_o$. This is called the net cloth area.

A standard bag size (diameter by length) is also selected from manufacturers specifications. The number of bags then is equal to A divided by (πDL). The bags are usually divided up and grouped into compartments so that a compartment may be shut off-stream for cleaning or repairs without disturbing the filtering operation. For this reason it is necessary, particularly in the case of the shaker/reverse-air cleaning mode, to allow additional area. The gross total cloth area required is computed by multiplying the net cloth area by an appropriate factor as given in the following table [43].

Net area-ft^2	Factor
1-4,000	2
4,001-12,000	1.5
12,001-24,000	1.25
24,001-36,000	1.17
36,001-48,000	1.125
48,001-60,000	1.11
60,001-72,000	1.10
72,001-84,000	1.09
84,001-96,000	1.08
96,001-108,000	1.07
108,001-132,000	1.06
132.001-180,000	1.05
>180,000	1.04

d. Determination of Filtering/Cleaning Cycle. The length of the filtering period is set on the basis of the changes in pressure drop and flow rate which may take place as the cake build-up proceeds. This must be done with regard to the characteristics of the fan being used to move the gas through the system, and to the total power consumption.

The fan must provide not only the ΔP of the filter as given by Eqn. (8.24) or (8.25), but also a ΔP_{duct} for the ductwork of the system. The ΔP_{duct} is proportional to Q^2 as the gas is always in turbulent flow. The total ΔP_{sys} at any instant t is therefore of the form

$$\Delta P_{sys} = AQ + BQ^2 \tag{8.26}$$

and ΔP_{sys} must be provided by the fan.

It is characteristic of fans to have an operating relationship between ΔP_{sys} and Q which may be represented by

$$\Delta P_{sys} = C\left(1-Q^2/Q^2_{max}\right)^{1/2} \tag{8.27}$$

in the operating range. These relationships (8.26) and (8.27) interact as shown in Fig. 3.

At the end of cleaning and beginning of filtering (t = 0) $\Delta P_{sys} = \Delta P_r + \Delta P_{duct}$ and the fan delivers flow rate Q_o. As filtration proceeds, added resistance enters the system increasing ΔP_{sys} by ΔP_f. The fan cannot continue to deliver Q_o, so the flow rate decreases along the fan curve. As it does so ΔP_{duct} and ΔP_r decrease somewhat, as ΔP_f continues to build up. THe process moves along the fan curve from the point marked t=0 in the direction of the arrow toward any later time t. If all of the constants in Eqns. (8.24), (8.26) and (8.27) are known, this path may be calculated and the amount of cake collected at t, as related to the ΔP_{sys}, may be determined.

The length of the filtering period (t) is usually determined on the basis of an allowable upper limit to ΔP_{sys}, based upon experience showing when an excessive value is reached. As a first

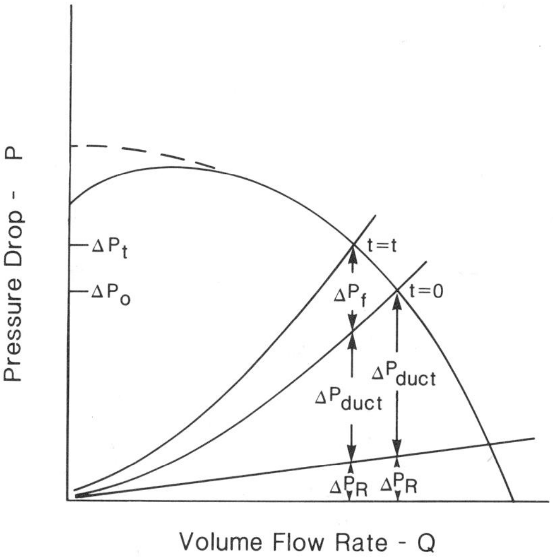

FIG. 3 *Variation of pressure drop with flow rate and time in fabric filter cycle. Equations (8.26) and (8.27).*

approximation, if the fan curve (8.27) is not known, Q_t may be taken as constant at Q_o.

The above remarks apply to a single bag, or to one compartment of bags operating together. For the complete analysis of performance of a number of compartments operating sequentially, see Crawford [41]. He also gives a detailed discussion of fan characteristics.

e. Cost Estimation. The total purchase cost of a baghouse filter may be reckoned as the sum of the cost of the bags themselves and the cost of the housing structure or compartments. The following data for the estimation of these costs are excerpted from Vatavuk [43] to illustrate briefly the basic features. The figures quoted are in 1981 dollars, for standard design, continuous duty and pressure construction. The costs of cooling systems, pumps, blowers and other accessories are not included in this estimation. For data relating to them see reference [43].

The cost of the bags is directly proportional to the gross total cloth area, according to the table below. The gross cloth area is related to the net area by the table given under section c. above.

Bag Prices for Selected Materials

Material	Price ($/ft^2) Shaker (<20,000 ft^2)	Shaker (>20,000 ft^2)	Pulse Jet	Reverse Air
Dacron	0.57	0.50	0.86	0.50
Orlon	0.93	0.71	1.36	0.86
Nylon	1.07	1.00	Not used	1.00
Nomex	1.64	1.50	1.86	1.50
Fiberglass	0.71	0.64	Not used	0.64
Polypropylene	0.93	0.78	1.00	0.78
Cotton	0.64	0.57	Not used	0.57

The cost of the compartments is a linear function of the net cloth area, that is the area which must be on-line for filtering at all times: cost = a + b (net area), where a and b are given in the table below.

Purchase Costs for Continuous Baghouse Compartments

Cleaning Mechanism[a]	Applicable Range, ft^2	Component	Parameter a	b
I. Mechanical shaker	0-70,000	Basic baghouse	9,500	4.99
		Insulation add-on	3,250	2.53
II. Reverse-air	0-100,000	Basic baghouse	36,600	4.28
		Insulation add-on	16,000	2.37
III. Pulse-jet	0-20,000	Basic baghouse	7,660	10.8
		Insulation add-on	7,010	3.42

[a] Mechanical shaker and reverse-air parameters are based on pressure construction, while those for pulse-jet units can apply to either pressure or suction construction.

C. *DESIGN EXAMPLE*

An open-hearth steel plant operates six furnaces each having a capacity to produce 320 tons of finished steel in a nine-hour operating period. The charge to the furnace is 7% over the finished steel weight. The oxygen lance [42] is used, for a maximum of 30 min. each hour. Source emission tests indicate that the emission rate maximum (while lance is used) is 7.2 lbm/hr per ton of melt. At the same time the effluent from one furnace is equivalent to 62,000 std cfm (dry, at 70° F and 29.92" Hg) and leaves the duct at 1350°F. It is desired to limit the emission to 0.010 grains/acf at 450-600°F and 8% humidity by volume. The feasibility of using a fabric filter system is to be considered.

A cursory search for relevant case studies reveals that fabric filters have been used in such an application. The work of Herrick [42] is pertinent and may be drawn upon for this study. Other data may be found in references [2,3,30,33,34,35].

Basic performance calculations are done on the basis of one furnace:

$$\text{Stack flow rate } Q = 62{,}000\ \frac{1350 + 460}{70 + 460} \times \frac{1}{(1-0.08)} = 230{,}150 \text{ acfm}$$

$$\text{Emission flux rate } E = 7.2 \times 320 = 2304 \text{ lbm/hr maximum}$$

$$\text{Maximum grain-loading} = \frac{2304 \times 7000}{230{,}150 \times 60} = 1.17 \text{ gr/acf}$$

Maximum efficiency required ≈ minimum penetration allowed

$$P = 0.010 \times \frac{450 + 460}{1350 + 460}/1.17 = 4.29 \times 10^{-3}$$

$$\eta_M = 0.9957$$

Although conditions fluctuate during this process, the design should be based upon the "most difficult" set of conditions and not upon average conditions.

Selection of Fabric: Because of the high temperatures involved a precooling system will be required, but a fabric to permit operation as hot as possible should be sought. It must also be established that the fabric selected can collect this kind of dust at 99.6% efficiency.

No data on the particle size and composition of this dust is given, but a good deal may be inferred from the literature. The material is primarily iron oxides, with small amounts of impurities, 45-50% in the 0-5 μm size range.

The Filter Handbook [2], Table 2.31, lists Fe_2O_3 cupola dust, 0.7 gr/acf, collected by fiber glass filter using shake cleaning, at 450°F, with an air/cloth ratio of 2.1 ft/min. The Particulate Control Document [33], Sec. 4.7 and 5.4, indicates that fiberglass bags may be used at 500°F, using reverse flow cleaning and sonic horns, with a ratio of 2 ft/min on a grain loading as high as 20 gr/acf. Table 1 shows fiberglass operating steadily at 500°F, and intermittently up to 550°F. The Herrick work [42] confirms these statements, indeed is probably the source of them.

We may therefore confidently and conservatively make the following selection: Fabric: Fiberglass bags; Operating temperature: 500°F; Air/cloth Ratio: 2 ft/min; Reverse-flow cleaning: Efficiency and grain-loading: as required.

Gas Cooling: A system to cool 230,150 acfm of gas at 1350°F to 500°F must be designed. For the sake of illustration let us look at all of the possible cooling systems discussed above. The gas is essentially N_2 and O_2 with some CO_2, CO and 8% vol. H_2O. It may be taken to have the properties of air in lieu of precise data on its composition: ρ_f = 0.0808 lbm/ft^3, C_p = 0.25 Btu/lbm°F. For Fe_2O_3 at 400°F, C_p = 0.263 Btu/lbm °F. The cooling required is as follows:

Mass of gas = 230,150 x 0.0808 x 492/1810 = 5050 lbm/min

Mass of dust = 2304/60 = 38.4 lbm/min

Heat removal rate:

For gas only = 5050 x 0.25 x (1350-500) = 1.073×10^6 Btu/min

For dust = 38.4 x 0.263 x (1350-500) = 8584 Btu/min.

Total = (1.073 + 0.0085) x 10^6 = 1.082×10^6 Btu/min

Dilution with ambient air at say 100°F will require: 1.082 x 10^6/0.25 (500-100) = 10,816 lbm/min which will add 10,816 x 1960/0.0808 x 492 = 261,200 cfm of volume flow to the gas at 500°F which would enter the baghouse. The total Q for the baghouse would thus be:

Q = 261,200 + 230,150 x 960/1810 = 383,300 acfm

against a volume of only 122,100 acfm undiluted at 500 °F. Dilution will more than triple the filtering surface required, which is clearly excessive.

A convective/radiative heat exchanger conservatively estimated to have an overall heat transfer coefficient of 10 Btu/hr ft^2°F, operating with a log-mean temperature difference of (1250-400)/(ln 1250/400) = 746°F will require a

$$\text{Heat Exchange Surface} = \frac{1.082 \times 10^6 \times 60}{746 \times 10} = 8700 \text{ ft}^2$$

This corresponds to a very large heat exchange unit. However, the volume of gas to be filtered would remain at 122,100 acfm.

For an evaporative spray cooler it may be assumed that the water will remove 1000 Btu/lbm evaporated. This will require 1.082×10^6/1000 = 1082 lbm/min or 130 gal/min of water. This will add 1082 x 38.8 = 42,000 ft^3/min of vapor volume to the gas at 500°F, making Q = 42,000 + 122,100 = 164,100 acfm. While this would increase the filter surface by about 35%, an equally significant consequence would be the effect on the humidity. This would increase from 8% by volume to (42,000 + 0.08 x

122,100)/164,100 or 31.5%. The absolute humidity will be approximately:

$$\text{A.H.} = \frac{(122{,}100 \times 0.08/38.8 + 1082}{5050 - (122{,}100 \times 0.08)/38.8} = 0.278 \text{ lbm } H_2O/\text{lbm dry gas}$$

Examination of a psychrometric chart shows that this will correspond to a dew-point of about 150°F. Obviously the baghouse must be insulated.

A final selection among these cooling processes cannot be made without a detailed economic analysis, which should consider the possibility of a waste-heat boiler as well. However, on the basis of the above calculations only, it seems that the evaporative cooler may be the most practical and the baghouse will be sized accordingly.

Baghouse Design: Taking v_o = 2 ft/min, Q = 164,000 acfm, the net filter surface required will be A = 82,050 ft^2. The gross total cloth area will be 82,050 x 1.09 = 89,430 ft^2. A baghouse configuration may be based upon bags each 11 ½" diameter by 34 ft. long, 80 bags per compartment [42]. This will correspond to 89,430/0.958 x 34 x 80π = 10.92 compartments. A reasonable design might use 11 compartments, assuming one at a time is off-stream for cleaning. Gas will enter the baghouse at 500°F and should not leave below 200°F.

Filtering Cycle: Some relevant data on the pressure drop constants for this kind of a system may be found in sources mentioned above.

From the Handbook, Table 2.31 [2], for iron oxide cupola dust on fiberglass the following values were obtained: S_E = 2.9 in. H_2O per ft/min; $\Delta P_{filt}/v_o$ = 4.3 in H_2O per ft/min, <u>maximum</u>; (K_2) = 121 in. H_2O x ft x min/lbm, all at v_o = 2.1 ft/min, c = 0.7 gr/ft^3, and 450°F.

Equation (8.24) may therefore be set up to read

$$\Delta P_{filt} = 2.9\ v_o + 121\ c\ v_o^2\ t/7000$$

for the present case (v_o = 2 ft/min at 500°F) in which the value of c will depend upon the actual gas flow according to the method of precooling used. The values of c, and corresponding equations for ΔP_{filt} are as follows.

Dilution: $$c = 1.17 \times \frac{230,150}{383,300} = 0.703 \text{ gr/acf}$$

$$\Delta P_{filt} = 5.8 + 4.86 \times 10^{-2}\ t$$

Exchanger: $$c = 1.17 \times \frac{1810}{960} = 2.21 \text{ gr/acf}$$

$$\Delta P_{filt} = 5.8 + 0.153\ t$$

Evaporative: $$c = 1.17 \times \frac{230,150}{164,100} = 1.64 \text{ gr/acf}$$

$$\Delta P_{filt} = 5.8 + 0.113\ t$$

If we accept the value of $\Delta P_{filt}/v_o$ = 4.3 in H_2O as the maximum tolerable, then ΔP_{filt} = 8.6 in H_2O and the length of filtration time is, in each case

Dilution: t = 2.80/0.0486 = 57.6 min

Exchanger: t = 2.80/0.153 = 18.3 min

Evaporative: t = 2.80/0.113 = 24.8 min

In lieu of information about a specific fan performance curve, (Fig. 3) these calculations assume that Q remains constant during the filtering cycle. The ΔP_{duct} has not been considered either.

The cost of this baghouse will be estimated as:

Bags: \$0.64 x 89,430 = \$57,235.

Compartments: allowing for insulation add-on,
(36,600 + 16,000) + (4.28 + 2.37) 82,050 = \$598,232

Total Cost: \$655,467 (as of 1981).

Note that this figure does not include the cost of the cooling system, blowers or other accessories.

All of the above constitutes only a preliminary survey of the problem. But it is sufficient to indicate that a baghouse filter may be feasible for the process and warrants further study and consideration.

REFERENCES

1. C.N. Davies, Air Filtration, Academic Press, London, 1973.
2. C.E. Billings and J. Wilder, Handbook of Fabric Filter Technology, Vol. IP.B. 200-648, Vol. II PB 200-649, U.S. Public Health Service, 1970.
3. J.H. Turner and J.D. McKenna in Handbook of Air Pollution Technology, S. Calvert and H.M. Englund, eds., Chapter 11, Wiley Interscience, New York, 1984.
4. A.C. Payatakes, AIChE Journ., 26: 443 (1977).
5. A.C. Payatakes and L. Gradon AIChE Journ., 26: 443 (1980).
6. H.C. Yeh and B.Y.H. Liu, Aerosol Sci., 5: 191 (1974).
7. H.C. Yeh, Ph.D. Thesis, University of Minnesota (1972).
8. H.C. Yeh and B.Y.H. Liu, Aerosol Sci., 5: 205 (1974).
9. C.T. Billiet, and J.I.T. Stenhouse Filt. & Sep., Nov./Dec.: 375 (1985).
10. L.A. Clarenburg and J.F. Vander Wal, Ind. Eng. Chem. Proc. Des. Dev., 5: 110 (1966).
11. D.C. Freshwater and J.I.T. Stenhouse, AIChE Journ., 18: 786 (1972).
12. M.W. First and W.C.Hinds, Journ. Air Poll. Control Assoc., 26: 119 (1976).
13. M.J. Ellenbecker, D. Leith, and J.M. Price JAPCA, 30: 1224 (1980).
14. K. Okuyama and A.C. Payatakes, in Particulate Systems, Technology and Fundamentals, J.K. Beddow ed., pg: 95 Hemisphere Publ., New York, 1983.
15. H. Emi, C-S Wang, and C. Tien AIChE Journ., 28: 397 (1982).
16. C.N. Davies, Proc. Inst. Mech. Eng., IB: 185 (1952).
17. L.A. Clarenburg and R.M. Werner, Ind. Eng. Che. Proc. Des. Dev., 4: 293 (1965).

18. T.E. Wright, R.J. Stasny and C.E. Lapple, WADC Tech., Rep. 55-457, Donaldson Co., Inc. (1957) PB-131, 570.

19. N. Kimura and K. Iinoya, Kagaku Kogaku, 23: 792 (1959).

20. S. Kuwabara J. Phys. Soc. Japan, 14: 527 (1959).

21. K.W. Lee, and B.Y.H. Liu Air Poll. Cont. Assn. Journ., 30: 377 (1980).

22. S.C. Yung, R.G. Patterson, S. Calvert, and D.C. Drehmel, Paper 77-32.6, Air Poll. Control Assoc., 70th Annual Meeting, Toronto, 1977.

23. H. Pendse and C. Tien AIChE Journ., 28:677 (1982).

24. G.I. Tardos, N. Abuaf, and C. Gutfinger Journ. Air Poll. Contr. Assn., 28: 354 (1978).

25. S. Ergun, Chem. Eng. Prog., 48:89 (1952).

26. S.A. Walata, T. Takahashi, and C. Tien Aerosol Sci. and Tech., 5: 23 (1986).

27. C.H. Black and R.W. Boubel, Ind. Eng. Chem. Proc. Des. Dev., 8: 573 (1969).

28. D. McCarthy, A.J. Yankel, R.G. Patterson, M.L. Jackson, Ind. Eng. Chem. Proc. Des. Dev., 15:266 (1976).

29. M.H. Peters, L-S Fan, and T.L. Sweeney AIChE Journ., 28: 39 (1982).

30. C.E. Billings, AIChE Symp. Ser. No. 137, 70: 341 (1974).

31. D. Leith and M.W. First, Journ. Air Poll. Control Assoc., 27: 534, 636, 754 (1977).

32. M. Fraser and G.J. Foley, Paper 74-99, Air Poll. Control Assoc., 67th Annual Meeting, Denver, 1974.

33. Control Techniques for Particulate Air Pollutants: AP-51, U.S. Dept. of Health, Education and Welfare, 1969.

34. R.P. Donovan, Fabric Filtration for Combustion Sources, Marcel Dekker Inc., New York, 1985.

35. J.H. Turner, A.S. Viner, J.D. McKenna, R.E. Jenkins, and W.M. Vatavuk JAPCA, 37: 749 (1987).

36. W. Strauss, Industrial Gas Cleaning, 2nd Ed., Pergamon Press, New York, 1975.

37. D. Leith, and M.J. Ellenbecker Air Poll. Cont. Assn. Journ., 30: 877 (1980).

38. D. Leith, and M.J. Ellenbecker Aerosol Sci. and Tech., 1: 401 (1982).

39. R. Dennis, and H.A. Klemm Report No. EPA-600/7-79-043a, NTIS No. PB 293 551/AS Feb. 1979.

40. D. Leith and M.J. Ellenbecker Atmos. Environ. 14: 845 (1980).

41. M. Crawford, Air Pollution Control Theory, Sec. 5-8, 10-8, McGraw-Hill, New York, 1976.

42. R.A. Herrick, J.W. Olsen, F.A. Ray, Journ. Air Poll. Control Assoc., 16 7 (1966).

43. W.M. Vatavuk, in Handbook of Air Pollution Technology, S. Calvert and H.M. Englund, eds., Chapt. 14, Wiley Interscience, New York, 1984.

PROBLEMS

1. (a) Repeat the calculations for the conditions of the example of fiber-bed filtration given in the text, except use velocities of 60 fpm, and of 80 fpm. Note the interplay between the face velocity values and L, A, and ΔP for each filter.

(b) For the conditions of this same example, assume that there is also present an image force brought about by the presence of 90 electronic charges per particle, and that the dielectric constant of the fibers is rather large. Estimate what effect this would have upon the filter dimensions for the case of 40 fpm face velocity.

(c) Again with reference to the same example, how would the results of the original case for 100 fpm be affected if the required efficiency were to be 95%?

2. A dust-laden air stream of 10,000 acfm at 70 F and dust concentration of 2.3 gm/m^3 is passed through a fabric filter consisting of 49 bags in parallel, each bag 20 ft long and 1 ft in diameter. Cleaning is by mechanical shaking of all the bags at the same time. Tests indicate that the pressure drop is 3.28" H_2O twenty minutes after shaking, and 3.58" H_2O forty minutes after shaking. Determine:

(a) air/cloth ratio in use during filtration;

(b) the values of S_E and K_2;

(c) time required to reach a ΔP = 4.0" H_2O;

(d) amount of dust collected when ΔP reaches 4.0";

(e) the time required to reach ΔP of 4.0", if an identical arrangement of 49 bags is added in parallel with the present arrangement.

3. Based upon available case history data (which you can find in the references), make a preliminary recommendation for the design of a baghouse filter system to collect flour dust from a flour mill. Grain-loading will be about 1 gr/acf and total flow 10,000 acfm. Assume bags are to be 8" diam. by 12 ft. long. Consider as many items of the system analysis as you can and explain what further information would be needed to complete a final design. Do you think an emission limitation of 0.12 lb/hr can be attained?

4. Refer to Problem 7 of Chapter 7. For the conditions stated there, consider the design of a baghouse filter system to collect the fly-ash. Assume that the fabric will be fiberglass. Make a preliminary design and compare it with that for the electrostatic precipitator design done for that problem.

9
Wet Scrubbing

I. SCRUBBER CHARACTERISTICS

A. *GENERAL FEATURES*

Scrubbing a particle-laden gas by bringing it into contact with a liquid can be an effective means of removing small particles. It also provides the possibility of dissolving gaseous components from the stream, thus providing for two kinds of purification simultaneously. The particulate matter in a wet state and the spent scrubbing liquor must be disposed of, and this obviously presents special problems not encountered in dry methods of collection.

There are many kinds of scrubbing devices. In an extensive study Calvert [1] has classified them into ten types. A handy summary of their designs and performance is also available [2]. To these may be added more recent concepts such as the reversejet [9] scrubber and the catenary grid [6]. However for purposes of modelling, these may be grouped broadly into two main classes: (a) those in which an array of liquid drops (sprays) form the collecting targets; (b) those in which wetted surfaces of various kinds are the collecting medium. In the first class are to be found preformed sprays, gas atomized sprays, and self-induced sprays. In the second class are a variety of plate towers and packed towers, including massive packing, fibrous packing, moving bed packing and irrigated fiber filters. The treatment in this chapter will be devoted to the first class.

One unique feature of wet collectors is the necessity for providing energy to move the liquid through the device as well as the gas. The total scrubber power required is an important parameter. It is well established that the higher the scrubber power, per unit volume of gas treated, the lower will be the cut diameter of the particles emitted. Calvert [2] uses this to correlate and compare the performance of the several kinds of devices. He shows that cut diameters of about 1 micron may be obtained with gas atomized spray devices, at about 1 hp per 1000 acfm of scrubber power. To obtain a cut diameter of 0.5 micron will require over 6 hp/1000 acfm. As shown by the sample calculation of Example 5 in Chapter 3, this is substantially higher than typical values for other devices. For the collection of sub-micron particles, the scrubber is a "high energy" device.

B. *SPRAY SCRUBBERS*

The term spray scrubber will be used to refer to a device in which the liquid is broken into drops, roughly in the range 100-1000 μm diameter, which are introduced into the particulate-laden gas stream. There the array of moving drops becomes the set of targets (regarded as spherical) upon which the particles may be collected by any of the mechanisms of aerodynamic capture. Of special interest are the possibilities of capture by diffusiophoresis and thermophoresis, as well as by "electrified" drops. (See Chapter 5)

Models for collection are based upon the concept of single-target efficiency for an individual sphere, by assuming that this efficiency may be calculated from the parameters of aerodynamic capture, as presented in Chapter 5. Inertial impaction (II) remains an important mechanism of capture for super-micron particles. For this to occur there must be relative motion between drops and particles. This would also be true of direct interception (DI). For any of the diffusion mechanisms or electrostatic forces, however, relative motion is not required except to bring the particles sufficiently close to the drops so that these kinds of forces may become effective. They are, of course, of primary importance for the sub-micron particles.

Relative motion is provided in all types of scrubbers. The principle kinds are:

(a) Pre-formed spray: drops are formed by atomizer nozzles and sprayed into gas stream: (i) Counter current gravity tower: drops settle vertically against rising gas-particle stream; (ii) Cross-current tower: drops settle through horizontal gas stream; (iii) Cocurrent: spray is horizontal into horizontal gas stream; (iv) Reversejet: spray is counter current to gas stream, usually vertically upward; (v) Ejector Venturi: liquid sprayed at high pressure into converging section ahead of throat pulls gas in as well.

(b) Gas-atomized spray: liquid is introduced more or less in bulk ahead of Venturi throat and shattered into drops by high-velocity gas: (i) Spray Venturi: liquid squirted in around periphery of Venturi, or flows over weir into gas; (ii) Annular orifice: liquid and gas flow in parallel into annular space between a movable disk and Venturi walls, ahead of throat; (iii) Rod bank: liquid and gas flow in parallel into constricted space formed between rods arranged parallel to each other in a row. From the point of view of basic principles it is better to treat all of the Venturi types together as a class, whether pre-formed (a, iv) or gas atomized (b,i; b,ii; b,iii). The catenary grid is also an example of gas-atomized spray, not a Venturi.

II. NON-VENTURI SCRUBBERS

A. *GRAVITY CHAMBERS*

1. *Counter current flow* Liquid is sprayed in at the top of a tower through atomizing nozzles. Drops then fall freely at their terminal settling velocities relative to the gas. The gas/particle stream is introduced at the bottom, flows upward against the rain of drops and clean gas issues from the top. Liquid containing particulate collects in a pool at the bottom and is pumped out for treatment to remove the solids. Cleaned liquid is usually recirculated to the sprays. The tower is simply an empty chamber.

The approach to modelling the collection process is similar to that used for fibrous filters. Consider a cross section of such a tower of differential thickness as shown in the sketch:

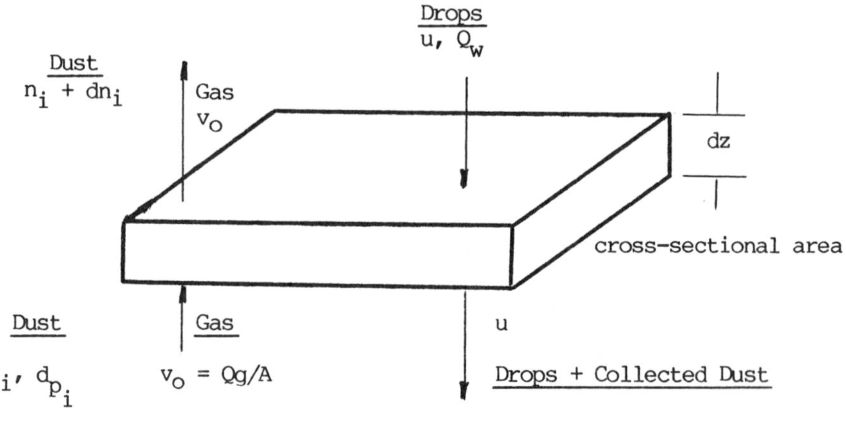

The model is developed on the assumptions that: (1) The drops are all of the same size, diameter = D, and are falling at their terminal velocity u_{s_D} relative to the gas. Thus, the velocity of the drops relative to the tower is $u = (u_{s_D} - v_o)$. (2) The drop concentration is sufficiently dilute that each drop may be regarded as acting independently as though it were present alone.

(3) The total quantity of scrubbing liquid, flowing at volume rate Q_ℓ is distributed uniformly over the cross-sectional area A. The total number of drops entering the section per unit time is equal to $6Q_\ell/\pi D^3 A$. (4) The individual dust particles are being carried upward in the tower at a velocity of $(v_o - u_{s_i})$ relative to the tower, where u_{s_i} = terminal settling velocity of dust particles of ith grade (in many cases $u_{s_i} << v_o$, and $u_{s_i} << u_{s_D}$ also). (5) The concentration of dust of grade i is uniform across the section and equal to n_i (number of particles/volume of gas) i.e., there is complete lateral mixing of uncollected particles across any level. (6) According to the definition of single target total efficiency η_{T_i}, each drop collects, per unit time, a number of particles equal to: $\eta_{T_i} (u_{s_D} - u_{s_i}) n_i \, \pi D^2/4$, wherein $\eta_{T_i} = f$ (Stk_i, Re_D, Pe_i, R_i, etc.). (7) The time required for a drop to fall through the section of thickness dz is equal to $dz/u = dz/(u_{s_D} - v_o)$.

The total collection of particles taking place in the section is

$$-dn_i \frac{Q_g}{A} = \frac{6Q_\ell}{\pi D^3 A} \eta_{T_i} \frac{\pi D^2}{4} (u_{s_D} - u_{s_i}) \, n_i \frac{dz}{u} \tag{9.1}$$

or

$$\int_{n_{o_i}}^{n_i} \frac{dn_i}{n_i} = \frac{3}{2} \int_0^H \frac{(u_{s_D} - u_{s_i})}{(u_{s_D} - v_o)} \frac{Q_{\ell T}}{DQ_g} \eta_{T_i} \, dz \tag{9.2}$$

Integration on the assumption that the quantities on the right are constant and independent of z, gives the overall efficiency:

$$\eta_{s_i} = 1 - \exp\left[-\frac{3}{2} \frac{(u_{s_D} - u_{s_i})}{(u_{s_D} - v_o)} \frac{Q_\ell}{Q_g} \frac{H}{D} \eta_{T_i} \right] \tag{9.3}$$

where H = total height of tower. This is a grade-efficiency relation because η_{T_i} depends upon d_{p_i}.

In case the drops are not all of the same size it is customary to take D as the Sauter mean drop diameter in (9.3). It is important to know the drop size distribution produced by the spray nozzles in this case. A more rigorous approach would be to divide the drop sizes into intervals, represented by D_j, and to determine a value of $\eta_{T_{i,j}}$ for each of the combinations of (d_{p_i}, D_j) pairs. Then equation (9.2) would have to be applied to each pair, and the total collection for all pairs summed.

The role of the drop size may be explored in part through its effect upon η_{T_i}. In the usual range of operation for gravity spray towers, it is found that II is the predominant mechanism of capture. Using a relationship recommended by Calvert [3]

$$\eta_{T_i} = [\mathrm{Stk}_{i_2}/(\mathrm{Stk}_i + 0.35)]^2$$

and

$$\mathrm{Stk}_i = \frac{\rho_p d_{p_i}^2 \left(u_{s_D} - u_{s_i}\right)}{18\mu D} = \tau_i \frac{u_{s_D}}{D} \tag{9.4}$$

so that

$$\eta_{T_i} = \left[\frac{\tau_i u_{s_D}}{\tau_i u_{s_D} + 0.35D}\right]^2 \tag{9.5}$$

For small drops $u_{s_D} \propto D^2$ (Stokes law) and η_{T_i} increases as D increases. At intermediate sizes $u_{s_D} \propto D$ and η_{T_i} is constant. For larger sizes u_{s_D} increases less rapidly than D and η_{T_i} decreases. Hence there is a value of D for which η_{T_i} is a maximum. Stairmand [4] has calculated that the peak value of

η_{T_i} occurs at about D ≈ 600μm regardless of particle size, is larger for larger particles, and is a rather flat peak extending for two or three hundred microns on either side of D ≈ 600μm.

The total effect of drop size on overall efficiency involves the interplay of u_{s_D}, η_{T_i}, and D as they appear in (9.2) or (9.3). Since η_{T_i} is relatively constant with D over a central range in which $u_{s_D} \propto D$, the net effect on the factor $u_{s_D}\, \eta_T (u_{s_D} - v_o)$ D is to make it a maximum at the low D end of this range, say in the neighborhood of 300-400 μm.

2. *Cross-flow* Proceeding in an analogous fashion, a model may be developed for the case of horizontal cross-flow of particle-laden air through a vertical settling spray. Similar assumptions are made, except that $v_o = 0$ in the vertical direction. The result is the same as (9.2) with $v_o = 0$. Then, also taking $u_{s_i} \ll u_{s_d}$

$$\eta_{s_i} = 1 - \exp\,(-\,3\,Q_\ell\,H\,\eta_{T_i}/2\,Q_g\,D) \tag{9.6}$$

The same equation as above may be used for calculating η_{T_i} but the value of Stk_i must be based upon v_o: $Stk_i = \tau_i\, v_o/D$. Now the ratio η_{T_i}/D in (9.6) clearly has no maximum value with respect to D, but continues to increase as D decreases.

The case of a spray nozzle operating horizontally in a duct with a cocurrent flow of gas, type a (iii), has been analyzed by Cheng [5]. He obtains equation (9.6), modified by the insertion of a factor λ ahead of Q_ℓ and replacement of H by a length ℓ. These modifications take into account the fact that the spray issues from the nozzle (assumed to be placed at the center of the duct of radius R) in a conical shape, with half-angle α, which does not fill the duct completely for some distance downstream, and that the spray drops have all settled to the bottom of the duct by a distance S ("projective distance" of the spray) downstream. Letting $r = S/R$, λ and ℓ are obtained:

$$\lambda = \frac{3r - 2 \cot \alpha}{2r^3 (1 - \cos \alpha)} \tag{9.7}$$

$$\frac{\ell}{R} = r - \frac{2}{3} \cot \alpha$$

and S is estimated according to the exit velocity of drops from the nozzle v_D and their terminal settling velocity:

$$S \approx (v_D - v_o)\, R/u_{s_D} \tag{9.8}$$

The model is shown to give good agreement with experiment, when η_{T_i} is based upon Stk_i in which the relative velocity is taken as the arithmetic mean of v_D and the final axial drop velocity, when the drop reaches the duct.

B. *THE L/G RATIO*

These models introduce the liquid/gas flow ratio, which is of basic importance in the performance of all kinds of scrubbers. This ratio should be expressed fundamentally in terms of mass flow rate of liquid (M_ℓ) divided by mass flow rate of gas (M_g). For convenience however this is usually expressed in terms of the equivalent volumes taken at the actual conditions of entrance. Thus

$$\frac{M_\ell}{M_g} = \frac{\rho_\ell Q_\ell}{\rho_g Q_g} \tag{9.9}$$

For the air-water system at 20°C, $\rho_\ell/\rho_g = 1/1.205 \times 10^{-3} \approx 830$. It is also common practice to express Q_ℓ/Q_g in units of gallons of liquid per thousand actual cubic feet of gas, at entrance conditions. This is commonly called the L/G ratio. Experience shows that L/G will typically fall in a range of roughly 2-20 gal. per 1000 cu ft, or 0.27-2.7ℓ/m^3. (1 gal/Mft^3 = 0.1337 ℓ/m^3.)

This ratio is directly proportional to the number of drops per unit volume of gas: $6Q_\ell/\pi D^3 Q_g = 2.553 \times 10^8$ L/(G D^3) drops/ cm^3, where D must be in microns. The following tabulation shows the number of drops per cubic centimeter to be obtained at various values of L/G and D.

	Drops per cubic centimeter		
L/G gal/Mft^3:	2	10	20
D = 100 μm	511	2553	5106
200 μm	64	319	638
400 μm	8	40	80
800 μm	2	5	10

It is clear that at combinations of low D and high L/G, the drops could become so crowded that collision and coalescence would be likely to occur. Assumption (2) of the model would not be valid. An operation attempted at too small D would automatically move toward a larger D. The feasible range of operating conditions is thus roughly delineated.

Example 1. Develop a set of grade-efficiency curves for a counter current gravity scrubber which is to remove particles of 2.60 gm/cm^3 density ranging in size from 0.6 μm to 20 μm. Consider L/G ratios of 5 and 10 gal/Mft^3 respectively, water drop sizes of 200 and 500 microns, and tower heights of 2 ft. and 10 ft. Assume the gas to be saturated with and at the same temperature as the liquid. Discuss what differences there would be if the liquid is colder than the saturated gas. (See Chapter 5.)
Take v_o at 0.20 u_{s_D}, and also at 0.50 u_{s_D}.

Figure 1 displays these grade efficiency curves for five cases. Case I may be regarded as the base case for which L/G = 10 gal/Mft^3, D = 500 microns, H = 2 ft, and v_o = 0.2 u_{s_D}. The other cases show that the grade efficiency is increased by: smaller

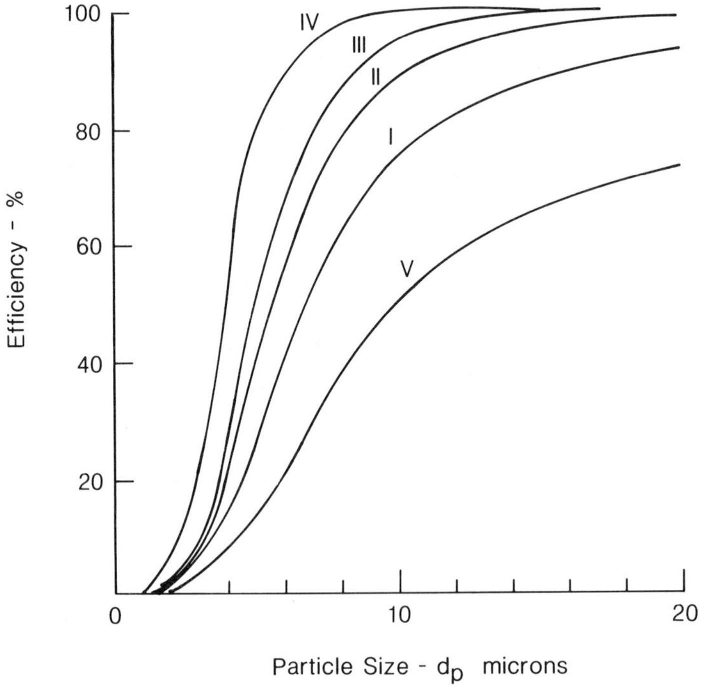

FIG. 1 *Grade efficiencies for spray tower.*

Case	*L/G*	*D*	*H*	v_o/u_{s_d}
I	*10*	*500*	*2*	*0.20*
II	*10*	*200*	*2*	*0.20*
III	*10*	*500*	*2*	*0.50*
IV	*10*	*500*	*10*	*0.20*
V	*5*	*500*	*2*	*0.20*

drop size D (Case II: D = 200 µm); larger ratio v_o/u_{s_D} (Case III: $v_o = 0.5\ u_{s_D}$); larger tower height H (Case IV: H = 10 ft.); larger L/G (Case V: L/G = 5 gal/Mft^3). These results are typical: the greatest efficiency would be obtained (out of these possible values) by the combination D = 200 µm, $v_o = 0.5\ u_{s_D}$. L/G = 10 gal/Mft^3, H = 10 ft. The calculations are straightforward using Eqns. (9.3), (9.4), and (9.5) together with the methods given in Chapter 4 for determining u_{s_D}. It is clear that, at best, the spray tower is not very efficient, especially for fine particles.

C. *TOWER DESIGN*

The above example focuses attention on some of the questions arising in the design of towers and other scrubbers. As in the case of other collectors discussed in previous chapters, the value of Q_g is a specified requirement. However v_o, Q_ℓ/Q_g (L/G), and D may be selected by the designer. Selection of v_o determines the tower cross-sectional area $A = Q_g/v_o$. Obviously $v_o < u_{s_D}$, but the larger v_o the longer the residence time of drops in the tower and the greater the overall efficiency.

The height H of the tower may be determined by applying Eqn. (9.3) to the particle size distribution in the same manner as was done for precipitators, and fiber filters according to Eqn. (3.12). An equation is set up to express the desired overall efficiency (see, for example, the Design Example in Chapter 7.) This equation is then solved by iteration, to give H. The terms in the equation depend upon the values selected for v_o, L/G, D, as well as the particle-size grade interval for g_i values.

Thus a range of possible tower sizes (A and H) exists in which all will be technically satisfactory solutions. The one combination to be selected must, as usal, be determined by an optimizing procedure. The applicability of the model is limited, as always, by the accuracy with which the conditions in a real scrubber are described by the set of assumptions made in deriving the equations.

The interplay among v_o, L/G, D, and scrubber geometry will be found to have a precisely analogous counterpart in the design of Venturi scrubbers. With obvious modifications, the system chart Fig. 3 may be applied to any scrubber.

D. *OTHER TOWER/SPRAY SCRUBBERS*

1. *Catenary grid scrubber [6,7,8]* An important modification of the tower scrubber is one in which a curved grid (of patented design, in the sahpe of a catenary) is placed a short distance below a central liquid inlet pipe, and convex to the upward flowing gas. The shape closely approximates the mirror image of the parabaloidal gas front, thus balancing liquid and gas forces. The action of the flowing gas, in passing through the grid, creates a zone of droplets and bubbles which has been likened to a "fluidized bed" in appearance. It is a uniform region of extreme turbulence and intense mixing, which permits use of L/G ratios as low as 1 or less, with gas velocities up to 30 ft/s.

The grid is made of a wire-mesh-type cloth (which may be made from a variety of materials) with approximately 3/8" holes and an open area of 50-80%. This results in a pressure drop of about 1.5" H_2O per grid. Two-grid units are typical, and are said to produce efficiencies of up to 98% of theoretical. The hydrodynamics and description of the "fluidized" zone have been published [7], but there has been no model of the collection process, which is obviously rather complex. The device has also been found to be very effective as a cooling tower [8].

Details of the design appear to be proprietary. However it is claimed that catenary scrubbers may be much smaller and cost less than other types [6,8]. Catenary Grid Scrubber is a registered trademark, and is marketed by the Chem-Pro Division of Otto H. York Co.

2. *Reversejet scrubber [9,10,11]* This is a counter current spray scrubber in which the liquid is sprayed (from a nozzle) vertically upward against a flowing gas stream in a large pipe. The spray drops decelerate due to the drag of the gas, and to gravity, until they reach zero velocity relative to the pipe, at a distance called y_s above the nozzle. A dense zone of droplets is

formed. Subsequently the droplets reverse direction, accelerate, flow concurrently with the gas stream and are collected. It is claimed that this operation results in a pressure drop that is half that of other scrubbers and produces an efficiency of 85-95% for particulate collection. The device was invented by Low [9]. The patent is licensed for exclusive marketing by the Koch Engineering Co.

A model of the collection process has been developed by Holmes et al [10]. In principle it is much like the one for spray towers as outlined above. Each drop is treated as a single target having a capture efficiency (primarly by inertial impaction) given by Eqn. (9.4), as in Chapter 5, and is assumed to behave independently. Their result is:

$$P_i = \exp\left(-\eta_i \, a \, y_s/4\right) \tag{9.10}$$

for the penetration P_i of particles of size d_{p_i}, , where η_i is the single target collection efficiency of drops of size D. Here a, the "interfacial contact area" is given by

$$a = (6/D)\left[Q_\ell/(Q_\ell + Q_g)\right](U_r/v_o) \tag{9.11}$$

where U_r is the relative velocity between gas (at v_o) and spray droplet. Both η_i and a have been assumed to be constant, but in fact they both depend upon U_r which varies considerably during the flight of the drop.

Obviously the value of y_s is of key importance in this model. It depends upon the trajectory of the drop, as determined by its size, the gas velocity, and the angle and velocity with which it leaves the nozzle. This has been investigated in detail, both theoreticallly and experimentally, by Faris [11]. A method of predicting the pressure drop through the scrubbing zone is given in Low's patent [9].

III. VENTURI SCRUBBERS

A. *BASIC ASPECTS*

The essential feature of a Venturi scrubber is the presence of a constricted cross section or "throat" through which the gas is forced to flow at high velocity. Drops introduced ahead of the throat at very low axial velocity are accelerated by the drag of the high speed gas. This provides the relative velocity needed for inertial and other collection mechanisms.

The cross section may be either circular or rectangular. The circular shape is not used in larger sizes (say above 1 ft. diameter) because it is too difficult to get a uniform distribution of drops across the entire section. The rectangular shape is therefore used with a rather large aspect ratio, in order to provide for a short side across which a good distribution can be obtained. Especially, this is the case in type b(iii) the parallel rod bank.

The cross-sectional area of the throat A_t determines the maximum velocity of the gas: $v_t = Q_g/A_t$. Ahead of the throat, there is a converging section in which acceleration of the gas occurs. Following the throat is a more gradually diverging section at the end of which the gas has returned to its initial velocity. Although the collection process may take place throughout the length of the Venturi, wherever there is a difference between gas and drop velocities, much of it will occur in the throat. The length of the throat must then be sufficient to accomplish the desired collection. For the parallel rod bank, the drops are introduced upstream and the "throat" itself is very short. Some acceleration of the drops will occur prior to the rods and the final acceleration, relatively brief, in between the rods. Also, in this case there are a number of "Venturis" operating in parallel.

The modelling of the Venturi must include the following features: (i) geometry of sections: throat velocity; (ii) introduction and acceleration of drops: drag on accelerating drops; (iii) relative velocity vs. axial distance: deceleration of drops beyond throat; (iv) drop size: pre-formed sprays, gas-atomized sprays; (v) the capture process: collection efficiency; (vi) the pressure drop: power required.

B. *MODELLING*

1. *Venturi geometry* Figure 2 shows a typical outline for types a(iv), b(i), b(ii), with the various sections designated. The cross-sectional area, with reference to that of the throat is given by:

$$A = A_t\left(1 + \frac{z_2 - z}{r_t}\tan\beta\right) \quad \text{converging:} \quad z < z_2$$

$$A = A_t\left(1 + \frac{z - z_3}{r_t}\tan\beta\right) \quad \text{diverging:} \quad z > z_3$$

$A_t = \pi r_t^2$ for circular sections

$A_t = 2r_t.W$ for rectangular ($2r_t$ by W) section

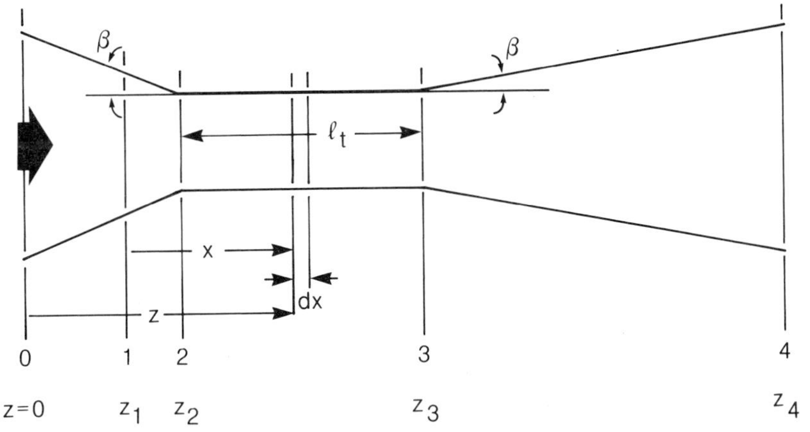

FIG. 2 *Venturi geometry. 0-2 Converging Section, 2-3 Throat, 3-4 Diverging Section; at 1 drops introduced; may coincide with 2;* $x = z - z_1$.

Drops are introduced at z_1, which often nearly coincides with z_2. In case $z_1 < z_2$, acceleration of drops must be considered in the converging section as well as the throat.

The gas velocity v_g at any point is given by

$$v_g = M_g/\rho_g A = \frac{v_t}{1 + \frac{(z_2 - z)}{r_t} \tan \beta} \qquad z < z_2 \tag{9.12a}$$

or

$$v_g = M_g/\rho_g A = \frac{v_t}{1 + \frac{(z - z_3)}{r_t} \tan \beta} \qquad z > z_3 \tag{9.12b}$$

where M_g is the mass flow rate of gas, taken as steady, and v_t is the gas velocity in the throat.

It is generally assumed that the flow through the Venturi is incompressible, i.e., that ρ_g remains constant, independent of z. However, in some operations ΔP along the throat may be as high as 100 in. H_2O (≈ 3.6 lb/in^2). If the initial upstream pressure is relatively low, then it may be necessary to make a correction for compressible flow of the gas according to the standard Venturi equations.

2. *Introduction and acceleration of drops* Drops may be introduced into the gas stream in a variety of ways. The location of introduction will depend in part upon the method. The method will also be a predominant factor in determining the size and size distribution of the drops. These matters will be taken up below.

Consider an individual drop, introduced with some initial axial velocity, which may often be taken as zero. It will be accelerated according to Newton's Law, by the drag force exerted by the gas (See Chapter 5).

$$F_D = \frac{\pi D^3}{6} \rho_D \frac{dv_D}{dt} = C_{DA} \frac{\pi D^2}{4} \rho_g \frac{(v_g - v_d)^2}{2} \tag{9.13}$$

or

$$\frac{dv_D}{dt} = \frac{3}{4} C_{DA} \frac{\rho_g}{\rho_D} \frac{(v_g - v_D)^2}{D} \tag{9.14}$$

Replacing

$$\frac{dv_D}{dt} = \frac{dv_D}{dx}\frac{dx}{dt} = v_D \frac{dv_D}{dx}$$

the differential equation relating v_D to x (distance downstream from z_1) becomes

$$\frac{dv_D}{dx} = \frac{3}{4} C_{DA} \frac{\rho_g}{\rho_D} \frac{(v_g - v_D)^2}{v_D D} \tag{9.15}$$

Solution of (9.15) depends upon selection of an appropriate representation for C_{DA}. This question was raised in Chapter 4 when it was shown that in general $C_{DA} \neq C_D$. There are several possibilities for dealing with it. In the range of Reynold's Number encountered in Venturis the results of Ingebo [12] may be applicable:

$$C_{DA} = 27\, Re_D^{-0.84} \qquad \text{where } Re_D = D\,(v_g - v_D)\,\rho_g/\mu$$

Not only is this relationship awkward to use, however, but also it has been criticized on experimental grounds by Boll [13].

Another approach has been taken by Hollands and Goel [14], to represent C_{DA} in terms of C_{D_1}, based upon conditions at z_1. Thus

$$C_{DA} = C_{D_1} \left[\frac{Re_{D_1}}{Re_D} \right]^{0.5} = C_{D_1} \left[\frac{(v_{g_1} - v_{D_1})}{(v_g - v_D)} \right]^{0.5} \tag{9.16}$$

Here C_{D_1} is taken from the standard relationship for steady motion, as represented in Chapter 4, based upon $Re_{D_1} = D\,(v_{g_1} -$

$v_{D_1}) \rho_g/\mu$. Appropriate expressions for C_D, are those of Schiller & Naumann, or Dickinson and Marshall, covering the proper range of Re for Venturi performance. Some workers have simply chosen to use one of these two for C_{DA} without making any correction for the acceleration aspect. To proceed we will use (9.16) and substitute in (9.15) to obtain:

$$\frac{dv_D}{dx} = \frac{3}{4}\frac{\rho_g}{\rho_D}\frac{(v_g - v_D)^{1.5}}{v_D D} C_{D_1} (v_{g_1} - v_{D_1})^{0.5} \tag{9.17}$$

Integration of (9.17) depends upon the conditions at drop introduction. For the case where the drops are introduced at the beginning of the throat by overflow through a weir or through nozzles, type b(i), $z_1 = z_2$, $v_{D_1} = 0$ at $x = 0$, and $x = \ell_t$ at the end of the throat. Then $v_g = v_1 = Q_g/A_t = v_t$, and (9.17) leads to

$$\frac{1}{v_t^{0.5}} \int_0^{v_D} \frac{v_D \, dv_d}{(v_t - v_D)^{1.5}} = \frac{3}{4}\frac{\rho_p C_{D_1}}{\rho_D D} \int_0^x dx \tag{9.18}$$

This has been integrated by Yung, Barbarika, and Calvert [15] to produce, in dimensionless form

$$u = v_D/v_t = 2\left[1 - X^2 + X\sqrt{X^2 - 1}\right] \tag{9.19}$$

Wherein

$$X = 3x\, C_{D_1} \rho_g / 16\, D\, \rho_D + 1 \tag{9.20}$$

The extent of acceleration of drops attained at the end of any throat length may simply be found by setting $x = \ell_t$ in (9.20) and obtaining u_t as a function of X_t from (9.19). For example, in order for $u_t = 0.90$, $X_t = 1.74$; for $u_t = 0.99$, $X_t = 5.05$; but for $u_t = 1$, $X \rightarrow \infty$.

For a set of representative conditions, we may explore the value of u_t. Taking air and water at °20 C, ρ_g = 1.205 x 10^{-3} gm/cm^3, ρ_D = 1.0 gm/cm^3, D = 100 microns, with Re_{D_1} = 400, C_{D_1} = 0.608, and ℓ_t = 1 ft:

$$X_t = \frac{3 \times (12 \times 2.54)\ 0.608 \times 1.205 \times 10^{-3}}{16 \times (100 \times 10^{-4}) \times 1.0} + 1 = 1.42$$

and u_t = 0.83. The gas velocity in the throat here is v_t = 197 ft/s, a typical value. For u_t = 0.99 in this case, the throat length would have to be ℓ_t = 9.64 ft.

In case the drop velocity at the point of introduction cannot be taken as zero (9.18) would be written:

$$\frac{1}{(v_t - v_{D_1})^{0.5}} \int_{V_{D_1}}^{V_D} \frac{v_D\, dv_D}{(v_t - v_D)^{1.5}} = \frac{3}{4} \frac{\rho_g}{\rho_D} \frac{C_{D_1}}{D} \int_0^x dx \qquad (9.21)$$

These results cannot be applied outside the throat because of the limitations placed upon the conditions of integration. To cover the cases where drops are introduced ahead of the throat, types a(iv), b(ii), where $z_1 < z_2$, integration would be done in two steps. First from z_1 to z_2, taking v_{D_1} at z_1 and using (9.12a) to represent v_g as a function of z between z_1 and z_2. This would determine v_{D_2} at z_2, which would become the starting point of the second integration. This would be through the throat, as above, from z_2 to z_3 except that $v_{D_2} \neq 0$, and (9.17) (or (9.21)) would become:

$$\frac{1}{(v_t - v_{D_2})^{0.5}} \int_{vD_2}^{V_D} \frac{v_D\, dv_D}{(v_t - v_D)^{1.5}} = \frac{3}{4} \frac{\rho_p}{\rho_D} \frac{C_{D_2}}{D} \int_{z_2}^{z} dx \qquad (9.22)$$

In case it is desired to carry the integration into the diverging section, still another calculation would have to be done using (9.12b) to relate v_g and z beyond z_3. The lower limits of the integration would be $z = z_3$ and $v_D = u_t v_t$.

These possibilities have been considered by Hollands and Goel [14] and by Chongvisal [16] and the integrations performed by computerized numerical integration. Results are presented in the form of non-dimensionalized charts, for the case of typical Venturi dimensions and operating conditions. They show that the drops continue to accelerate in the diverging section for some distance downstream of the end of the throat. However, since the gas velocity is decreasing in that region, the drop velocity eventually reaches a maximum when it becomes equal to the gas velocity at some point in the diverging section. Thereafter the drop velocity is greater than the gas velocity, but both continue to decrease and approach one another as the flow through the diverging section is completed.

These calculations reveal that, for drops of given size, the profile of drop velocity against distance downstream depends upon the geometry of the Venturi, the size of drop, and the gas velocity in the throat. The liquid-to-gas ratio (L/G) has no effect. Larger drops are accelerated more slowly and reach a lower peak velocity at a point further downstream, for a given geometry and gas velocity. At lower gas velocity, the drops are initially accelerated to a higher velocity which is reached more quickly.

It has been suggested by Crowder et al [17] that the point at which gas and drop velocities become equal be called the "minimum contactor length". They studied the drop velocity profile for drops injected just ahead of the throat whose size was determined by the shattering effect of the gas (see below). They found that for a given gas throat velocity, the minimum contactor length moved downstream as the L/G ratio increased (larger drop sizes). However, for a fixed L/G this length moved upstream as throat velocity increased, provided L/G was less than about 2 ℓ/m^3. Otherwise, this length moved downstream as L/G increased.

All of the above treatment is based upon the equation of motion applied to individual drops. Where drops are flowing in a group, as it were, their behavior may be modified, especially if the drop concentration is high. Crowe [18] studied the acceleration of solid spherical particles fairly uniformly distributed across the entire cross section of a shock tube. He obtained values of C_{DA} which lie parallel to and just above the "standard" curve for C_D in the range of Re_1 for 400 to above 1000. Hesketh, Engel, and Calvert [19] studied acceleration of droplets and found that a "cloud" type of atomization could occur where water is injected in a stream from nozzles greater than about 1 mm in diameter. Particles making up the clouds were much finer and accelerated more slowly than would otherwise be expected. However, in an extensive study of liquid atomization in Venturi scrubbers, carried out by high-speed photography, Roberts and Hill could not find this type of behavior [20].

3. *Drop size and size distribution* a. Gas Atomized Sprays. The size of drops is determined by the shattering impact of the high speed gas upon a stream of liquid injected from a nozzle radially into the gas stream. Many Venturis have this mode of liquid introduction. This process is governed by a critical value of a dimensionless group called the Weber number:

$$We = \rho_g v_t^2 D/\sigma \tag{9.23}$$

This number expresses the magnitude of the ratio of the inertial stress produced by the gas, to the surface tension stress opposing deformation.

Values of the We_{crit} have been reported to range from 6-11 for various liquids, when the stress is applied slowly. Radhakrishnan and Licht [21] found that a value of We = 5 fitted the experimental data of Boll et al [22] where D is the Sauter mean diameter, at conditions in a type b(i) Venturi: L/G = 10 - 20 gal/Mft^3, v_t = 150 - 300 ft/s.

Boll and his group measured the Sauter mean diameter of drops produced by injection through a series of nozzles located 12" upstream of the throat of a rectangular section Venturi. The value of r_t = 7", ℓ_t = 12", throat width = 12", β = 12.5° converging, β = 3.5° diverging. This represents a one foot slice taken out of a full-scale Venturi in which the width would be several feet. Their results were correlated by

$$\bar{D}_{1,2} = \left[283{,}000 + 793\ (L/G)^{1.932}\right] \Big/ v_{g_1}^{1.602} \tag{9.24}$$

where $\bar{D}_{1,2}$ (the Sauter mean diameter), is in microns and v_g, in ft/s. L/G ranged from 4.5-18 gal/Mft^3, v_t from 100-300 ft/s., and $v_{g_1} = 0.725\ v_t$ from (9.12a).

The study of Roberts and Hill [20] shows how drops are formed as the transverse jet is broken up. Determination of drop sizes and size distribution was not a primary objective, but some such data were produced. These data do not lead to any method of predicting size distribution, but they do show that the existing equations seem to underestimate the value of $\bar{D}_{1,2}$.

No data have been found for types b(ii) and b(iii) Venturis, nor any data on size-distribution for any of the type b's. Some authors have used data from studies on pre-formed sprays (as reported below), but conditions are not comparable [21] and it seems to be a very questionable practice. Estimates of $\bar{D}_{1,2}$ may be made using We_{crit} of 6-10 in these cases.

b. Pre-formed Sprays. This term is taken to mean sprays produced from atomizing nozzles introduced with the gas stream ahead of the throat, as in type a(iv). The nozzles may or may not be pneumatic. Such sprays and nozzles may also be used in gravity chambers.

(1) Pressure nozzles. Liquid under high pressure is forced into a whirl chamber through a tangential passage (hollow-cone) or directly through an orifice (full-cone). Experiments by Nelson and Stevens [23] on drop-sizes produced have been correlated in terms of the volume-median (same as MMD) diameter $\tilde{x}$ by defining

$$Y = \log_{10} \tilde{x}/D_o \tag{9.25}$$

and

$$Z = \log_{10} Re_o \, (We_o/Re_o)^a \, (u_t/u_a)^b \tag{9.26}$$

Then

$$Y = AZ^2 + BZ + C \tag{9.27}$$

In these dimensionless expressions: D_o = diameter of nozzle orifice; $u_a = Q_\ell/A$ = average axial liquid velocity through orifice; $u_t = u_a \times \tan \theta/2$ = average tangential velocity through orifice; θ = maximum cone angle of spray; $Re_o = D_o u_a \rho_D/\mu_D$; $We_o = u_a^2 D_o \rho_g/\sigma$. For water sprays: a = 0.2; b = 1.2; A = -0.144; B = 0.702; C = -1.260. For organic liquids: a = 0.55; b = 1.2; A = -0.0811; B = 0.124; C = -0.186.

Drop-size distribution has been found to fit the upper-limit function of Mugele and Evans [24], as well as the Rosin-Rammler distribution [25], both of which are discussed in Chapter 2. However, a square-root normal distribution is also useful here [23, 26]. This consists of using

$$A = \log_D s(We_o/D_o)^{0.5} \tag{9.28}$$

$$B = \log_{10} We_o/\sqrt{Re_o} \tag{9.29}$$

and

$$A = 0.150 B^2 - 0.359 B + 0.986 \tag{9.30}$$

They also give a graph from which the Sauter mean drop size can be determined from s and $\tilde{x}$.

(2) Pneumatic nozzles. The usual design consists of a central core of liquid surrounded by a concentric coverging nozzle for air. In addition to the geometry and dimensions of the nozzle, the relative velocity of air to liquid at the nozzle and the ratio of the mass flow rates are basic parameters which govern the nozzle performance. This flow rate ratio in the nozzle must not be confused with the L/G ratio in the Venturi when such nozzles are used.

A classical study by Nukiyama and Tanasawa [27] is often cited for performance of these nozzles. Their correlation for the Sauter mean drop diameter produced is

$$\bar{D}_{1.2} = \frac{585}{v_{rel}} \sqrt{\frac{\sigma}{\rho_D}} + 597 \left[\frac{\mu_D}{\sqrt{\sigma \rho_D}} \right]^{0.45} \left(\frac{1000\ Q_\ell}{Q_g} \right)^{1.5} \tag{9.31}$$

where $\bar{D}_{1,2}$ = Sauter mean drop diameter, microns; v_{rel} = $(v_g - v_o)$ at nozzle, m/s; σ, dynes/cm; ρ_D, gm/cm^3; μ_D, poises; Q_ℓ, m^3/min; Q_g, m^3/min. These studies covered nozzles ranging from 0.2 to 1.0 mm diameter for liquid and 1-5 mm for air. Relative velocities ranged from 260 ft/s to sonic, Q_ℓ/Q_g from 0.85 to 15 in gal/Mft. The liquids studied were gasoline, water, alcohol and heavy oils.

A more recent study by Kim and Marshall [28] employed an improved technique for measuring the drop sizes. For a single-air nozzle they found

$$D_m = 160.8 \frac{\sigma^{0.41} \mu_D^{0.32}}{(v_{rel}^2\ \rho_g)^{0.57} A^{0.36} \rho_D^{0.16}} + 1573 \frac{\mu_D^2}{(\rho_D\ \sigma)^{0.17}} \left(\frac{M_g}{M_\ell} \right)^n \frac{1}{v_{rel}^{0.54}} \tag{9.32}$$

where D_m = mass median drop diameter, microns; A = cross-sectional area for air flow, cm^2; M_g = atomizing air mass flow rate, g/s; M_ℓ = liquid mass flow rate, g/s, and the other symbols and units are the same as for (9.31). Here n = -1 if $M_g/M_\ell < 3$, and n = -0.5 if $M_g/M_\ell > 3$. These experiments included nozzles ranging from 1.83-2.54 mm diameter for liquid and 3.05-6.9 mm for air. Relative velocities ranged from 250 ft/s to sonic and relative flow rates from 0.75 to 75 in gal/Mft^3. The liquids were limited to molten wax and wax-polyethylene mixtures.

The drop size distribution did not follow the log-normal, square-root-normal, or Rosin-Rammler forms. Kim and Marshall correlated it by

$$\Phi_v = \frac{1.15}{1 + 6.67 \exp(-2.18\ D/D_m)} - 0.150 \tag{9.33}$$

where Φ_v = volume fraction-less-than size D, and $\bar{D}_{1,2} = 0.83\ D_m$. Licht [29] found that the upper-limit function fitted well with a maximum drop size always equal to 2.97 times the mass median, and expressed as

$$\Phi_v = 1/2 \left[1 + \text{erf}\ \frac{\ell n\ u/0.52}{\sqrt{2}\ \ell n\ 2.94} \right] \tag{9.34}$$

where $u = D/D_{max}$.

For corresponding values of L/G and v_{rel}, applied to the air-water system, (9.32) leads to $\bar{D}_{1,2}$ values which are less than half of those obtained from (9.31).

4. *Particle capture efficiency* The model for capture efficiency of the stream of accelerating drops is developed according to the same principles as have been used previously for spray chambers. A material balance over a control volume of length dx, taken anywhere downstream of the point of drop introduction, for the particulate removed by collision with drops is:

$$-v_g A\ dn_{p_i} = \eta_{T_i} \left(\frac{\pi D^2}{4}\right) (v_g - v_D) n_{p_i}\ n_D\ A\ dx \tag{9.35}$$

where n_{p_i} = number of particles of size d_{p_i} per unit volume of gas n_D = number of drops (targets) per unit volume of Venturi. This assumes that both particles and drops are uniformly distributed over the cross-sectional area of the Venturi, and that each drop collects particles independently and in the same manner as every other drop of the same size.

Assuming further that no drops are lost either by coalescence with each other or by transfer to the duct wall, that none are created by further shattering or pick-up from the wall (in other words that the wall is dry), the value of n_D is given by

$$n_D = \frac{\text{number of drops in control volume at any instant}}{\text{volume of gas in control volume at any instant}}$$

$$n_D = \frac{M_\ell}{v_D \rho_D} \frac{6}{\pi D^3} \bigg/ \frac{M_g}{v_g \rho_g} = \frac{6}{\pi} \frac{Q_\ell \, v_g}{Q_g \, v_D \, D^3} \tag{9.36}$$

Combining (9.35) and (9.36) leads to

$$-\frac{dn_{p_i}}{n_{p_i}} = \frac{3}{2} \frac{\eta_{T_i}}{D} \frac{Q_\ell}{Q_g} \frac{(v_g - v_D)}{v_D} dx \tag{9.37}$$

which must be integrated along the axis of the Venturi in order to obtain the grade efficiency of collection of particles d_{p_i} by drops of size D. This integration depends upon the interrelationship among v_g, v_D, and x based upon (9.15) as discussed above. The exact nature of it will vary according to the conditions of drop introduction and Venturi geometry. A comprehensive simulated computerized solution has been carried out by Placek and Peters [30]. With the aid of this, the effect of a number of operating variables has been studied. The results seem to check well against such experimental data as were available for testing. However it is desirable to have simplified methods to use and several special cases have been proposed.

a. Calvert Model - A fairly simple case uses an approximation to Ingebo's data [12] for C_{DA} in (9.15), combined with (9.37). The approximation is $C_{DA} = 55/Re_D$, so that

$$dv_D = \frac{165\ \mu}{4\ \rho_D D^2} \frac{(v_g - v_D)}{v_D} dx \tag{9.38}$$

and (9.37) becomes

$$-\frac{dn_{p_i}}{n_{p_i}} = \frac{2}{55} \eta_{T_i} \frac{Q_\ell}{Q_g} \frac{\rho_D D}{\mu} dv_D \tag{9.39}$$

To allow for the variation of η_{T_i} along the throat, the basic relationship for η_{T_i} for spherical targets is used:

$$\eta_{T_i} = \left[\frac{Stk_i}{Stk_i + 0.35} \right]^2 \tag{9.40}$$

where

$$Stk_i = \frac{C_i \rho_{pi} d_{pi}^2}{18\mu} \frac{(v_t - v_D)}{D} = \left[\frac{C_i \rho_{pi} d_{p_i}^2 v_t}{18\ \mu\ D} \right]^{(1-u)} = K_i (1-u)$$

These relationships are substituted into (9.39)

$$-\frac{dn_{p_i}}{n_{p_i}} = \frac{2}{55} \frac{Q_\ell}{Q_g} \frac{\rho_D D}{\mu} v_t\ K_i^2 \frac{(1-u)^2}{[K_i(1-u) + 0.35]^2} du \tag{9.41}$$

This may be integrated for whatever limits are appropriate, for example, namely $u = 0$ at $z_1 = z_2$. In case diffusiophoresis and/or thermophoresis is contributing to the capture process, the relationship (9.40) will have to be modified along the lines described in Chapter 5. Placek and Peters have incorporated this into their general model [31].

Calvert [1] has produced an integrated version of (9.40) which allows for reconciliation with actual performance for grade penetration. He assumed that particle collection did not begin until the drops had attained some relative velocity $(v_t - v_{D_1}) = fv_t = (1 - u_1)\, v_t$ which should fix the lower limit of integration of (9.40) at $u_1 = 1 - f$. Further, he assumed that acceleration was complete at the end of the throat, i.e., $u_t = 1$, which becomes the upper limit. In effect this approach regards the Venturi geometry as simply a straight tube of the throat diameter, and takes no account of what happends in the diverging section. Calvert expressed the result in terms of f as follows:

$$P_i = \exp\left[\frac{2}{55}\frac{Q_\ell}{Q_g}\frac{\rho_D D}{\mu_g} v_t\, F(K_{pt_i} f)\right] \qquad (9.42)$$

Here

$$K_{pt_i} = \frac{d_{pa_i}^2 v_t}{9\mu D} = \frac{2\,\mathrm{Stk}_i}{(1-u)}$$

$$d_{pa_i}^2 = C_i \rho_p d_{p_i}^2 = (\text{aerodynamic diameter})^2$$

and

$$F(K_{pt_i} f) = \frac{1}{K_{pt_i}}\left[-\,0.7 - K_{pt} f\; 1.4\, \ell n\left(\frac{K_{pt_i} f + 0.7}{0.7}\right) + \frac{0.49}{0.7 + K_{pt_i} f}\right] \qquad (9.43)$$

Regarding the factor f, Calvert [32] states

> The factor f is an empirical factor which absorbs the influence of various parameters not explicitly included

in (9.42). These parameters include collection by means other than impaction, particle growth due to condensation or other effects, drop sizes other than those predicted, loss of liquid to the Venturi walls, maldistribution, and other effects.

Data available at that time indicated that: $f \approx 0.25$, for hydrophobic particles; $f \approx 0.4$-0.5, for hydrophillic particles; $f \approx 0.5$, for large scurbber tests; and that f increases for $L/G <$ $0.2\ \ell/m^3$, or 1.9 gal/M ft^3. In a later study by Rudnick et al [33] it was found that $f = 0.31$ gave a much better fit than $f = 0.25$. Apparently it is not possible to predict f, but if it is determined in pilot-scale tests for a given system this value could be used for scale-up calculations. From the strictly mathematical point of view values of f must be interpreted in terms of u. Thus if $f = 0.25$, $u_1 = 0.75$, which means collection occurs while the drops accelerate from 0.75 v_t up to v_t.

In using (9.42) and (9.43), D is taken as $\bar{D}_{1,2}$ from the Nukiyama and Tanasawa relation (9.31), in which Q_ℓ/Q_g is taken for the Venturi as a whole and v_{rel} is based upon velocities at z_2. It has been pointed out above that this is not a proper application for gas-atomized sprays. Furthermore, it ignores the size distribution of drops which should be taken into account in the Stk values.

For the air and water system the Scrubber Handbook [1] presents a number of graphs showing P_i as a function of d_{pa}, for various values of Q_ℓ/Q_g and v_t. From these are derived other graphs giving the aerodynamic cut diameter (value of d_{pa} for which $P_i = 0.50$) as a function of Q_ℓ/Q_g, and v_t. All of these graphs are prepared for $f = 0.25$.

Calvert's design method consists in finding (with the aid of such charts) a set of operating conditions which will yield an aerodynamic cut diameter equal to the required performance cut diameter. The latter is determined from a knowledge of the inlet particle size distribution and the required overall penetration (efficiency) according to the methods described in Chapter 3. The equations, as stated above, do not specifically link the throat length with the performance. In other words an integrated form of (9.15) or (9.38) is not used.

b. Yung Model. A more elaborate but perhaps more accurate approach is to combine (9.17) with (9.37), obtaining

$$-\frac{dn_{p_i}}{n_{pi}} = 2\,\frac{\eta_{T_i}}{C_{D_1}}\,\frac{Q_\ell}{Q_g}\,\frac{\rho_D}{\rho_g}\,\frac{1}{v_t^{0.5}(v_t-v_D)^{0.5}}\,dv_D \tag{9.44}$$

Using (9.40) again for η_{T_i} the form to be integrated becomes

$$\frac{dn_{p_i}}{n_{p_i}} = \frac{2\,Q_\ell\,\rho_D\,K_i^2}{C_{D_1}\,Q_g\,\rho_g}\,\frac{(1-u)^{1.5}}{[K_i(1-u) + 0.35]^2}\,du \tag{9.45}$$

This has been integrated by Yung et al [34] over the length of the throat only, to give

$$\begin{aligned}\frac{\ln Pi}{B} = {} & \frac{1}{K_{il}(1-u) + 0.7}\Bigg[4K_{il}\,(1-u_t)^{1.5} + 4.2\,(1-u_t)^{0.5} \\ & - 5.02\,K_{il}^{0.5}\,(1-u_t + \frac{0.7}{K_{il}})\,\tan^{-1}\left(\frac{(1-u_t)\,K_{il}}{0.7}\right)^{0.5}\Bigg] \\ & - \frac{1}{K_{il} + 0.7}\left[4K_{il} + 4.2 - 5.02\,K_{il}^{0.5}\left(1 + \frac{0.7}{K_{il}}\right)\tan^{-1}\left(\frac{K_{il}}{0.7}\right)^{0.5}\right]\end{aligned} \tag{9.46}$$

In this expression u_t is evaluated by (9.19) using $x = \ell_t$ in X, and K_{il} is evaluated similarly to (9.40): $K_{il} = C_i\rho_p d_{p_i}^2(v_t - v_{D_1})/9\mu D$; and $B = Q_\ell\rho_D/Q_g\rho_g C_{D_1}$. It is assumed that drops are introduced at the throat entrance (z_2) and v_{D_1} may if desired be taken as zero. Note

rison and critique of models. The most recent and most
sive testing of all of the above models is that of Rudnick
, already mentioned above. In a series of over a thousand
nts of penetration covering twenty-three particle size
, collected over a wide range of Venturi designs and
conditions, they compared the models on a statistical
ney found the Yung model to give the best overall agreement,
edominating tendency to overestimate the penetration
The Calvert model with f = 0.25 gave poor agreement, but
).31 ranked just below the Yung with, however, a greater
to overestimate penetration. The Boll model gave much the
greement, with a pronounced tendency to underestimate
on.

he basis of this study, Eqn. (9.46) must be recommended for
e. It is somewhat cumbersome, but can be programmed for
ulator. Furthermore, Yung et al have provided a chart for
sionless correlation of P_i with K_{i1} and X, which will assist
lculations.

of these models suffer from certain faults. None of them
o account the well-known fact [35,36] that the walls of the
re not dry, that at any one time as much as 20% of the
present in a layer on the wall instead of in the form of
d that there may be a continuing interchange of liquid
rops and wall. This would lead to an underestimation of
on. Likewise, the use of the Nukiyama-Tanasawa formula, in
s, greatly underestimates the Sauter mean drop size [33].
, would tend to underestimate the penetration. Furthermore,
rt model and the Yung model both take into account only the
n process taking place in the Venturi throat. None of the
al with the reality of the atomization process, which has
rved to be quite complex and to occur over some distance
ection [20]. Boll, for example, assumed that atomization
immediately, hence would tend to over-estimate the amount of
n. In many respects the Boll model is more realistic and
a better starting point for improving predictive models [33].

that for Venturis with long throats, $u_\ell \to 1$ and right side approaches zero. Equation (9.46) gav [33,34] with experimental tests on large scale \ Sauter mean diameter $\bar{D}_{1,2}$ as given by (9.31) fo

For application to design purposes, it is (9.46) as a relation between three dimensionles K_{i1}, and a dimensionless throat length L, defir as

$$L = \frac{3\, C_{D_1}\, \rho_g}{2\, D\, \rho_D}\, \ell_t$$

so that x_ℓ = (L+8)/8, and u_t is related to L al have published [34] a graph of $(-\ln P_i)/B$ parameter. From a study of this chart they L = 2-3 would determine an optimal throat le based upon making a suitable compromise betw penetration and still having an acceptable values of L would correspond to the values 0.82 according to (9.19).

c. Boll Model. Boll [13] carried out the numerically without attempting to make any mentioned. He used (9.16) for the represe coefficient, and Dorsch's data for η_{T_i}, ea by a series of linear approximations. He estimate $\bar{D}_{1,2}$. Stepwise integration of (efficiency relationships for any given se These could then be compared with experim agreement was obtained in some cases. Ir be explained as due either to the presen or to maldistribution of liquid. He con contradict the model, as used, which cou study of optimum design and operating co procedure, however, this approach is ra

d.
comp
et a
meas
inte
opera
basis
with
somew
when
tende
poore
penet
C
design
desk
the di
in the
A
takes
Ventur
liquid
drops,
between
penetra
all mod
This, t
the Cal
collect
models
been ob
after i
occurre
collecti
might be

None of the above equations for grade efficiency attempts to take into account the size distribution of the drops. All assume, without proof, that the Sauter mean diameter $\bar{D}_{1,2}$ is a satisfactory representation of the array of drop sizes. In the absence of data on the drop size distribution existing in operating Venturis, Chongvisal [16] explored the effect of assuming various reasonable size distributions and found that better agreement with the experimental overall penetration could be obtained than by using the Sauter mean.

Subsequently some experimental data on liquid distribution in a Venturi have been gathered by Leith et al [36]. Injection of water was through a set of parallel tubes coaxial with the Venturi and even with the beginning of the converging section. They sampled the stream of drops at four locations: inlet and outlet of the throat, and midpoint and outlet of the diverging section. Droplet size distribution, fraction of injected liquid in the gas, and distribution of liquid across the scrubber cross section all varied from point to point. These observations establish the ever-changing nature of the liquid condition in the scrubber, and indicate how difficult it will be to model it. No generalized way is available to describe drop size distribution in a simple and yet realistic manner. If one should appear, there are methods available for the computation of collector efficiency [16,30,31,36].

5. *Pressure drop* a. Inlet Velocity Heads. A momentum balance is written over the control volume of length dx, in a manner analogous to the material balance (9.35).

$$AdP + M_g\,dv_g + M_\ell dv_D + (M_g + M_\ell)\,\frac{f_t v_g^2}{2D_e}\,dx = 0 \tag{9.47}$$

where the last term represents wall friction based upon equivalent diameter D_e of the Venturi, and upon turbulent friction factor f_t. To obtain the total pressure drop, (9.47) should be integrated from z_0 to z_4. However the estimation of the net loss ΔP can be simplified by applying (9.47) to the throat section only ($dv_g = 0$) and neglecting the wall friction, which is assumed to be offset (at least to some extent) by the pressure recovery in the diverging section.

Yung, Barbarika, and Calvert [15] have taken this approach using (9.47) transformed to:

$$dP = -\rho_D v_t (Q_\ell/Q_g)\, dv_D \tag{9.48}$$

For the usual conditions: $v_{D_1} = 0$ at $z_1 = z_2$, and v_{D_2} at $x = \ell_t$

$$\Delta P = P_3 - P_2 = -\rho_D v_t (Q_\ell/Q_g) v_{D_2} \tag{9.49}$$

or

$$\Delta P_{3,2} = -\rho_D v_t^2 (Q_\ell/Q_g) u_t < 0$$

Combining this with Eqns. (9.19) and (9.20) leads to

$$\Delta P_{3,2} = -2\rho_D v_t^2 (Q_\ell/Q_g) \left[1 - X^2 - X\sqrt{X^2 - 1} \right] \tag{9.50}$$

This may be written in the form of N_H (number of inlet velocity heads) lost, as in Chapter 3, by dividing:

$$N_H = \frac{|\Delta P_{3,2}|}{\rho_g v_t^2/2} = 4 \frac{\rho_D Q_\ell}{\rho_g Q_g} \left[1 - X^2 + X\sqrt{X^2 - 1} \right] = 2 \frac{\rho_D}{\rho_g} \frac{Q_\ell}{Q_g} u_t \tag{9.51}$$

Note that as X increases the term in brackets will never exceed 0.5, i.e. no matter how long the Venturi throat becomes. This equation was tested [15] against experimental data of Boll [13] using again the Nukiyama and Tanasawa expression (9.31) for $\bar{D}_{1,2}$ to give D in calculating X from (9.20), and was shown to fit quite well.

For example, in an air-water system $\rho_D/\rho_g = 830$, $\mu = 1.81 \times 10^{-4}$ gm/cm s with throat cross section (35.6 x 30.5) cm^2, ℓ_t = 30.5 cm, v_t = 91.5 m/s, Q_ℓ/Q_g = 1.0 ℓ/m^3; σ = 70 dynes/cm:

$$\bar{D}_{1,2} = \frac{585}{91.5}\sqrt{\frac{70}{1}} + 597\left[\frac{0.01}{\sqrt{70 \times 1}}\right]^{0.45}\left(1.0\right)^{1.5} = 82.4\ \mu m$$

$$Re_{D_1} = \frac{82.4 \times 10^{-4} \times 91.5 \times 10^{2} \times 1.205 \times 10^{-3}}{1.814 \times 10^{-4}} = 500;\ C_{D_1} = 0.55$$

$$X = \frac{3 \times 30.5 \times 0.55}{16 \times 82.4 \times 10^{-4} \times 830} + 1 = 1.46;\ u_t = 0.843$$

$$N_H = 4 \times 830 \times 1.0 \times 10^{-3}\ [1- (1.46)^2 + 1.46\sqrt{(1.46)^2 - 1}] = 1.40$$

$$\Delta P_{3,2} = -1.40\text{x}1.205\text{x}10^{-3}(9150)^2/2 = -70{,}620\ \text{dynes/cm}^2 = -72.0\ \text{cm}\ H_2O$$

Note that N_H is not quite linear in Q_ℓ/Q_g as this ratio enters into the determination of $\bar{D}_{1,2}$ which is involved in C_{D_1} and in X. Note also that N_H is dependent upon v_t as well, as it enters independently into $\bar{D}_{1,2}$ and Re_{D_1}, hence into C_{D_1}. For example, in the above case when v_t = 45.8 m/s, the value of N_H drops to 1.30. If longer throat values ℓ_t are considered, note that the maximum value of N_H will be 1.66.

Boll [13] used the complete equation (9.47) and performed a numerical integration, step-by-step in order to construct the entire curve of pressure vs. distance downstream from z_1. This reproduced the experimental pressure measurements very well, and may be carried out when a complete detailed study of pressure variation is desired.

The simple model, Eqn. (9.50), may be improved by allowing for the pressure regain in the diverging section in addition to the loss in the throat. To do this in a rather simple way, Leith et al [37] assumed that:

(a) the gas expands abruptly in the diverging section; that is that $v_g = v_{g4} = v_{go}$.

(b) drops decelerate and reach $v_D = v_{g4}$ in this section.

Then, Eqn. (9.48) applied to this section will give the pressure regain from point 3 to point 4 as

$$\Delta P_{4,3} = \int_{P_3}^{P_4} dP = -\rho_D (Q_\ell/Q_g) v_{g4} \int_{u_t v}^{v_{g4}} dv_D \tag{9.52}$$

$$\Delta P_{4,3} = -\rho_D \left(\frac{Q_\ell}{Q_g}\right) v_t^2 \left[\left(\frac{v_{g4}}{v_t}\right)^2 - \left(\frac{v_{g4}}{v_t}\right) u_t \right] \tag{9.52a}$$

Noting that $(v_{g4}/v_t) \approx (A_t/A_4) = (A_t/A_o)$ this may be written as

$$\Delta P_{regain} = \rho_D \left(\frac{Q_\ell}{Q_g}\right) v_t^2 \left[u_t \left(\frac{A_t}{A_o}\right) - \left(\frac{A_t}{A_o}\right)^2 \right] > 0 \tag{9.53}$$

Combining this with Eqn. (9.50) for the loss in the throat, $\Delta P_{loss} = \Delta P_{3,2} < 0$, will give the value of $\Delta P_{total} = \Delta P_{3,2} + \Delta P_{4,3}$, the net ΔP of loss and regain:

$$\Delta P_{total} = \rho_D \left(\frac{Q_\ell}{Q_g}\right) v_t^2 \left[u_t \left(\frac{A_t}{A_o} - 1\right) - \left(\frac{A_t}{A_o}\right)^2 \right] < 0 \tag{9.54}$$

This is the improved expression to be used for estimating ΔP_{total}. The ratio of ΔP_{total} to ΔP_{loss} may be written as:

$$\frac{\Delta P_{total}}{\Delta P_{loss}} = \frac{u_t \left(\frac{A_t}{A_o} - 1\right) - \left(\frac{A_t}{A_o}\right)^2}{-u_t} \tag{9.55}$$

A comprehensive set of experiments was carried out to measure total ΔP and compare the results with values calculated from the Yung model, Eqn. (9.50), the Boll approach based upon Eqn. (9.47) and the new model (9.54). They found that all methods tended to overestimate ΔP_{total} but (9.47) the least, as would be expected. Although the Boll method gave comparable results, it is much more cumbersome to use. This new model is, therefore, recommended for general use.

Continuing the example started above, assume that the ratio of throat cross-sectional area to approach area is 0.5. Then

$$\Delta P_{regain} = \frac{1 x 1 x 10^{-3} \ (9150)^2}{980}\left[0.843\left(\frac{1}{2}\right) - \left(\frac{1}{2}\right)^2\right] = 14.7 \text{cm } H_2O \qquad (9.53)$$

$$\Delta P_{total} = \frac{1 x 1 x 10^{-3} (9150)^2}{980}\left[0.843\left(\frac{1}{2} - 1\right) - \left(\frac{1}{2}\right)^2\right] = -57.3 \text{ cm } H_2O \qquad (9.54)$$

and

$$\frac{\Delta P_{total}}{\Delta P_{loss}} = \frac{57.3}{72.0} = \frac{0.843\left(\frac{1}{2} - 1\right) - \left(\frac{1}{2}\right)^2}{-0.843} = 0.797 \qquad (9.55)$$

As is the case in the grade-efficiency calculations discussed above, these pressure-drop calculations also ignore the effect of drop-size distribution and use only the Sauter mean size to represent the array. Chongvisal also studied this question and showed that by assuming certain reasonable size distributions (in lieu of actual data) better agreeement with experimental pressure-drop values could be obtained. It would seem that a mean diameter based upon mass would be more appropriate than the Sauter mean as the pressure-drop depends upon a momentum transfer to the accelerating drops.

b. Relationship Between Power and Efficiency. Since pressure drop is seen to depend primarily on the energy expended to accelerate the drops, and since this relative motion between drops and gas provides the principal capture mechanism, it is evident that pressure drop and overall collection should be closely linked. As was shown in Chapter 3, ΔP represents power consumption per unit of volumetric gas flow rate Q_g. A relationship between power consumption and overall collection efficiency has been recognized empirically for a long time [38], well in advance of the theoretical developments just described. It forms the basis of another design method, based upon the concept of "contacting power," due to Semrau [39].

The basic relationship may be developed by combining (9.37) with (9.47), using (9.15), for the usual case of drops introduced at the beginning of the throat. Then

$$\frac{dP}{d\ln n_{p_i}} = \frac{\rho_g v_t}{2} \frac{C_{DA}}{\eta_{T_i}} (v_g - v_D) \tag{9.56}$$

This might be integrated, for example, using (9.16) or (9.38) for C_{DA} and (9.40) for η_{T_i} to give

$$\frac{dP}{d\ell n\; n_{p_i}} = C_{D_1} \frac{\rho_g v_t^2}{2} \frac{[K_i(1-u) + 0.35]^2}{K_i^2 (1-u)^{1.5}} \tag{9.57}$$

The integration would have to be done numerically because of the complex relation between $\ell n\; n_{p_i}$ and u (such as given by (9.42) or (9.46), but it is clear that the kinetic energy of the gas in the throat $\rho_g v_t^2/2$ is directly involved.

The term "contacting power," as defined by Semrau [39], refers to power, per unit of Q_g, that is dissipated in contacting gas and liquid, and is ultimately converted into heat. In general it may include three parts: gas-phase contacting power P_g (due to the effective friction loss as measured by Δp), liquid-phase contacting power P_ℓ (due to the pumping of the liquid stream), and mechanical contacting power P_m (due to a mechanically driven rotor, not involved in Venturis). Total contacting power is the sum of these: $P_T = P_\ell + P_g + P_m$. The interrelationships are as follows:

	ΔP - in. H_2O	ΔP - cm H_2O
Pg:	$0.1575\ \Delta p$ hp/Mft^3/min	$0.02724\ \Delta p$ kWh/Mm^3
P_ℓ:	$0.583\ p_f(L/G)$ hp/Mft^3/min	$0.02815\ p_f(Q_\ell/Q_g)$ Mm^3
P_f:	lb/in^2	atm
L/G:	gal/ft^3	Q_ℓ/Q_g: ℓ/m^3

where P_f = liquid feed pressure.

The empirical relationship is to the effect that

$$\text{N.T.U.} = \alpha(P_T)^{\gamma} \tag{9.58}$$

where NTU = number of transfer units overall, as defined in Chapter 3, and α, γ are emipirical constants that depend upon the specific aerosol collected. They are little affected by the scrubber size or geometry, or the manner of applying the contacting power. Note that α and γ automatically take into account the particle size distribution of the specific aerosol, so that no grade-efficiency relation is developed explicitly. Values for α and γ are determined by running pilot scale tests on the actual gas stream, for which portable small size scrubbers may be used conveniently. Tabulations and graphs of these data are available [39].

C. *DESIGN PROCEDURES*

1. *The system* The interaction of the independent design (Venturi geometry, liquid introduction) and operating (L/G ratio, gas velocity v_t) parameters is shown in Fig. 3. The mathematical interrelationships have been detailed above, at least in principle. It is evident that the system is sufficiently complex that no comprehensive design method of calculation can be set up without making simplifying approximations, or without resorting to fairly elaborate computerized routines.

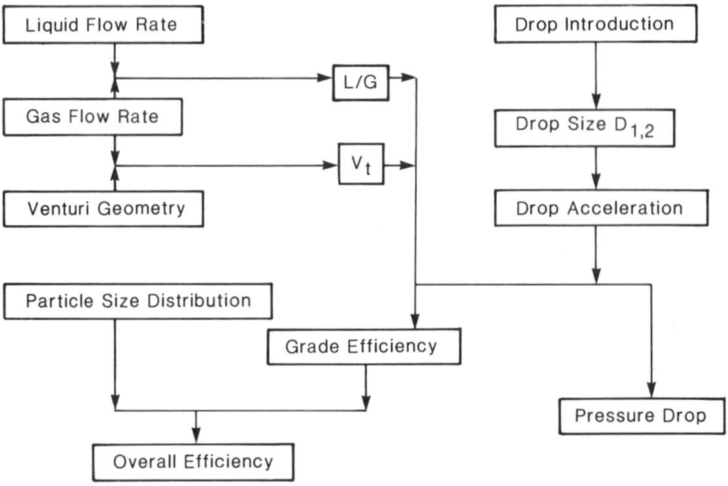

FIG. 3 *System analysis for Venturi scrubber.*

Usually a particular type of Venturi, and liquid introduction system, is assumed. Then for this situation, combinations of L/G and v_t are determined which will produce the required overall efficiency. As usual there will be a range of these combinations which will "work." For each, the corresponding pressure drop, energy requirements, and throat length will also be determined. An optimum set of conditions may then be found. Finally, the dimensions of the Venturi throat may be calculated.

2. *General method* The following outline is a general step-by-step method by which to implement the design calculations for a selected type of Venturi and liquid introduction system.

1. Select a pair of values for Q_ℓ/Q_g (or L/G) and v_t. These may be chosen from the range of commonly used values.
2. From these calculate $\bar{D}_{1,2}$, using (9.31), or (9.24).
3. For the v_t chosen, and $\bar{D}_{1,2}$ calculate $Re_{\bar{D}}$ and C_{D1}.
4. Using the above determined values, calculate B.
5. Select a particle size (d_{p_i}) and find the C_i for it.

6. For this particle size, calculate K_{i1} as defined for Eqn. (9.46).
7. Select a value of L (say L=3) and calculate u_t from (9.19).
8. Using (9.46), calculate P_i.
9. Repeat steps 5-7 for a range of particle sizes.
10. Determine overall penetration from these P_i values.
11. Repeat the entire process using different values of Q_ℓ/Q_g and/or v_t. Also consider other values for L, i.e. u_t.
12. Determine throat length, ΔP_{total} (9.54), and throat area.

Obviously this can be a cumbersome process even though computerized. Methods of shortening it are desirable, at least for preliminary design screening puposes. Some short-cuts to selection of optimum combinations of parameters have been suggested [40,41]. The Calvert method, which may only be used when a value of f has been determined, will be illustrated below.

The cost of the entire Venturi set-up including scrubber, cooler, separator, pumps and controls may be estimated by

$$C = 10{,}200 + 0.582\ Q_g - 1.21 \times 10^{-6}\ Q_g^2$$

where Q_g is in acfm ($0 \leq Q_g \leq 200{,}000$) and C is in 1981 dollars [42]. This is for 1/8 in. carbon steel construction.

Example 2. The author is indebted to Dr. R. L. Byers for proposing this problem, and for some of the calculations.

Consider a dust stream which is the outlet from a cyclone (primary collector) and is to be collected with at least 99.5% overall efficiency by a secondary collector, possibly a Venturi scrubber. The design of such a scrubber is to be investigated. Operating conditions will be: gas flow rate: Q_g = 13,300 acfm; gas temperature: 20°C (same as water temperature); gas pressure: 1 atm; dust properties: ρ_p = 1.5 gm/cm^3, MMD = 5.9 μm, σ_g = 2.4,

c_o = 2.63 gr/acf; inlet mass rate = 300 lbm/hr; outlet mass rate = (1-0.995) x 300 = 1.5 lbm/hr.

A type b(i) Venturi will be studied: gas-atomized spray, liquid introduced through tubes at the throat, rectangular cross section r_t = 6", design as illustrated in Fig. 1. It will be assumed that pilot-scale testing has indicated a value of f = 0.25.

The recommended procedure for the Calvert method would be as follows: 1. select possible values of Q_ℓ/Q_g as 1 ℓ/m^3 (or L/G = 7.48 gal/Mft^3), and v_t as 150 m/s (or 492 ft/s). 2. Calculate the Sauter mean drop size, preferably using (9.24), but alternatively using (9.31) simply because this has been used in the evaluation of f. Here $\bar{D}_{1,2}$ = 15.7 μm from (9.24), and $\bar{D}_{1,2}$ = 62 μm from (9.31). Note the wide difference. 3. Set up the calculation of Stk_i or K_i for use in (9.42) in terms of particle size d_{p_i} with the corresponding Cunningham factors C_i. Here, in general

$$K_i = \frac{C_i d_{p_i}^2}{\bar{D}_{1,2}} \frac{1.5 \times 15{,}000}{18 \times 1.814 \times 10^{-4}} = 6.891 \times 10^6 \frac{C_i d_{p_i}^2}{\bar{D}_{1,2}}$$

and for d_{p_i} = MMD = 5.9 μm, C_i = 1.03:

$$K_i = 398 \text{ for } \bar{D}_{1,2} = 62 \ \mu\text{m}$$

4. Calculate grade efficiencies for values of d_{p_i}, using (9.42) and (9.43) with $K_{pt_i} = 2K_i$, and f = 0.25. Since this is to be based upon Calverts' determination of f, $\bar{D}_{1,2}$ must be taken as 62μm. As an example, for d_{p_i} = MMD, (9.43) gives

$$F(K_{pt} f) = F(796 \times 0.25) = 0.2412 \text{ and (9.42) gives}$$

$$P_i = \exp\left[-0.2412 \times \frac{2}{55} \times \frac{0.001 \times 1 \times 62 \times 10^{-4}}{1.814 \times 10^{-4}} \times 1500\right] = 0.0112$$

5. With the grade efficiencies calculated in Step 4, and the particle size distribution of the inlet aerosol, calculate the overall efficiency as explained in Chapter 3. If this exceeds the required value, (here 99.5%) the Q_ℓ/Q_g and v_t selected are a workable combination. If not, new value(s) must be selected and the entire calculation repeated. This repetitious procedure is shortened by the Calvert method described below.

An alternative would be to continue with the general approach outlined above, using the Yung model. Thus calculating

$$Re_D = \frac{62 \times 10^{-4} \times 150 \times 10^{2} \times 1.203 \times 10^{-3}}{1.814 \times 10^{-4}} = 617$$

$$C_{D1} = \frac{24}{617} + \frac{3.60}{(617)^{0.313}} = 0.521 \qquad \text{(Schiller \& Naumann)}$$

$$B = \frac{830 \times 1 \times 10^{-3}}{0.521} = 1.593$$

Then selecting L = 3 (u_t = 0.816), and using d_{p_i} = MMD, calculate P_i from (9.46). Here K_{i1} = 2 x 398 = 796 and P_i = 0.025. Since the penetration for MMD is so great, it seems that the overall efficiency will not be satisfactory, as estimated from this model.

The corresponding throat length would be:

$$\ell_t = \left(\frac{2 \times 62 \times 10^{-6} \times 830}{3 \times 0.521}\right) \times 3 = 0.197 \text{ m} = 7.8''$$

and the net total pressure drop (taking A_t/A_o = 0.5):

$$\Delta P = -\frac{1 \times 1 \times 10^{-3}\left(150 \times 10^{2}\right)^{2}}{980 \times 2.54}\left[0.816\left(\frac{1}{2} - 1\right) - \left(\frac{1}{2}\right)^{2}\right] = -151 \text{ cm } H_2O$$

3. *Calvert method* In case the inlet particle size distribution is log-normal, as in the above example, the calculation procedure outlined may be greatly facilitated. The concepts of "required cut diameter" and "performance cut diameter," together with a set of charts, presented by Calvert [1] are utilized.

The "required cut diameter" $d_{p_{50}}$ is the particle size (in an aerosol of given MMD and σ_g) which must be collected with 50% efficiency in order to produce a specified overall (integrated) efficiency. It is a function of ($K_{pt}f$). A series of three charts, Figs. 8.3-5, 6, 7 in [1], for values of σ_g = 2.5, 5, 7.5 respectively are available from which $d_{p_{50}}$ may be found. In the example, σ_g = 2.4; $K_{pt}f$ = 199. For 99.5% collection efficiency required, P = 0.005. Using Fig. 8.2-5 (for σ_g = 2.5) the value of $d_{p_{50}}$/MMD is read as 0.055. Hence, $d_{p_{50}}$ = 0.055 x 5.9 = 0.32 μm. The corresponding aerodynamic diameter (C_i = 1.51) will be d_{p_a} = $\sqrt{1.51 \times 1.5}$ (0.32) = 0.48.

The "performance cut diameter" is that particle size which will be collected with 50% efficieny under the operating conditions assumed. Only one chart, Fig. 5.36-13 in [1] is available for this, based upon the air-water system and f = 0.25 only. For other systems, or other values of f, this performance cut diameter could be calculated by (9.42) and (9.43) following the procedure outlined. In the example case, the chart gives the aerodynamic diameter of 0.47 μm at Q_ℓ/Q_g = 1ℓ/m^3 and v_t = 15,000 cm/s. Note that σ_g and MMD are not involved here.

Where the "performance" aerodynamic cut diameter equals the "required" aerodynamic diameter, the set of values for Q_ℓ/Q_g and v_t is correct, i.e., these conditions will work. The agreement of 0.47 vs. 0.48 in the example is taken to be satisfactory.

For these operating conditions, the cross-sectional area of the throat must be A = Q_g/v_t = 13,300/492 x 60 = 0.451 ft^2. For a rectangular shape of height of 1 ft, this would correspond to a width of only 5.4", or a circular shape of diameter 9.1 inches could be used. The maximum pressure drop (regardless of throat length) estimated from (9.50) would be

$$\Delta P = \frac{-2 \times 1.0 \times (15{,}000)^2 (0.001) \times 0.5}{980} = -\ 229 \text{ cm } H_2O$$

This is a very high pressure drop. These conditions, therefore, are probably far from optimum and other values of Q_ℓ/Q_g and v_t must be explored. With the Calvert charts this can be done rather quickly.

The problem of determining a precise value of throat length remains unresolved by this approach. As a guideline, ℓ_t would have to be close to that used in the pilot scale tests which gave the value of f. This value should be used to find X and then a more precise estimate of ΔP . The basic theoretical difficulty here lies in the fact that the integration of (9.18) used in (9.50) to get ΔP goes from $v_{D_1} = 0$ up to v_{D_t} at end of throat, while the integration of (9.42) to produce P_i, in the Calvert scheme, goes from $v_{D_1} = fv_t$ up to $v_D = v_t$ at the end.

REFERENCES

1. S. Calvert, J. Goldschmid, D. Leith and D. Mehta, Scrubber Handbook NTIS, PB 213-016, 1972.
2. S. Calvert, Chem. Eng., 54 (Aug. 29, 1977).
3. S. Calvert, Amer. Inst. Chem. Eng. Journ., 16: 392 (1970).
4. C.J. Stairmand, Trans. Inst. Chem. Eng. (London) 28: 130 (1950).
5. L. Cheng, Ind. Eng. Chem. Proc. Des. Dev., 12: 221 (1973).
6. Chem. Eng. pg. 39, Oct. 13, 1986.
7. H.E. Hesketh, K.C. Schifftner, R.P. Hesketh, and B. Powell Atmos. Environ., 19: 1565 (1985).
8. Y.Y. Liu, and H.E. Hesketh Journ. Eng. Gas Tub. & Power, __: (1986).
9. D.N.Low U.S. Patent 3,803,805, April 16, 1974.
10. T.L. Holmes, C.F. Meyer, and J.L. Degarmo Chem. Eng. Prog. pg. 60 (Feb. 1983).
11. D. Faris M.S. Thesis, Chemical Eng., Univ. of Cincinnati, 1986.
12. R. Ingebo, NASA Tech. Note 3762 (1956).
13. R.H. Boll, Ind. Eng. Chem. Fundam., 12: 40 (1973).
14. K.G.T. Hollands and K.C. Goel, Ind. Eng. Chem. Fundam., 14: 16 (1975).

15. S.C. Yung, H.F. Barbarika and S. Calvert, Journ. Air. Poll. Control Assoc., 27: 348 (1977).

16. V. Chongvisal, Ph.D. Dissertation, University of Cincinnati, Dept. of Civil and Environ. Eng. (1979).

17. J.W. Crowder, K.E. Noll, and W.T. Davis Atmos. Environ., 16: 2009 (1982).

18. C.T. Crowe, Ph.D. Thesis, University of Michigan, Ann Arbor, Mich., 1961.

19. H.E. Hesketh, A.J. Engel and S. Calvert, Atmos. Environ., 4: 639 (1970).

20. D.B. Roberts, J.C. Hill Chem. Eng. Commun., 12: 33 (1981).

21. E. Radhakrishanan and W. Licht, A.I.Ch.E. Symp. Ser., No. 175, 74: 28 (1978).

22. R.H. Boll, L. R. Flais, P.W. Maurer and W.L. Thompson, Journ. Air. Poll. Cont. Assoc., 24: 934 (1974).

23. P.A. Nelson and W.F. Stevens, A.I.Ch.E. Journ., 7: 80 (1961).

24. S. Katta and W.H. Gauvin, A.I.Ch.E. Journ., 21: 143 (1975).

25. W.R. Marshall, Jr., Chem. Eng. Prog. Monograph Series 2, 50 (1974).

26. E.H. Taylor and D.B. Harrison, Jr., Ind. Eng. Chem., 45: 1455 (1954).

27. S. Nukiyama and Y. Tanasawa, Trans. Soc. Mech. Engrs. (Japan), 5: 63 (1939).

28. K.Y. Kim and W.R. Marshall, Jr., A.I.Ch.E. Journ., 17: 575 (1971).

29. W. Licht, A.I.Ch.E. Journ., 20: 595 (1974).

30. T.D. Placek and L.K. Peters A.I.Ch.E. Journ., 27: 984 (1981).

31. T.D. Placek, and L.K. Peters A.I.Ch.E. Journ., 28: 31 (1982).

32. S. Calvert, in Air Pollution (A. Stern, ed.) 3rd Ed., Vol. IV: Chap. 6, Academic Press, New York, 1977.

33. S.N. Rudnik, J.L.M. Koehler, K.P. Martin, D. Leith, and D.W. Cooper, Environ. Sci. and Tech., 20: 237 (1986).

34. S.C. Yung, S. Calvert, H.I. Barbarika and L.E.Sparks, Environ. Sci. and Tech., 12: 456 (1978).

35. B.J. Azzopardi, A.H. Govan Filt. and Sep., : 196 May/June (1984).

36. D. Leith, K.P. Martin, and D.W. Cooper Filt. and Sep.: 191, May/June (1985).

37. D. Leith, D.W. Cooper, and S.N. Rudnick Aerosol Sci. Tech., 4: 239 (1985).

38. C.E. Lapple and H.J. Kamack, Chem. Eng. Prog., 51: 110 (1955).

39. K.T. Semrau, Journ. Air Poll. Control Assoc., 10: 200 (1960); 13: 587 (1963).

40. H.E. Hesketh, K. Mohan JAPCA, 33: 854 (1983).

41. D.W. Cooper and D. Leith Aerosol Sci. Tech., 3: 63 (1984).

42. W.M. Vatavuk in Handbook of Air Pollution Technology, S. Calvert and H.M. Englund, eds, Chap. 14, Wiley Interscience, New York, 1984.

PROBLEMS

1. Refer to Example 1 and Figure 1 in this chapter. Select any two of the cases and verify the calculated value of grade efficiency for any two particle sizes for each case.

2. A gravity spray tower 3 m high is operating at a liquid-to-gas ratio of 1 ℓ/m^3 with a drop diameter of 400 μm. The gas velocity is 0.1 of the drop terminal velocity which is 157 cm/s. The operation is at 20 C. What is the grade efficiency for particles of 1 μm diameter, having a density of 2.0 gm/cm^3?

3. The solutions obtained for Example 2 in this chapter are different for the two different models used. How do you account for the difference? What value of f would be required to make them agree?

4. A Venturi scrubber is to be designed to collect dust from an asphalt stone drier. The dust has a mass median diameter of 1.8 μm and a density of 2.6 gm/cm^3. The uncontrolled emission rate is 2310 kg/hr, but state regulations require that this be reduced to a maximum of 25 kg/hr. The air flow is 20,000 acfm at 250 F. No additional data are available. Assuming a throat velocity of 150 ft/s, make a preliminary determination of the necessary L/G value, and of the maximum pressure loss in the

throat. Discuss the additional data and calculations which would be required to make a final design of the Venturi.

5. A Venturi scrubber having a throat 9" high by 20" wide by 30" long is to be used under the set of operating conditions described for Example 2 in the text, and with a flow of 10 gal water per 1000 actual cubic feet of gas. Determine:

(a) the mass rate of collection (kg/min) particles lying in the size range between 1 μm and 2 μm;
(b) the velocity of the drops at a point 15" downstream from the beginning of the throat;
(c) the pressure drop (net total) and the power consumption.

Would you recommend any changes in the dimensions of the throat?

Appendix—
Answers to Selected Problems

Chapter 2.

1. (a) 16.5 μm; (b) 24 μm; (c) 22.4 μm; (d) 69 %;
   (e) Yes. NMD = 16.5 μm, σ_g = 1.42.
3. 11.6 %.
6. (a) 4.5 %; (b) 493 μm; (c) 3.0 x 10^{-3} μm^{-1}
   (d) 148 μm.
7. (a) 128 kg; (b) 65 μm; (c) 4.2 μm.

Chapter 3.

1. (a) 1.48 μm; (b) 83.9 %; (d) 16.1 kg.
2. 0.047.
4. No-mix model; No. Complete mix model.
5. (a) 9.4" H_2O; (b) 21.2.
7. (b) 10.1 gr/cu. ft.

Chapter 4.

1. 188 cm/s.
2. 671 μm.
3. Upward; 2.84 s.

4. (a) 0.015 cm/s; (b) 458 cm/s.
5. 66.4%
6. 1.64 cm/s.
7. (a) 980 cm/s^2; (b) 6 x 10^{-7}.
8. (a) 0.51 s; 23,100 C/cm.
9. 0.67 C/cm.
10. 12 (for Th ≈ 0.50).
12. n/n_o = 0.99 by coalescence, 0.82 by deposition.
13. (b) 1.24 x 10^6 cm^{-3}; (c) 0.054 cm^3/min x 10^{-6}.

Chapter 5.

1. @ 1 µm: 5.3 %; @ 10 µm: 63.2 %: >12 µm: 100 %.
2. (a) 3.5 cm; (b) 64 %.
4. (a) 26.0 m/s; (b) 0.020 s; 11.2 cm/s;
   (d) 0.65 < Re < 4.9
6. (b) $\eta_{II} \approx 0$, $\eta_{DI} \approx 1.5 \times 10^{-4}$, $\eta_{BD} \approx 7.6 \times 10^{-4}$.
7. $Stk_A = 6.76 \times 10^{-4}\ v_o/D$; $Stk_B = 1.05 \times 10^{-4}\ v_o/D$.
8. (a) Stk_A = 0.50C, Stk_B = 0.062C;
   (b) $\eta_{BD} \approx 0.22$ (Ranz).
9. 75 %.
10. (a) 0.82; (b) 0; (c) 0.
11. $\eta_{BD} \approx 0.009$, others 0, for 100 µm.
12. 0.625 gm/cm^3.
13. $c/c_o \approx 0.91$.
14. @ 75 F: for d_p = 0.1 µm, ACE = 0.0065;
    for d_p = 10 µm, ACE = 0.99.
17. (a) for d_p = 1.0 µm: @ 1 cm/s, 0.030,
    @ 100 cm/s, 0.107; (b) for d_p = 1.0 µm:
    @ 1 cm/s, 0.18, @ 100 cm/s, 0.513.

Chapter 6.

1. (a) 8.3 ft/s, 2.78 times larger.
   (b) 2.25 µm, 1.77" H_2O.
2. 42.5 %.
4. Approximately 0.05 mm high, 8 mm wide.

Chapter 7.

1. For $d_p = 0.1$ μm: 82.2 coulombs x 10^{18}/μm.
4. (a) 0.173 ft/s; (b) 0.20 μm; (c) 1.79 μm: $\bar{D}_{1,1}$ or $\bar{D}_{1,0}$; (d) 99.16 %.
7. 43,800 ft^2.

Chapter 8.

2. (a) 3.25 cfm/ft^2; (b) $S_E = 1280$ "H_2O min^2/lbm, $K_2 = 14{,}000$ "H_2O min^2 ft^2/lbm^2; (c) 68 min; (d) 669 min.
3. Wool fabric; reverse jet; 8.6 cfm/ft^2; 46 bags.

Chapter 9.

1. See Fig. 1.
2. 0.051.
3. (a) 0.161 kg/min; (b) 43.7 m/s; (c)
   (c) -27.8 cm H_2O, 23.0 hp

Index

L

M

N

O

P

T

U

V

W